INDIA'S ARMY

INDIA'S ARMY

MAJOR DONOVAN JACKSON

Low Price Publications
2012

First Published 1902
Reprinted in **LPP** 2001, 2006, 2012

ISBN 10: 81-7536-438-6
ISBN 13: 978-81-7536-438-7

Published by
Low Price Publications
A-6, Second Floor, Nimri Commercial Centre,
Ashok Vihar Phase-IV, Delhi 110052 (INDIA)
tel.: +91.11.27302453 fax.: +91.11.47061936
url: www.Lppindia.com e-mail: info@Lppindia.com

Printed at
Salasar Imaging Systems
Delhi

PRINTED IN INDIA

TO ALL THE SOLDIERS OF INDIA

Foreword

by

Field-Marshal The Lord Birdwood, G.C.B., G.C.S.I., G.C.M.G., G.C.V.O., C.I.E., D.S.O., L.L.D.

TO any old soldier of the Indian Army—after forty-five years of Indian Service, I feel I can describe myself as such—it must be a pleasure to have an opportunity of reading Major Jackson's—what I may call great work—book *India's Army*, and not only to read it but to see the excellent photographs which cover all units and classes enlisted in India's Army. Such has been wanted for a long time, as though individual histories have as we all know been published, it has not been possible up to now, to obtain any volume which contained all, this, Major Jackson has now provided for us.

Individual portions of this have already appeared in *The Statesman*, and the very favourable—indeed great reception they have received speaks for itself, and shows that the complete volume will meet a real want, and I well know how popular it will be. I trust that it will meet with great demands both in England and in India, and I heartily wish Major Jackson all success with it.

Deal Castle.

Author's Foreword

THE contents of this book originally appeared in more or less the same form as articles in *The Statesman*, India's principal newspaper, in 1938–39. They were inspired by a series of articles which told of the glories of the British soldier, and by the fact that little has been done to make known to the Indian public something of the history and glory of their own regiments who, as the succeeding chapters will show, have played down the years an almost world-wide part in the protection and advancement of the association of nations now known as the British Commonwealth.

As Lord Birdwood has said, individual regimental histories have from time to time been published, but he would add, largely for private circulation and written generally from as different angle of view to that adopted by the writer, who, while making no claims whatever to literary style or skill, has attempted to present the stories of the regiments of India's Army up to the outbreak of the present war, in a way which he hopes will appear to her citizens and the non-military mind. If he succeeds in that, then he has at least achieved his object.

It is not feasible to give a bibliography, but all the histories were compiled from official or regimental sources, and in most instances with the generous assistance of the regiments themselves it has been impossible to include all that they would have wished. To them, and to the writers of the published unit histories upon which I have drawn so freely, as well as the individual officers who offered help and advice or supplied photographs and other details which assisted the writer to paint the illustrations, my gratitude. All of the latter were

compiled from old records or material supplied by the units. The full dress depicted is that of 1914 or used since upon occasions when full dress has been worn.

In particular, thanks are due to Lieut.-Colonel Sir Patrick Cadell, C.S.I., V.D., for his assistance and advice over the stories of the regiments which had their genesis in the old Bombay Army, to Army Headquarters and the India Office, for permission to use official photographs, the Military Adviser-in-Chief of the Indian States Forces, to *The Statesman*, and to Field-Marshal Lord Birdwood for his early encouragement and later his kindness in writing the Foreword.

The very many who wrote to me during the fifteen months the series were appearing in India, and to whom I was unable to answer, will understand my apparent lack of courtesy when I disclose that during that time I was suffering from a serious illness. In fact, the stories would never have been completed had it not been for the greatest helper of all, my Wife. And, let me add, my doctors all aver that during my enforced idleness their completion was better medicine than they were able to prescribe.

Now, war has come upon the world again and the peace-time army strength of 150,000 Indians of all ranks is now being increased, as a first step, to 500,000 men of all arms, trained, equipped and mechanized in accordance with modern requirements and modern tactics. Recruiting has been thrown open to all castes and creeds, and is no longer confined to the so-called "martial classes"; Indianization of the officer cadre is proceeding at speed, and Indian King's Commissioned Officers are being posted to all regiments. Before these lines appear in print, it is certain that the Indian soldier will again be adding fresh glories to the story of his Army.

The war has delayed publication, and military considerations, as well as military duties, prevent any revision of the manuscript to accord with present times; so the writer craves

indulgence for leaving the stories of the regiments of India's Army in the style they were written just before this Greater War began.

Finsbury Barracks,
London, E.C.1.

Contents

Contents

Contents

Contents

INDIA'S ARMY

CHAPTER I

Note on the Evolution of the Indian Army

THE stories of the Indian regiments may perhaps be better followed if something of the evolution of the army which they comprise is understood.

The first purely Indian troops of the British era had their foundations in the formation of watch-keepers employed by the East India Company to protect their trading stations. These evolved into battalions, and eventually, as military responsibilities of the Company grew, into the three great Presidency Armies of Bombay, Madras, and Bengal; each more or less separate entities which were made necessary by conditions of communication and terrain. They were, however, from 1748 all under one Commander-in-Chief, the first father of the Indian Army being Major Stringer Lawrence.

Clive, in Bengal, in 1757, conceived the organization of sepoy battalions, armed and dressed and trained on the European model and commanded by a nucleus of British officers. Prior to this, Indian troops of the Company adopted their own weapons and indigenous dress, and were officered by their own kin. Certain features of Clive's system still remain in the Indian Army of to-day.

Madras adopted the system with six battalions in 1759, and Bombay in 1767. Cavalry was not a strong feature of the early forces of the East India Company.

A regimental system on a two battalion basis was introduced between 1796 and 1804. It was not satisfactory, the battalions though theoretically linked, were not mutually interdependent, and had no common *esprit de corps*. Control by the regimental commander was excessive and exasperating to the battalions, while owing to the increase in the number of British officers, twenty-two per battalion, the authority and dignity of the Indian officers was diminished. In this system was the foundation of the Great Mutiny, a crime with which both the Madras and Bengal Armies had not been unfamiliar, including Madras European officers.

Single battalion regiments were reverted to in 1824 and the units numbered according to the dates upon which they were raised—the most conspicuous defect of the earlier system; the high proportion of British officers, was, however, continued.

About this time, corps recruited for a particular locality, as well as irregular cavalry, came into being. The latter were organized on the indigenous system of the local rulers known as the *sillidar*, whereby for a lump sum down and a maintenance grant, the man provided his own horse, weapons, and such military attire as was considered necessary; while the number of British officers was kept at a very small minimum.

Regular regiments of cavalry were formed about 1784, none but three survived the Great Mutiny.

Then came vast territorial expansion, the armies of one Presidency were often employed on service beyond the Presidency frontiers with the troops of another, and with the establishment of peace, left there in garrison. As a consequence, the allowances for field service were withdrawn. Men detailed for service outside their own Presidency not only objected to serving so far from their homes, but considered such service entitled them to special compensatory treatment. Bengal

troops were the principle sufferers. Thus, the Presidency army system, with its mistaken ideas of vested rights which it fostered, must take its share of the many causes which led to the Great Mutiny, principally the Bengal Army, in 1857.

After the Mutiny, all the Company's troops were transferred to the Crown. And 1861 saw a general reorganization of the army in India. All the cavalry, save the remnant regiments of the Madras cavalry, were based on the *sillidar* system. Indian artillery was abolished. And the number of British officers reduced to six per battalion.

The Presidency Armies were still continued. The total units of the Indian Army proper, including the Punjab Frontier Force and the Hyderabad Contingent—the two principal local forces—being 42 cavalry regiments, 142 infantry battalions and 3 corps of engineers; approximately some 135,000 men.

A commission of enquiry set up after the Afghan War of 1878-80, recommended the abolition of the Presidency Armies. But although the ordnance, supply, transport, and pay branches had been or were then unified, the main unification for a variety of reasons, did not take place.

But in 1886 a further step was taken, when the Punjab Frontier Force, then under the control of the civil authority, was transferred to the Commander-in-Chief. While the same year, infantry battalions were grouped, generally in pairs, and given permanent centres at which one of the two could be located to serve as a draft-finding unit for the other in war. Recruits were thenceforth enlisted for the group, though the battalions were not made components of a regiment. A certain bond of mutual interest was thus created, but the defect was that one battalion could only be reinforced at the expense of the other.

At the same time, a reserve for the fighting units was created, service therein being voluntary.

At last, in 1895, after many alternatives had been considered, the Presidency Armies with their separate organizations were

A unique view of the Indian Military Academy, Dehra Dun, with the riding school in the foreground, main buildings in the centre, and other buildings—quarters, swimming baths, sports fields, etc.—in the background, with the Himalaya mountains in the distance.

abolished, and geographical area commands established. But the units were still to be localized in their peace garrisons, and the area commands were still to be as separate from each other as the Presidency Armies had been.

More drastic measures were wanted to properly complete the unification of the Indian Army. The great Lord Kitchener, who assumed the Commander-in-Chiefship in November, 1902, wasted no time. On the 1st January, 1903, the regiments were renumbered consecutively on an all India basis, all traces of the Presidency Armies deleted, save in the names of the regiments; the Hyderabad Contingent units were delocalized, and the Punjab Frontier Force as such abolished—all units were to have experience of the North-west Frontier. Troops were to be distributed and trained together in peace in the formations to which they would belong in war. All ranks, officers and men, were to belong to one corps, the Indian Army.

Lord Kitchener's scheme had not been completely carried out by 1914—finances would not permit, since the abolition of some thirty-four stations was to be effected in the idea of concentrating troops, not only in their tactical formations, but in areas to the north, from whence aggression might come.

But Lord Kitchener's principles by 1914 had been carried far enough into effect to enable India to promptly more than pull her weight in the team of the Armies of the British Commonwealth in the world war.

On August 1st, 1914, her total fighting forces were 155,000. By November, 1918, they had reached 573,000. But it must be stated that India only bore the cost of her normal army, the expense of the increase being met entirely by Britain.

The war brought out defects. The linked battalion and reserve system failed to stand the strain. The lack of unification in equipment in the *sillidar* cavalry, as well as the replacement of horses, placed great difficulties in maintaining units efficiently in the field.

So in 1921, evolution was carried a stage further. *Sillidar*

cavalry were abolished; thirty-six existing regiments were paired, and from the marriage, one new regiment was produced, while three, the 27th and 28th Light Cavalry (already non-*sillidar*) and the Guides Cavalry were left as separate entities; thus producing twenty-one regiments, which was considered sufficient for post war requirements. The regiments were grouped in threes, the racial class composition of each group being identical, so that a pool of trained reservists would be available to serve in any regiment of the group on mobilization, when group depots would be formed. This was carried a stage further in 1937, when the groups were reduced to three, of seven regiments each, one of which became permanently located as a training regiment to serve as a depot unit for the whole group. Two regiments were mechanized in 1938. Mechanization of more has been announced.

The 1921 reorganization introduced a proper regimental system into the infantry (save Gurkha regiments which had all consisted of two battalions since 1908). An average of six battalions were regimented, one of which was to be a training battalion to supply and train recruits for the active battalions which in war would proceed on service. Training Battalions were numbered the 10th, the intervening numbers being left for new active or field units in war. Further reductions in the infantry—principally the old Carnatic regiments turned into the 3rd Madras Regiment in 1921—were effected in 1923: the Pioneer Regiments suffered a similar fate in 1932. In 1939 there were eighteen regiments of infantry, each of an average of six battalions, and ten Gurkha regiments of two battalions each.

Two cavalry regiments and six infantry battalions were selected for complete Indianization of the officer cadre in 1923. A further cavalry regiment and six infantry units, as well as components of the engineers and signals, being added to the scheme in 1933.

The Indian Military Academy for the training of Indian

gentlemen as officers independently of the Royal Military College at Sandhurst, was established in 1931.

Indian artillery, absent from the army since the Great Mutiny, save for certain mountain batteries forming part of the Royal Artillery, was revived as Field Artillery in 1935.

The Indian soldier is fed, clothed, housed, and horsed entirely by centralized services, in direct contrast to the unit arrangements, often inadequate, before the Great War. Pay and Pensions are on a liberal scale, which brings readily forth the number of recruits required and secures their efficiency and contentment. The medical treatment and hospital services compare favourably to those of any army in the world. His arms and equipment are identical with those of his British comrades, and now no longer one step behind, as was formerly the case.

In support of the regular army there are the battalions of the Indian Territorial Force, started in 1920, to provide for military aspirations of those classes not enlisted into the regulars, and for those who wished to serve their country in their spare time on an amateur basis.

There is also the Auxiliary Force, descendants of the old Indian Volunteers, comprised of Europeans or persons of mixed descent, available for service locally in case of emergency and to assist in home defence.

Last but not least, there are the forces of the Indian princes, trained in varying degrees, some to the standard of the regular Indian army, which their rulers place at His Majesty's disposal when danger threatens his Imperial Realm.

India's Army, under normal conditions, has responsibilities to be found nowhere else in the British Commonwealth. In addition to the maintenance of the field army, troops have to be allotted for internal defence duties in support of the civil power, and, most important duty of all, a covering force must be kept on the North-west Frontier to control the ever restless war-like tribes.

The lessons of 1914-20 have been well learnt, and the King-Emperor now has at his disposal an army of the highest quality, well instructed in the hard school of the Frontier, and trained in the modern tactics of attack and defence; with mechanization steadily progressing, it is a most efficient fighting machine.

CHAPTER II

Ranks in the Indian Army

TO those unacquainted with India, a note on the ranks in the Indian Army may be helpful. British rank names are not used except for King's commissioned officers, and the Indian styles adopted when the East India Company first formed armies are still retained.

The principal Indian ranks and their badges are:

	Badge
CAVALRY	
Officers	
Risaldar-Major	A crown
Risaldar	Two stars
Jemadar	One star
Other ranks	
Duffardar-Major	Royal Arms
Duffardar	Three stripes
Lance-Duffardar	Two stripes
INFANTRY	
Officers	
Subadar-Major	A crown
Subadar	Two stars
Jemadar	One star
Other ranks	
Havildar-Major	Royal Arms
Havildar	Three stripes
Naik	Two stripes
Lance-Naik	One stripe

A sowar is the *trooper* in the cavalry, and a sepoy the *private* in the infantry.

CHAPTER III

The Body-Guards

THE Body-guards of India are now the only bodies of select cavalry maintained in the Armies of the Commonwealth for the purpose of escorting and guarding the person and establishments of the personal representatives of the King. The three Governors of the old Presidencies, Madras, Bombay, and Bengal, each have such units as part of their household establishments, the other Governors having to be content with escorts from Indian cavalry on occasions of state. Above all, the Viceroy and Governor-General has his own separate corps, which has enjoyed an unbroken existence for nearly one hundred and seventy years, and which are household troops of picked men in every sense of those words.

The first body-guard raised in India was composed of Europeans recruited in 1762 from the East India Company's Infantry. There being no cavalry then with the army, two troops of Dragoons and one of Hussars were formed, the latter becoming the personal guard of the Governor.

Great difficulty appears to have been experienced in procuring suitable horses, while the men having been trained to fight on foot, did not take kindly to an equestrian career. The troops were perfectly useless cavalry, as well as being very expensive, in consequence their disbandment was ordered; one troop of sixty alone being retained for the duties of reconnaissance and

patrol, in addition to providing the necessary escorts for the head of the Company's government.

Clive, on his return in 1765, re-organised the army, with the result that the European troop of horse was further reduced to one subaltern and twenty-four other ranks, purely for the personal use of the head of the Presidency. All the three Hindustani cavalry regiments were similarly treated, and by the end of 1772 no mounted force remained in the Army of Bengal.

The depredations of the Senaissies—bandits who in the garb of gipsies or religious mendicants roamed the then North-West Frontier plundering and destroying—led Warren Hastings, when in Benares in 1773, to raise, in conjunction with Raja Cheyt Singh, a body of Indian horse for the purpose of keeping the raiders in check. Of this unit, fifty were to be used by Hastings as his personal guard in peace and to accompany the Commander-in-Chief when campaigning.

The records show that Hastings advised the Court of Directors he desired in no way to be ostentatious, but he considered a guard necessary for the dignity and safety of the Governor, as during his journey from Calcutta to Benares, not more than eight horsemen could be raised to escort him.

As a result of Hastings' decisions, in September, 1773, the present Governor-General's Body-Guard came into being. The first Commandant was one Captain Sweeny Toone, whose successful leading of the new corps in the guerilla operations against the Senaissies, resulted in the raiders being driven from the Company's territory and the cessation of their depredations for many years.

Under Toone, the Body-Guard saw service in the Rohilla War (1774), but it is not definite as to whether they were present at the then much talked of Battle of St. George—so called because it was fought on April 23rd, St. George's Day.

The official designation of the Corps at this time was "The Governor's Troop of Moghuls", Hastings then being merely Governor of Bengal, the title "Governor-General" not being

conferred until the 20th October, 1774. Subsequent titles during the first decade were "The Troop of Body-Guards", "The Troop of Horse Guards", "The Troop of Black Cavalry", "The Body Troop", and lastly, "The Governor-General's Body-Guard", by which title it now takes its place in the Army List as the senior Corps of the Indian Army. The title was, curiously, never officially conferred, and the Corps to-day is more often than not authoritatively styled "The Viceroy's Body-Guard".

The Corps was the only mounted unit of the Bengal forces until 1777, when two regiments of cavalry, raised in the preceding year by the Nawab of Oudh, passed into the "John Company's" service—neither of these units survive to-day.

In January, 1777, Toone, on proceeding to furlough, was succeeded by a Captain Briscoe, a nominee of Hastings'. Briscoe being an infantryman, great concern was shown by the General, Clavering, who could not reconcile himself to the idea of a select body of cavalry being commanded by such a person of little consequence as an infantry officer; the polite, but corrosive, minutes upon the subject make amusing reading to-day. Hastings, of course, won; but Briscoe had to give up the command in the following April, for, having been promoted Major in the February, the Honourable Directors decided that his appointment to the Body-Guard was incompatible with field officer's rank.

The internal economy of the Corps must have been somewhat loose at this time, for we find entries in December, 1784, to the effect that the pay of "The Governor-General's Detachment" was Rs.42,000 in arrear—the trooper's pay was then Rs.10 per mensum, Captain Rs.124, Subaltern Rs.62. Toone also appears to have been a careless book-keeper, for on his return to India he was presented with a bill for Rs.18,190 by the Military Paymaster-General. Toone's explanations were apparently eminently satisfactory, for he was subsequently promoted to Lieut.-Colonel for his services. He was held in great esteem by Hastings, and accompanied the Governor-General to England as one of his personal staff in 1785.

With Hastings' departure, his temporary successor—McPherson—wishing doubtless to ingratiate himself with the authorities at home, at once started to make sweeping economies, and as a result reduced the Body-Guard to fifty men. Here it may be mentioned that in 1782 an infantry section of the Body-Guard, 600 strong, had been formed, also from a regiment in Oudh's service. These McPherson also disbanded.

But let us return to 1790, when Lord Cornwallis, who had relieved McPherson four years previously, decided that war upon Tippoo Sahib, Sultan of Mysore, was necessary for the safety of the Honourable Company's territories. The operations were in the first instance entrusted to General Medows, Governor and Commander-in-Chief of Madras, but so dissatisfied was the Governor-General with the conduct of the operations, that he himself proceeded to the field and assumed the command in person. The Body-Guard were ordered on active service. Embarking at Calcutta on Christmas Eve, 1790, they took twenty days on the passage to Madras. Before Bangalore, in the March of the following year, they were instrumental in preventing an attempted assassination of the Governor-General. By the July, the cavalry of the line with the Company's forces in the field had become completely *hors de combat*, due in no small measure to inefficient horse-management, in complete contrast to that of the Body-Guard, that the latter, together with the small Body-Guard of the Governor of Madras, were for some time the only effective cavalry in the field.

As is usual after any successful war, demands for reductions in the armed forces are always to the fore. And in 1796 the Court of Directors ordered the complete disbandment of the Body-Guard and directed that its duties be performed by ordinary cavalry. The order was, however, never carried out, and 1801 saw the acquisition of the land at Ballygunge, Calcutta,

which became the normal winter station of the Household troops of the Governor-General for over one hundred years.

1801 is also important to the Indian Army, for that year saw a force of Indian troops despatched overseas for the first time. Wellesley, the Governor-General, sent 5,000 men under General Baird to assist Abercromby in expelling the French from Egypt. With this force went an experimental troop of six guns of horse artillery (which is now "F" ("The Sphinx") Battery, R.H.A.), but the horses, drivers, and other non-gunner personnel were provided by the Governor-General's Body-Guard. Baird, unable to reach Suez owing to the monsoon, landed at Kosseir on the Egyptian Coast, and marched, in the height of summer, across the intervening 120 miles of desert to the Nile. All the horses died on the march and the guns had to be placed on camels. Alexandria had capitulated when the Indian troops arrived, but their conduct while with the army of occupation, together with their smartness and discipline, brought forth great praise and admiration. On their return in 1802, they were publicly received by the Governor-General at Fort William, each man being presented with a medal. Why the honour of "The Sphinx superscribed Egypt" was not awarded, as was done to other regiments of the forces who took part in that campaign, is a mystery; similarly the corps do not bear the honours "Seringapatam" or "Carnatic", although they took part in the campaign for which these honours were awarded. The authorities at this time, be it noted, were pleased to record that the Body-Guard was not kept for mere show, but were capable of rendering useful service in the field; so much so, that in 1803, they were again on service, against the Rajah of Berar and the reduction of his stronghold at Cuttack.

1804 brought the establishment for a time of a mounted band—the first ever formed in India. The same year the Mahratta War developed, which resulted in the Corps' departure for service under Lord Lake against Scindia. On their return, pistols were introduced as a side arm for the first time.

As a result of the appearance of French men-of-war in the Bay of Bengal in 1809, the Corps were embarked as Marines in the Company's ships, and they thus share with the British 17th Lancers the distinction of being the only cavalry of the King-Emperor ever to perform that duty.

The continued interference with the Indian trade by the French, which was facilitated by their bases in the Dutch East Indies, led to the necessity for an oversea expedition against the island of Java. The forces were led in person by Lord Minto, the then Governor-General, and were found from the Madras and Bengal armies. Assembling at Malacca, they landed on the western coast of Java in July, 1811, and on the 8th August, H.M.'s 22nd Dragoons and the Governor-General's Body-Guard captured Batavia. The Body-Guard took part in all the subsequent actions against the French and Dutch, thus earning the first battle honour on their standards and appointments of "Java".

Lake's operations against the Mahrattas having had little lasting effect, further sustained efforts to break the power of the Pindaris were considered necessary, and from 1817 to 1818 the Body-Guard were on active service against the Mahratta States. They suffered severely in this campaign from cholera, and at one time had half their number down.

Minor activities between 1818 and 1824 ranged from an expedition against the Larka Kols—a fearless, war-like tribe, armed with bows, arrows, and battle-axes, inhabitants of Orissa which no Raja had been able to subdue, but against which a successful cavalry charge was made; to the suppression of a mutiny at Barrackpore in the 47th Native Infantry, who had refused to embark for Burma.

Owing to the unprovoked aggressions by the Burmese Governors of Arracan, and the silence of the Burmese King of Ava to the British representations, Lord Amherst, the then Governor-General, declared war; and an attack on Rangoon was ordered. No cavalry at first went with the expedition,

TROOPER, THE GOVERNOR-GENERAL'S BODY-GUARD, 1815

but the advisability of attaching some mounted men soon became evident. The campaign necessitated an oversea journey, then difficult to get Indian troops to undertake, but the Body-Guard volunteered their services and were duly embarked. One of the transports was completely wrecked *en voyage*, resulting in the loss of brave men and many valuable horses. It is recorded that the gallantry of the troopers, both Brahmins and Mohammedans who had never before seen the sea, was superb. The Corps took part in all the important actions of the first Burmese War, one squadron successfully charging a column of seventeen war elephants supported by infantry columns, a feat requiring great courage and perfect horse control. On another occasion their prompt action resulted in the capture of several guns. The war ended in 1826, after which the honour "Ava" was granted. It is noteworthy that the Commandant of the Corps returned from Rangoon to Calcutta in the *Enterprize*, the first steam-ship to reach India from England.

Between 1826–1845 the Corps was employed in escorting the Governors-General of the time in the lengthy tours they made through India; tours which sometimes lasted two to three years. Lord Amherst to Oudh, Agra, and Bhurtpore. Lord William Bentinck to Benares, Allahabad, Cawnpore, Meerut, Mussoorie, Saharanpur, Rupar, Karnal, Ajmere, Delhi, and Gwalior—Lord Bentinck was actually away from his capital at Calcutta for a little over three years. Remember there were no railways or other modern conveyances, nor were there telegraphs or wireless. Lord Auckland, the next Governor-General, made a similar tour lasting two years, as far as Lahore. Lord Ellenborough, his successor, to Ferozepore, where he reviewed the troops on their return from the first Afghan War. Here it was ordered that admission to the Body-Guard in future was to be an honourable reward for good service in the regiments of the regular cavalry. Whilst at Ferozepore, Lord Ellenborough's notice was attracted by the record for valour

during the war of a young officer, so much so, that as a mark of appreciation he appointed him to the Body-Guard. The officer afterwards became a well-known figure in the person of Field-Marshal Sir Neville Chamberlain. Lord Ellenborough subsequently said at a public dinner: "With Mayne and Chamberlain in the Body-Guard, I would face the devil"—Mayne was the Commandant.

The death of the Maharajah of Gwalior and the consequent troubles over the succession, resulted in intervention by the paramount power, and Lord Ellenborough—still on his tour—ordered a force into Gwalior, which he himself joined with the Body-Guard in December, 1843. The Maharani's forces opposed the British entry, and battle was offered at Maharajpore and Punniar. The Body-Guard took part in the former and earned the honour bearing that name; each man receiving a reward of Rs.18.

Lord Ellenborough returned to Calcutta in January, 1844, but it having long been realised that a trial of strength between the British and the Sikhs could not much longer be delayed, only two troops returned with him, the remainder of the Corps being left to watch the frontier.

Hostilities broke out in the December of 1854 and the Body-Guard was seen in the thick, taking part in the cavalry charges at Moodkee, Ferozeshah, Aliwal, Sobraon; for all of which honours are borne. The Body-Guard subsequently escorted the Governor-General to Lahore, where peace was signed with the Khalsa.

After the first Sikh war the Corps occupied for the first time its present summer quarters at Dehra Dun, and that war was the last major campaign in which the Corps as such, took a part. They were, however, employed in putting down the Santhal Rebellion in 1855, when they were ably commanded by one Captain Rattray, an infantry officer, and later the founder of the battalion of famous Sikh infantry which still bears his name.

During 1857 the Corps were employed in disarming the mutinous native infantry at Barrackpore, and from then on, until the end of the troubles in 1858, in training and taking up remounts for British cavalry employed in suppressing the mutiny; a duty, no light task, as the marches were long and through disaffected districts.

The Body-Guard escorted Lord Canning to the grand Durbar at Allahabad where, on the 1st November, 1858, was proclaimed that the future government of India would be by the Queen, the title of Viceroy being conferred on the Governor-General. The Viceroy returned to his capital by road, and for the first time in their history, the Corps were entrained at Raneegunj, and brought into Calcutta by rail.

At the outbreak of the Great War, Lord Hardinge, the then Viceroy, offered his Body-Guard as Divisional Cavalry for the Meerut Division, then under orders for France. But the proposal fell through, as it was considered that better use could be made of the Corps as a remount training centre. Their standard of equitation being of the highest class, they would have been wasted as divisional cavalry in continental warfare, and so they remained throughout the war as trainers of raw remounts for the cavalry and artillery. A detachment of men were, however, sent to Skinner's Horse, and served with distinction in that regiment in France from 1914 until 1916.

Recruitment to the Corps has varied throughout its history; originally Muslims, then Madrasis; while later nearly all the martial classes were at one time enlisted to its ranks. Transference to the Corps as a reward for good service from other cavalry units was not successful, as the men were either reluctant to leave the regiments, or the regiments loath to let their best men go. To-day, men are directly enrolled for the whole of their service. The establishment is one squadron of four troops, two being of Sikhs (first enlisted in 1884) and two of Punjabi Musalmans (enlisted since 1887). The Commandant and Adjutant are seconded from the Indian cavalry.

The present uniform of scarlet frock coat with blue facings, plain lancer girdle of yellow and red, white breeches, long Napoleon boots, and white gauntlet gloves, with a dark blue and gold pagri has not always been worn. The first uniforms were on the style of that of the European cavalry then in general use; tight-fitting hussar tunics, shakoes, leather stocks; uniforms not only unserviceable for India, but detested by the wearers. White frock-coats with blue piping are worn in the summer; and when on dismounted duty inside the Viceregal establishments, white *chust pyjamas* with brass-buckled black shoes. British officers wear a scarlet tunic with blue facings embroidered with lotus leaves in gold, gold-laced pouch belt, aigullette, white breeches, Napoleon boots, and a white helmet. The officers still wear the sabretache on their swords—an article of equipment after the style of the satchel introduced into the British Army during the Peninsula War for the purpose of carrying documents. The sabretache is heavily embroidered in gold with the regimental crest and battle honours, and they are the only Corps in the armed forces of the Crown who still wear this ancient accoutrement.

The horse furniture of the Corps is striking, they still retaining the old Hanoverian bossed bit, while the British officers have a blue cloth shabraque—a saddle and wallet cloth—edged with gold lace bearing the crest and battle honours, with a panther skin as a holster flounce. The other ranks have a blue cloth shabraque edged with yellow lace, and a saddle cover of white sheepskin after the pattern of the Life Guards. The Viceroy's Body-Guard is the only Corps in the army where the shabraque is used by the other ranks.

The lance was introduced in 1865, and this, with a curved sabre of Indian pattern, completes the arms worn in ceremonial dress. Contrary to popular thought the Body-Guard undergo regular training and are armed and equipped for field service on the same basis as the cavalry of the line. They are, however, primarily for the duty of guarding the Viceroy, and as their motto "Auspicio Regis et Senatus Angliae" implies, do so by authority of the King and Parliament.

The Corps has no standard to-day, but on State occasions silver clarion trumpets presented by Lord Reading are carried. The trumpet banners are blue, one bearing the devices of the Star of India, the other the coat of arms of the ruling Viceroy—each Viceroy presents his own. The trumpeters mount white horses, the remainder of the Corps, bays with black points.

Many distinguished officers have served in the Viceroy's Body-Guard, two reaching Field-Marshal's rank; Sir Neville Chamberlain already referred to, and Lord Birdwood, who was Adjutant of the Corps from 1893 to 1898.

The Body-Guards of the Governors of Madras and Bombay were originally raised for the protection of the Governors of these Presidencies. Madras is the oldest, having been founded in 1778, and like the Viceroy's Body-Guard, were originally a small body of European troopers. This Corps saw service on escort duty in Persia and in the Mysore War, though like the senior Body-Guard, they bear no honours for that campaign. They fought with distinction in the Mahratta War, their charge with that of the Bengal Cavalry, being the decisive factor in the battle of Seetabuldee, in relief of the Nagpur Residency (26th and 27th November, 1817). This is the only honour borne on their appointments.

Ten years later, when on tour, Sir Thomas Munro, then Governor of Madras, died from cholera, and on the infantry of his escort refusing to carry his remains to the grave on the grounds of caste, the troopers of the Body-Guard, men of the highest caste, volunteered to do so, they claiming it a privilege. The Madras Body-Guard was also a remount training centre in the Great War, and on several occasions were called out to control local disturbances; while the unit has the distinction of having been under German fire in India during the Great War, they patrolling the Madras beaches during the bombardment by the *Emden*. Their ranks are composed of

Rajputana Rajputs, and Jats from the United Provinces and the Punjab.

The Bombay Body-Guard was raised from men of the old Southern Mahratta Horse, when that famous body was disbanded. To-day the unit consists of Ranghars and Sikhs. The junior Corps of all, that of the Governor of Bengal, comprised of Punjabi Musalmans and Rajputs, was formed of volunteers from Indian cavalry regiments in 1912, consequent upon the transfer of the capital of India to Delhi, and the elevation of the Bengal to a Presidency Governorship. Both these units were remount centres in the Great War, as well as suppliers of drafts of men to cavalry regiments on service, an honour also shared by Madras. All the Provincial Body-Guards follow the Viceroy's in the matter of uniform, though none achieve the plain simplicity of the senior, they all wearing additional articles of dress, such as cummarbunds and plastrons.

The Presidency Governor's Body-Guards have a smaller establishment than the senior Corps; they are, however, all trained cavalrymen and form part of the military internal defence forces, and rank senior to the cavalry of the line.

CHAPTER IV

The Nepal Escort

NEXT in the Indian Army List after the Body-Guards come the little known Corps of the Escort to the British Envoy in Nepal, raised at Dinapore, Bihar, in April, 1816, on the establishment of a Residency in the Gurkha territories consequent upon the termination of the Nepalese War by the victories of General Ochterlony. The Governor-General-in-Council decided that an independent Escort of the same size as those attached to the Resident at Nagpore and at the Court of Scindia was necessary for Khatmandu. The Corps, with an establishment of one complete company, was thus formed from volunteers from the 18th, 21st, 22nd Native Regiments and the Champaran Light Infantry. Lieut. H. Boileau of the 18th Regiment was appointed to the command, which was at the time described as comprising the *élite* of the army of Bengal. To-day the Escort remains in the Army List as one of the oldest units of the Army of India. They thus form a link with the time which marked the earliest beginnings of the close relations with a country, the outstanding feature of which has been a loyal and lasting friendship, in whose cause Nepal and its Gurkhas never failed to make the greatest sacrifices.

The seventy-five men comprising the present establishment, are all recruited from British India, principally from Bihar, the United Provinces and Rajputana; all are Brahmins or Rajputs, a provision desirable in a purely Hindu country. Enlistment is for whole service, pensions on the Indian Army scale are granted upon termination, while all are eligible for the honours and rewards available to the Indian soldier.

The Corps comprises both cavalry and infantry, the former escorting the Minister—whose full title is His Britannic Majesty's Envoy Extraordinary and Minister Plenipotentiary at the Court of Nepal—on occasions of state; while the latter provides the Quarter-Guard over the Treasury, Armoury, and Magazine, with a Guard for the Legation at night. Owing to the friendliness of the Nepalese Government, single sentries only are posted, as opposed to the double posts usual in British India. The Minister tours considerably and escort duty in the Nepal Terai makes heavy work; thus the duties of the Corps are by no means light when it is considered that they undergo musketry, and all the similar parade duties of the average Indian sepoy. In fact, their musketry standard is of a very superior order—a factor alone which speaks for their high standard of training—and they have of recent years been winners or runners up in competitions of the Army Rifle Association of India with other regular units. The officers are all Indians, the Minister being ex-officio Commandant, training being supervised for six months of the year by an officer from one of the Gurkha Regiments.

In 1857–58, when a large portion of the land forces became disaffected, it was stated that: "a finer or better behaved set of men exists not in the Army, while their service has considerable drawbacks, such as slow promotion, high prices, want of society, and the necessity of the most irksome caution in all their intercourse with the people of the country." The only one of these drawbacks since alleviated, is that of dearness of food, for which they now receive compensation; while the relations which exist between the Escort and the Nepalese are of the friendliest nature.

The Corps earned the commendation of the Government of India by the excellence of their conduct during the revolution in Nepal in November, 1885, when certain members of the ruling family took refuge in the British Residency.

The Escort being a local corps, recruitment is only for service inside Nepal, and thus on the outbreak of the Great War in 1914 they, to a man, volunteered for foreign service. Although too small to be of use alone as a fighting unit, it was urged that they would creditably acquit themselves in a theatre of war. The Government of India, while thanking them for their keen sense of loyalty, decreed they were to remain in Nepal, and thus to their regret, they had no opportunity of attaining merit with their other comrades in arms on one of the many fields of service.

Like the Body-Guards they still wear full dress, which is red with blue facings, blue breeches, and a blue and gold pagri; uniforms, which have to last for twenty years.

CHAPTER V

Skinner's Horse
(1st Duke of York's Own Cavalry)

"Bhurtpore", "Ghuznee, 1839", "Afghanistan, 1839", "Khelat", "Candahar, 1842", "Maharajpore", "Moodkee", "Ferozeshah", "Aliwal", "Kandahar, 1880", "Afghanistan, 1879-80", "Punjab Frontier", "Pekin, 1900", "France and Flanders, 1914-16", "N.W. Frontier, India, 1915", "Baluchistan, 1918", "Afghanistan, 1919".

SKINNER'S HORSE—the old "yellow boys" of Scindia—are the only Corps in the Armies of the British Commonwealth to wear a canary coat.

The third year of the nineteenth century brought a nasty shock to the French-trained regular troops of the Mahratta princes. They were defeated—a novelty for them—and after Lord *Lik Sahib* had rubbed it in at the gory battle before Delhi, they decided to change their employers, and thus offered their services to the victorious general. Lord Lake gladly accepted them; and their conditions—never to be led against Scindia, and to be allowed to choose their own commander. Thus James Skinner, renowned among them for his prowess, famed for

his bravery, son of a Scot and a Rajputni prisoner-of-war, returned to lead his "yellow boys" as he had led them in a hundred fights before. Dismissed from his service when that Prince of Mahrattas, Scindia, decided to cross swords with the British trading company's forces; refused re-employment by the former's Latin generals when his great power of leadership might have saved them the day; in disgust Skinner offered his sword to Lord Lake; and the canary *risalahs* claimed him for their own.

Many then were the tales of old James; the saviour of Scindia's life; the boy put to the printer's trade to become a hero; the man left for dead with the bodies of 900 of his 1,000 troopers, who alone had remained loyal to their master. Many more are told of him after he entered the Company's service; this, however, is a brief narration of the illustrious regiment that bears his name, not the story of old *Sikandar* of Hansi, the builder of the ancient Church of St. James that overlooks the ruined ramparts of old Delhi.

Skinner's Horse, as they came to be known, soon covered themselves with glory under their new flag. At the fortress of Malaghur near Meerut, after twice being repulsed in the attack, a small band of the "Yellow Troop" once more returned to the fray, fell upon the enemy, captured his guns, and cut down to a man the opposition infantry. Eight hundred of them surprised the Sikh chiefs about to join hands with the Holkar's armies, by a sudden descent upon 5,000 of their followers at Saharanpur, with the capture by Skinner of all the leaders.

They harassed Holkar, and brought about his final defeat. A pursuit lasting 500 miles, when horses were rarely unsaddled, the men slept under arms; and the daily going averaged thirty miles, with men and horses subsisting on the standing crops.

Amir Khan, being the next competitor in the ring with the British, Skinner was again to the fore. He entered the enemy's camp in disguise, discovered their plan, and chased Amir

TYPES OF SKINNER'S HORSE
Hindu Rajput, Hindu Jat, Musalman Rajput

Khan 700 miles back to his own territory. In these operations, Skinner's Horse were the only Indian corps employed throughout.

During the course of these wars, Lake had assured the Corps of continuous employment in the forces; but, as usual, on a termination, what the Military Commanders promise the Civil fail to keep, and the temporary and effete Governor-General, Sir George Barlow, ordered their disbandment, greatly to Skinner's disgust. He was, however, permitted to keep on a few as a nucleus of personal retainers, and by 1809, when the Skihs threatened hostilities, the Corps was again raised to 800. For five years they patrolled incessantly the Sutlej frontier.

In 1815, the Corps had expanded into three, each a thousand strong, with Skinner's brother, Robert, as second-in-command. They then began to make themselves well known to those terrors of the countryside of middle India, the robber Pindaris.

Conditions by 1819 enabled the 3rd Corps to be disbanded; while to appease Robert Skinner, the 2nd Corps was detached and the command given to him. The two Corps were brigaded for a time under the elder Skinner in 1838, but from then on, each had a separate existence and went different ways, until their re-junction over one hundred years after.

The 1st Corps assisted Combermere at the second siege and capture of Bhurtpore, James Skinner receiving the Companionship of the Bath; but the rules of that honourable Order, however, required the recipient to be a Lieut.-Colonel in the British Army; *Sikandar* was thus made so, which settled the question of his rank, long a sore point; he was also promoted local Brigadier.

Major Robert Skinner's death in 1821, gave the vacant command of the 2nd to a Captain Baddeley, and the 2nd in consequence became known as "Baddeley's Frontier Horse". As such they marched to Kandahar in the war of 1839, participating in the captures of Ghuznee and Khelat on the way.

A detachment remained with the ill-fated army of occupation, the Regiment returning to take part in the Gwalior excursion; where at Mahrajpore a weak troop routed a Mahratta square, a standard and two guns falling to their charge.

Two years prior to this fight, a detachment of the 1st Corps had marched also to Afghanistan, earning an honour for their part in the fighting around Kandahar in 1842; the severity of which is evident from their casualties—108 to fall on the field for ever out of the 180 engaged.

James Skinner had died the year before; had he lived how he would have revelled in such glory. He was buried in great splendour at Delhi, attended by both Regiments of his Horse.

Moodkee, Ferozeshah, and Aliwal, battles of the first Sikh War gave employment to the 2nd Corps in 1845.

During the construction of the Ganges Canal in 1854, the 1st Corps were employed to patrol the banks in the prevention of sabotage, and the yellow coats seemed to impress the countryside, for when the civil department concerned took over the duty they dressed their own men similarly. The custom still continues, and this is the reason for the yellow garment still worn by minor officials in the Irrigation Department of the United Provinces to-day.

When revolt crept over the army of Bengal, the 1st were at Multan; they were employed in putting down insurrection along the banks of the Ravi, a party greatly distinguishing themselves at Chichawatni. The 2nd saw service in Oudh and Bundelkand.

Both were in Afghanistan, 1879-80. A trooper of the 1st saving the afterwards great Lord Roberts' life in the action at Kila Kazi. The 2nd accompanied the force in the Khyber to Kabul, from whence they marched with Roberts to Kandahar and victory.

No further campaigning came the way of either corps of Skinner's until the 2nd joined the fighting on the Frontier in 1897-98.

TRUMPETERS, SKINNER'S HORSE

Known to the world as the 1st Bengal Lancers, they made brilliant charges, cutting up the Tartar cavalry, with the capture of three standards and some cannon during the Chinese Expedition of 1900. They also made a joint mounted attack with American cavalry; the first time on a field of battle where the forces of India and the U.S.A. were allied together.

The 2nd had the luck in 1914, when they proceeded to France with the Meerut Cavalry Brigade, to take a part, sometimes great, sometimes small, but never unworthy, in all the actions of the 2nd Indian Cavalry Division, until their transfer back to India for service in Mesopotamia in the summer of 1916. The banks of the Euphrates never saw them, however, for they remained in India for frontier duty, whereat, in the defence of Gumbaz Post on the 19th and 20th February, 1918, eighty men of the Regiment showed such spirit against repeated attacks by tribesmen, that the action was merited by inclusion in the official lists of battles and other actions of the Great War.

The 1st remained throughout the war on the North-West Frontier and, despite disappointment at not seeing service in other spheres—a disappointment shared with Britain's junior cavalry regiment, the Empress of India's 21st Lancers—fully maintained its reputation as the premier cavalry regiment of the King-Emperor, in the many border actions of those years.

Both Corps "put up good shows" in the subsequent trial of strength between Amanullah and India in 1919.

The 2nd Corps after their separation, adopted at first a red uniform, subsequently changed to blue. The 1st had always stuck to "Sikandar's fancy", and the 2nd came back to wear the yellow coat and black facings of the 1st after over a century of separation, at Sialkot in the spring of 1921.

The subsidiary title of the Duke of York's Own arises from 1899, when His Royal Highness became Colonel-in-Chief of the 1st Corps, an office which he held until his death as His Majesty King George V. in 1936.

Skinner's Horse (1st Duke of York's Own Cavalry)

The motto of the Corps adopted since the amalgamation is "Himmat i mardan madad i Khuda" or "The bravery of man is by the help of God".

The Regiment was mechanized in 1939.

The modern Indian cavalryman—a sowar of Gardner's Horse in hot-weather active service order.

CHAPTER VI

The 2nd Royal Lancers (Gardner's Horse)

"Arracan", "Sobraon", "Punjaub", "Mooltan, 1857-8", "Afghanistan, 1879-80", "Tel-el-Kebir", "Egypt, 1882", "La Bassee, 1914", "Givenchy, 1914", "Neuve Chapelle", "Festubert, 1915", Somme, 1916", "Morval", "Cambrai, 1917", "France and Flanders, 1914-18", "Egypt, 1915", "Megiddo", "Sharon", "Damascus", "Palestine, 1918", "Tigris, 1916", "Mesopotamia, 1915-16", "Afghanistan, 1919".

THE 2nd Royal Lancers are a combination of two of the oldest regiments of the Bengal Army, which on their amalgamation in 1922, were the 2nd Lancers (Gardner's Horse) and the 4th Cavalry.

Gardner, originally an officer of Highlanders who had early left the Queen's service for a more adventurous career with the Mahratta Princes, when threatened with execution for refusal to proceed against the British, escaped, joined the Company's forces under Lake, and subsequently was ordered to raise, in 1809, the Corps which now bears his name.

Primarily required for police duties in the newly occupied territories around Agra, the Corps were irregular to the last degree. Colonels received a lump sum for the unit in those days, and Gardner paid out strictly according to ability, a first-class swordsman receiving as much as Rs. 150 per mensem, His reputation for fair dealing quickly spread, and the Corps attracted the best type of recruit from the surrounding states. A picturesque and medieval lot they must have been, dressed in silver-embroidered emerald green coats and red pyjamas. and armed with long matchlocks, curved tulwars, lances, and shields; a uniform which, save for the breeches which later became yellow, was used for over eighty years.

Gardner's Corps were first engaged by detachments in the Nepal War of 1815. A war rendered necessary by the frequent inroads by the Gurkhas into Indian territory. Gardner himself led the column which invested the stronghold of Almora, an advance of extreme difficulty, no communications other than village to village foot tracks existing in the densely wooded Himalayas. No coolies would face the Gurkhas, and every military necessity, food, and ammunition, had to be carried on the backs of soldiers.

For its meritorious services against the Pindaris, 1817-19, the Corps came on to the establishment and towards the end of the latter year, were employed on the eastern frontiers invading Arracan. In this Burma campaign they earned great fame, marching 2,000 miles without having a single case of desertion, while the infantry were leaving the ranks, it is said, by hundreds. After losing most of their animals they cheerfully fought with the line on foot, then considered very degrading by mounted men. They are the only regiment to bear the honour "Arracan".

Colonel Gardner, who had given up the command in 1828, died in July, 1836. He had married forty years before, a thirteen-year-old Princess of Cambay. Their marriage was both happy and ideal; their son wedded into the royal house of Delhi, and their descendants still form a little Christian community at Khasgunj, the family estates near Agra.

The 4th Cavalry have their origin in a cavalry regiment raised under the Company's orders for service with the King of Oudh in 1838. They were later transferred back to the Bengal Army proper and saw much service in Scinde and on the Baluch frontier in 1844, for which they received an Honorary Standard bearing the device of a lion. They were then the 6th Regt. of Bengal Irregular Cavalry; they became the 4th in 1861; from 1900 to 1903 they were "Lancers".

The 2nd fought with credit at Sobraon and in clearing the Jullundur Doab in the Sutlej and Punjab expeditions against the soviet armies of the Sikhs in 1846 and 1848. In 1855 they were in Bengal quelling the Santhal Rebellion.

Both Regiments were staunch and untainted when revolt crept over the Bengal Army. The 4th were employed against the renegades at Multan, several men earning Orders of Merit. Gardner's, who had previously been on several minor Frontier shows against the Mohmands and other clans, cleared the Gugera and Gurdaspur Districts, where many exciting small combats occurred.

A long period of quiet followed until 1882. Then the 2nd went off to Egypt, to fight dismounted at Kassasin and mounted in the decisive battle of Tel-el-Kebir. After their return from Egypt the old green uniform was changed to blue with light blue facings, and a blue pagri; the dress of the Regiment to-day. The 4th prior to the amalgamation wore red. The officers of the 4th, for a good many years *circa* the mutiny period, wore as a headdress an almost exact copy of the helmet of the Prussian Guard complete with the double eagle; a distinction conferred by the Prussian King, as Prince Waldemar of Prussia had been attached to them during their stay in Sind.

When what was thought to be Armageddon began, the 4th, of the two regiments, was the first to go; proceeding to France in 1914 as the Meerut Divisional Cavalry. At Festubert, being

THE STANDARD OF THE 2ND ROYAL LANCERS (GARDNER'S HORSE)

the last reinforcements available, they were rushed into the fray, and, with the 2nd Black Watch, held the line against all comers, fighting in the trenches hand to hand with their lances. They transferred to Iraq in the winter of 1915, taking part in the actions to relieve Kut, and finally arrived back in India in the winter of 1917 with their ranks greatly depleted by scurvy, owing to the unsuitability of the rations issued.

The 2nd were bemoaning, as an internal security unit, when the decision to send a complete Indian Cavalry Corps to the western front set them packing, and by the end of November, 1914, they were well in the Red Sea, *en route* to France as part of the Mhow Brigade of the 2nd Indian Cavalry Division.

How much importance was attached to the automatic gun as a cavalry weapon at that time is evidenced by the fact that those issued to Gardner's Horse for service in Europe were almost twenty-five years old; while their tripods, even when clamped down with a spanner, collapsed after firing twenty rounds. The men went to France with curved tulwars and lances, great faith being placed on the latter. They arrived at Orleans, dressed in cotton drill, to camp in snow.

The Somme actions and the hopes of a great break through gave them their first serious employment. At Cambrai in the Tank attack, and in the subsequent counter battle, they did sterling work, and were the only Indian Regiment to be mentioned by name in Lord Haig's Despatch. One of their number, Lance Duffadar Gobind Singh, received the Victoria Cross for the courage and resource shown in carrying messages mounted in full view of the enemy; three horses were shot under him that day, one being cut in half by the direct hit of a shell.

After Cambrai, France saw them no more. Reorganizing on their old battlefield at Tel-el-Kebir, where they changed back to drill with caustic comments on the efficiency of the rearward services in Egypt as compared to France, they proceeded to the Jordan Valley.

In the night advance through the Musmus Pass, September 19th, 1918, the success of which was the keystone of Allenby's plan, the 2nd Lancers had the honour of leading the way. Surprising the enemy piquet found around the camp fire on the summit, they continued to lead, and by daybreak were through, but emerged to encounter a Turkish battalion specially sent to hold the Pass. The enemy were promptly attacked, and the rapidity with which the operation was planned, and the yell which preceded the charge, completely unnerved the Turk. Hands went up, and with the loss of only one man and twelve horses wounded, a Turkish battalion over 500 strong had been put completely out of action.

The 2nd shared in the remaining victories of the greatest cavalry drive in history, and after the Armistice were employed in the pacifying operations in Palestine and Syria.

While at Beirut in 1919, Trumpeter Mangal Jain won the peace time "Victoria Cross" in shape of the Albert Medal for his wonderful daring in rescuing three drowning soldiers from the treacherous surf. He was the first Indian to earn this decoration.

The 2nd Lancers returned to India in 1920 and amalgamated with the 4th Cavalry in April, 1922, at Bombay.

The title "Royal" was bestowed on the occasion of the Silver Jubilee of H.M. King George V.

General H. H. The Maharajah of Bikaner is Honorary Colonel of the Regiment.

CHAPTER VII

The 3rd Cavalry

"Mooltan", "Punjaub", "Afghanistan, 1878-80", "Mesopotamia, 1917-18"

Skinner's Horse, Gardner's Horse, and the 3rd Cavalry are the only regiments remaining which had their beginnings in the old irregular cavalry of the East India Company's Bengal Army. The two partners of the marriage in 1921 were then the 5th and the 8th Cavalry, both of which numbers were original. The 5th were raised as a consequence of the Afghan War, at Bareilly in 1841; the 8th after the first Sikh War, at Sultanpur in Oudh, five years later.

The senior Regiment first saw serious fighting before Multan in 1848, a detachment being hurried into trenches before that fortress, when friendly Sikhs went over to the rebel defenders, the Regiment subsequently making a most gallant charge, capturing many guns and prisoners.

Both Regiments were at Peshawar in 1857, and, being of undoubted loyalty, were employed disarming and rounding up mutineers, and, but not least, keeping incessant watch on the frontier, assisting in the many minor operations which went on all through that critical time.

The 5th came south in 1864 to operate against the Bhutanese

A State Trumpeter of the Indian Cavalry at The Delhi Durbar, 1902

in the difficult and unhealthy tea-garden country of the Dooars; a brief campaign notorious for the fever and cholera casualties. The year previous they had also sent volunteers to complete the 12th Bengal Cavalry when that regiment was ordered on active service in Abyssinia.

About this time much humour was raised by the issue of an army order that cavalry horses must be trained to jump hedges, ditches, walls, and other obstacles. It is also of interest to read in the records that twelve years previously carbines were obtained from England, at a cost of Rs.26 each, to replace the old match-locks, and in consequence the Commanding Officer lost his "match" allowance of half an anna per mensem.

To retrogress, the years 1854–56 are noteworthy; the first seeing the outbreak of the Crimean War between Britain and France, and Russia, service for which the 8th Cavalry volunteered to a man. The Government, while thanking them, declined the offer; but nevertheless ordained in the latter year that all recruits in future must be sworn to serve in any part of the world. Previously when oversea expeditions were contemplated, volunteers had been called for.

The Afghan wars of 1878–1881 gave much difficult escort duty and mounted skirmish. The 8th operated through the Khojak Pass to Kandahar; the 5th in the Khyber column to Kabul. These campaigns were severe on cavalry; water was scanty, long marches had to be made over a countryside of sharp stones and boulders devoid of human habitation, rations were short, while snow and ice added often to the discomfort of horse and rider.

From the time of their return to India after their honours in Afghanistan, neither Regiment was privileged to see active service; nor were they lucky enough to be included in the first contingents from India into the war arenas of 1914. Nevertheless, they had, despite their long period of absence from active warfare, maintained the highest standard of efficiency, and they sent many hundreds of men, as well as officers, to units

in France, Egypt, and other theatres. Many distinguished themselves in these fields, which is a testimony to the training they received in the two Regiments destined as such to remain behind.

The 5th were eventually the more fortunate of the two; as at the end of 1917 they were ordered to Mesopotamia to take part in the operations protecting the oil fields, and subsequently in Persia itself. Two squadrons on one occasion covered 250 miles in ten days, through desert, without a man or horse falling out. The Arab Rebellion gave them brilliant opportunities of which they never failed to take advantage, capturing a standard and four hundred prisoners.

The 4th had their share of oversea service in 1920, being sent to garrison Palestine. No ordinary garrison work, but as arduous as any frontier campaign, and more so. One squadron charged 1,500 armed Arabs with decisive results.

The present uniform is blue, but it was not always so. The 5th on their raising, wore a collarless red coat, and a hussar busby of green cloth, with a red bag and yellow lines. The red coat continued until the amalgamation, when the blue worn by the 8th was the choice for the new Corps, the facings becoming primrose; the facings of the 8th were red.

Junction was effected at Secunderabad in 1921. Eleven years later the 3rd Cavalry, as they had then become, were selected for the experiment of Indianization, whereby all the officers commissioned to the Regiment in future will be Indian gentlemen who have passed through Sandhurst or Dehra Dun, and the troops and squadrons officered on the British model.

The 3rd Cavalry, like all the regiments of horse which remained in the post-mutiny Indian forces, the three Madras cavalry regiments excepted, were raised and maintained on the *sillidar* system. Under this almost age-long method of India, for a lump sum down the recruit brought his own horse, arms and equipment, which he maintained from a pay more generous than that given to the soldier clothed and equipped direct by the Government. The more wealthy *sillidars* were allowed to enlist countrymen not so well off as themselves, and for whom the *sillidar* provided the necessary horse and equipment. The latter men were known as *barghirs*, and since their maintenance in a regiment was at the expense of their patrons, the latter received a regulated share of the pay of their nominees.

As time went on, military requirements demanded a better standing of mounting and equippage than the *sillidar* was able to supply. The system was modified. The recruit no longer received the lump sum, but instead he himself had to produce money, known as the *assami*, in return for which horse and equipment were provided under regimental arrangements. A portion of the man's pay was taken for upkeep and replacement. On the soldier's discharge the *assami* was returned in full. The horse, etc., remained with the regiment.

This modification of the original system became very popular and was much cheaper to the country. Many regiments were managed in a fashion with which commerce could vie, while the men, being shareholders in a very going concern, took a lively interest in the administration.

In war the Government took the monthly deductions of pay and in return maintained the regiment in the field, and at the close of the campaign returned it to peace conditions, complete with serviceable horses and equipment.

Prior to 1914 the system had stood the test of past wars extremely well. But the Great War proved to be too much for it. Each regiment had been allowed to maintain its own pattern of saddlery, clothing and equipment. Replacement proved too great a problem. At the same time, men discharged, and the relatives of men killed, could not readily obtain a return of their *assamis*, owing to the fact that recruits enlisted during the War had not to find this money, with a result that the regimental funds were called upon for heavy expenditure

which they could not meet in the absence of fresh income, which the Government failed to provide.

The system had outlived its usefulness. It was abolished, after much discussion of other expedients, in 1920. Now all Indian regiments are maintained and administered on the European or regular model.

CHAPTER VIII

Hodson's Horse (4th Duke of Cambridge's Own Lancers)

"Delhi, 1857", "Lucknow", "Abyssinia", "Afghanistan, 1878-80", "Suakin, 1885", "Chitral", "Punjab Frontier", "Givenchy, 1914", "Somme, 1916", "Bazentin", "Flers-Courcelette", "Cambrai, 1917", "France and Flanders, 1914-18", "Megiddo", "Sharon", "Damascus", "Palestine, 1918", "Khan Baghdadi", "Mesopotamia, 1916-18".

THE audacity of the Regiment's first Commandant in obtaining the surrender of the escaping King of Delhi in 1857, and his personal execution the following day of the remaining members of the royal house thought to be the leaders of the revolt which swept the bulk of the Bengal Army out of the Indian Army List, which was for many years the subject of much bitter controversy, are doubtless the cause of "Hodson's Horse" being more widely known than probably any other cavalry corps of the Indian Army.

William Stephen Raikes Hodson, student of Rugby School and Cambridge University, son of a Christian Divine, subaltern

(*India Office Photo.*)

LIFE IN THE INDIAN ARMY—A PHYSICAL TRAINING DEMONSTRATION

of the Bengal Fusiliers, and possessor of a magnetic personality, was a man who did what he conscientiously conceived to be his highest duty, and who never at any time sought to avoid the responsibility or consequences of his actions. He died, as he would have wished, a soldier's death before Lucknow on the 13th March of the following year. He lies buried, near where he fell, in the grounds of La Martinere School.

The Regiment was actually raised in small *risalahs* or troops, principally in the Lahore and Amritsar Districts of the Punjab by prominent local Sirdars. As the troops were formed they were despatched to join the field force before Delhi, and in the camp behind the now historic Ridge were incorporated by Hodson into an irregular cavalry corps in the mid-summer of 1857. They were armed with matchlocks, swords, daggers or spears, just as each man fancied.

The credit of raising the first troop goes to an officer of the Punjab Police—Man Singh—who from thence for upwards of a double decade rendered to the Regiment services of great merit. As Sirdar Bahadur, Man Singh, C.I.E., he will probably be better remembered by the Sikh community, for he was manager of their principal shrine—the Golden Temple of Amritsar—for many years after his retirement from the Regiment until his death at a ripe old age.

After Delhi the Corps were marched to join the columns relieving the besieged Lucknow. There the gallant and intrepid Hodson laid down his life, while one of his lieutenants—afterwards Sir Hugh Gough—won a well-merited Victoria Cross.

A successor to Hodson was found in Major Daly of the Guides. Daly duly took over, to discover that his command, now some 1,200 strong, was without any form of pay roll, account books, or understandable records. Book-keeping had always been a bugbear to Hodson; and there had been serious trouble over similar matters, from which he had been afterwards exonerated, when he too had served with the Guides. Daly, it seems, found only Rs.16,000 in the Regimental chest,

while some Rs.1,20,000 appeared to have been drawn from the Government Treasury. Nobody exercised any supervision over the administration, and confusion reigned supreme. Nothing, however, was seriously wrong, and a complete account entirely satisfactory to the financial authorities was produced within one year.

The fables of rich rewards waiting on the Gangetic Plains for the stalwarts of the north had caused such an influx of recruits, that by 1858 the original command became so unwieldy that sub-division was necessary. Three regiments were thus formed. The junior was disbanded in 1860, the men nearly all going over to "Fane's Horse"—a forbear of the present 19th King George V's Own Lancers—then being formed for service in China.

The Regiment almost lost its identity the following year, the renumbering on the transfer to the Crown making the Corps the 9th and 10th Bengal Cavalry; these numbers were retained, although the titles were many times changed, until 1921. The two units were, however, more often than not known by their foundation name, which was in the end officially restored in 1901. It is as the 1st and 2nd Regiments of "Hodson's Horse" that their career will be traced, since, although from 1859 the two had maintained a separate existence, they were of common parentage, bore a common subsidiary name, and are to-day again one.

The Abyssinian Field Force of 1867, although found principally from units of the Bombay Army, comprised a brigade from Bengal. With the latter went the 2nd Regiment, they being towed in tiny craft by steamers from Calcutta to the Red Sea coast. The march of the Force from the port of Zula over most difficult country to Magdala—the then capital—supported by what were for the time almost perfect arrangements for supply and transport, justly deserves a high place amongst the achievements of the forces of India.

Both Regiments were also out of Hindustan during the

years 1878–80. The 1st with part of the 2nd, voyaging to Malta, to join an Empire force ready to act if Russia extended the activities of her then war with Turkey into Asia. The balance of the 2nd went on active service into Afghanistan, to take part in the second war with that country.

The rising in the Sudan, with the death of General Gordon in the summer of 1885, again necessitated an expedition to Africa. With the object of opening a route to the Mahdi's capital at Khartoum, an Indian force was sent to Suakin on the Red Sea. The 1st "Hodson's Horse" were the only cavalry employed; and although a great honour, it was also great hardship, as the force of some 18,000 men, and about half that number of animals, had to operate in a sandy desert of great heat, where every single drop of water had to be transported from the base; while the rations and fodder were both scant and poor. But the men of the 9th Bengal Lancers—their then official title—made no demur, and carried on with a cheerfulness which was a fine tribute to their regimental spirit.

The honours for the North-West Frontier outbreaks of 1895–97 were also earned by the senior Regiment. But they saw little serious fighting, being employed, as cavalry usually are in mountain warfare, by detachments.

Many men and large numbers of horses were sent for remount duties to South Africa, and many British officers for active service, during the 2nd Boer War, 1899–1902.

One non-commissioned officer—Dafadar Wadhawa Singh—became orderly to the great Lord Roberts, then Commander-in-Chief in the field, and remained with him throughout the campaign.

The Dafadar subsequently became State Mace Bearer to His Majesty the King-Emperor George V. He was also depicted as Lord Roberts' orderly in the pageant at the Royal Tournament in London during King George's Silver Jubilee year.

But the producers of the pageant made a mistake; for they showed the Dafadar as wearing a red *achkhan*. "Hodson's"—

except in the early days when a salmon coat brought them the nick-name of the "Flamingoes"—never wore red, but blue. The blue *achkhan*, or long coat, with red facings and cummarband, together with a blue *pagri*, is still the full-dress uniform, while the original *multani mutti* ochre-coloured breeches complete the kit.

Mention of the Royal Tournament also recalls that "Hodson's Horse" were the first Indian regiment ever to appear at what is one of the most popular entertainments in London, a detachment winning gold medals for tent-pegging and other military competitions in 1902.

To-day one hears much of officers being kept in command too long; thus one wonders what the present authorities would say to the case of Major-General Sir Charles Palliser. Appointed Commandant of the 2nd "Hodson's Horse" in 1859, he held a most successful tenure for twenty-two years—a record for the army.

Now to the deeds of the two Regiments in the conflicts of 1914–20.

The 1st were in France and Flanders' fields from November, 1914, until February, 1918. All the honours for those theatres were earned by them, they forming part of the Ambala Cavalry Brigade, whose mounted action at Cambrai was one of the effective cavalry feats of the western front. It was in this action that Captain Som Dutt, M.C., I.M.S., the medical officer of the Regiment, was decorated with the wearer's own Iron Cross—the highest German award for gallantry—by a badly wounded German colonel, whose injuries Captain Dutt had dressed whilst under heavy fire.

The honours for Allenby's great cavalry campaign in Palestine also were the 1st's; they having left France to join his command early in 1918.

Police work in Syria detained them after the *coup-de-grace*, and it was not until five years had passed since their departure did they again land on India's shores.

The 2nd Regiment's oversea exploits in the Great War did not begin until 1917, when the avengement of the disaster of Kut by the capture of Baghdad called for their services with Cassells' Brigade, the outstandingly dexterous tactics of which brought about the victory of Khan Baghdadi.

The general armistice of 1918 ,however, did not mean home for the 2nd; or for many other brave soldiers in the lands of the Tigris and Euphrates. Subversive forces and treachery soon caused the employment of the armed forces against the enemies of law and order.

The 2nd Regiment of "Hodson's Horse" fought many minor actions in the attempts to restore civil government in the country we now call Iraq. Of those frays, all will pass save one—the defence of the armoured train at Samawa on the 3rd September, 1920. Deserted in mistake by the escorting infantry, derailed and surrounded by thousands of hostile Arabs, the Dogra detachment of "Hodson's" manning the train fought on in those iron trucks in the Mesopotamian heat until the end. Why no relief came seems incredible, since the main camp of the column was just a mile and a half away. Yet due to some trick of tropical atmospherics, no sound of the fight, which lasted from early morning until late afternoon, was heard by any authoritative ear.

Later, when the almost unbelievable story was eventually proved unquestionably true, the grant of a posthumous Victoria Cross was recommended for the gallant officer-in-command—Captain D. P. B. Russell, M.C. But the grant was refused on the grounds of lapse of time; an excuse difficult to understand, since there had been ample precedent—a Cross earned in 1857 was not awarded until fifty years later. But reward or no reward, the memory of Russell and the gallant lancers of "Hodson's Horse" will for ever be remembered as one of those lonely epic defences which are the pride and glory of India's Army.

Order was at last restored, and the 2nd traced their steps

homeward, only to find that the peace and the need for retrenchment would require only one regiment to carry on the high ideals, and the by now much honoured name; the two thus re-united at Multan on the 3rd September, 1921, after individual existences for upwards of sixty years.

The subsidiary title of the Duke of Cambridge's Own dates from 1877 when H.R.H. became Colonel of the 2nd Regiment. As Commander-in-Chief of the British Army, he inspected them at Malta the same year. A first cousin of the Queen-Empress Victoria, he led a division in the Crimean War, 1854, and had a horse shot under him at Inkerman. A lovable creature of forcible language, whose inclusion of an umbrella with his full-dress uniform in inclement weather was always a source of amusement to the troops, he took a great interest in the Regiment until he died at the great age of eighty-five in 1904.

General Sir Arthur Power Palmer, G.C.B., G.C.I.E., Commander-in-Chief in India, 1900–1902, originally belonged to the Regiment.

It was an officer of "Hodson's Horse"—Major Osmund Barnes as Delhi Herald—who proclaimed Queen Victoria Empress of India, at the great Imperial Assemblage on the 1st January, 1877.

CHAPTER IX

Probyn's Horse (5th King Edward VII's Own Lancers)

"Lucknow," "Taku Forts", "Pekin, 1860", "Abyssinia", "Ali Masjid", "Peiwar Kotal", "Charasiah", "Kabul, 1879", "Afghanistan, 1877-80", "Chitral", "Malakand", "Punjab Frontier", "Mesopotamia, 1915-18".

WOULD the small force investing Delhi hold the insurgent garrison? Suppose the Bengal mutineers gain some spectacular success before law and order can be restored? Questions such as these one may be sure exercised the mind of the Punjab's Chief Commissioner in the summer of 1857. Realizing the situation unfavourable answers might bring, that officer—Sir John Lawrence—ordered south all the soldiery he dare spare. But that in his opinion would not be enough; new corps must be raised; and for them, what better material than the martial Sikhs to fill the fighting ranks? Thus, as a result of

(*India Office photo.*)

MECHANIZED INDIAN CAVALRY—A LIGHT TANK AND ITS CREW

Lawrence's orders, there were brought into being at Lahore the 1st and 2nd Regiments of Sikh Irregular Cavalry.

The command of the first regiment was given to a Captain Wale of the 18th Irregular Cavalry; that of the second to Captain Hockin of the 17th Irregular Cavalry. As soon as the units reached full strength they were at once sent down country, to the scene of active operations in what is now the United Provinces. The baptismal fight of the 1st, however, had already taken place in the September, between the revolting inhabitants of the Gugera tract between Multan and Lahore, who were endangering the communications between the two cities. The 1st covered the eighty miles from Lahore Cantonment in a single march, and beat off the rebels, causing them heavy losses. Many of the 1st had fought for the armies of the Khalsa against the British, and were thus not exactly on their tirocinium.

A description of the Sikh Irregular Cavalry at the time states them as being composed of every variety of horse, entire, mare, and gelding, from large chargers to ponies; with a similar miscellany of bits, bridles, saddles, and swords. Mounted drill was equally primitive, and it was all they could do to form *threes right or left*. No two of the men rode alike, and none had a *cavalry seat*; but they were undeniably horsemen, and there was never any difficulty in getting them to form some sort of line and to ride hard and straight at the enemy. Such was the 1st, or "Wale's Horse" as they had come to be known, when the Corps concentrated at Lucknow to take a prominent part in the operations before that city.

The 1st Regiment was not long under Wale; he, gallant soldier, being shot dead on March 1st, 1858, by a solitary rebel sepoy in ambush, whilst at the head of his men in pursuit of the fleeing e emy. To the command was appointed another brilliant officer who had won the Victoria Cross at Agra in the previous year, Major D. M. Probyn; it is his name the Regiment now bears, although such was not officially bestowed until 1904.

The Oudh operations gave much mounted work and minor action to the 2nd Regiment. At Keotee—a small village in Rewa State—one detachment covered themselves with glory, and captured a standard which is one of the Regiment's most prized possessions.

At Keotee, on March 4th, 1859, sixty-one tired and exhausted *sowars* of the 2nd Sikh Irregular Cavalry charged the centre of a formed battle line of twelve hundred fully armed, disciplined, desperate rebel soldiers, and after a most bloody hand-to-hand encounter, completely drove them from the field. The astonishing results achieved by such a small body of men, in complete disregard of the enemy's numbers, make the action at Keotee one which is an epic in the history of cavalry combat.

The seizure by the Celestial Government of a Hong Kong ship, led, in 1859, to hostilities against China, in which the French joined. An expedition was decided upon to comprise among other troops from India, a complete cavalry brigade, the two Indian regiments of which were to be the 1st Sikh Cavalry, now known unofficially as "Probyn's Horse", and a regiment raised from volunteers of other cavalry styled "Fane's Horse" now the 19th Lancers; the British regiment was the King's Dragoon Guards. The duty of serving abroad was voluntary, each man being required to attest individually his willingness. In "Probyn's" not a man said nay. They marched from Lucknow the six hundred miles to the rail-head at Raneegunge in eighteen days, and by April 1st, 1860, the whole regiment was *en voyage* to Hong Kong.

The allied forces—11,000 British Indian and 6,700 French—stormed the Taku forts. The Cavalry, led by Probyn himself, made many brilliant charges against the Chinese horse—Tartars mounted on hardy ponies and armed principally with bows and arrows, which they used from the saddle with telling effect. The Regiment in this expedition were armed with lances for the first time, and so effectively were they used, that in the

end no Chinese troops would stand up to the Indian Cavalry and fled upon their approach, and so Pekin capitulated.

The renumbering carried out as the result of the transfer of the East India Company's troops to the Crown caused the 1st Sikh Irregulars to become the 11th Bengal Cavalry, and the 2nd, the 12th. Both Regiments were not exclusively Sikhs, other martial classes of the Punjab being admitted.

A year later, 1862, the 11th took part in the desperate fighting in the Ambeyla Pass caused by the Swat tribal rising; the last encounters they were to see for fourteen years.

King Theodore III of Abyssinia, having threatened with death all foreign subjects in his territory, the British Government were forced in 1867 to send an expedition to that country, and with it went the 12th Bengal Cavalry. The Regiment were split up in detachments along the long line of communication from the base, and their posts were frequently attacked. The Commander-in-Chief at the conclusion of the campaign reported that—"the cavalry have had peculiar responsibility thrown upon their officers and soldiers. Left in isolated positions, far from control, entrusted with commissariat duties, while charged to be conciliatory to the people of the country, they have been firm in maintaining inviolate the respect due to their positions as soldiers. . . . Seldom or never have cavalry had such a variety of duties in maintaining communications for so many miles, climbing over mountains and through forest ranges, often benighted, where a false step would mean destruction, and in danger of treacherous attacks from the wild border tribes who are honoured among themselves for slaying without reason and without scruple". As was only to be expected, the wear and tear of horses, as well as equipment, on such duties was very considerable.

In the second Afghan War, 1878–80, the 11th, which had now been honoured by the Prince of Wales, afterwards King Edward VII, becoming their Colonel-in-Chief, were with Sam Browne before Ali Masjid and on through the Khyber. The

12th joined Roberts in the Kurram and were at the storming of the Peiwar Kotal; they were subsequently transferred to the Kabul Field Force, being present at the Shutar-Gardan Pass, Charasia and the engagements around Kabul. Thence for nearly forty years the 12th were not required for active service.

During the Afghan-Russian boundary settlement of 1885, 200 lancers of the 11th were sent to escort the representatives of India. It is recorded in the regimental archives that once, when far from shelter, the party was overwhelmed in a blizzard. The night spent in the open brought death to several followers, many mules, and to every canine member of the caravan. Although the horses' legs were practically coated with ice, none died; which exhibits the little known peculiarity that horses can stand extremes of cold when camels or mules cannot, and where dogs, almost without exception, die.

Squadrons of the 11th were employed in the Black Mountain country in 1891; the whole Regiment with the Chitral Relief Force, 1895—they and the Guides being the only cavalry engaged—and in the Malakand Field Force, 1897.

The year 1908 saw the institution of the Indian Distinguished Service Medal, and one of the first recipients was an Indian officer of the 11th Bengal Lancers, Jemadar Shahzadmir. The Jemadar's record as an explorer is unique in the history of the Indian Army. He accompanied Sir Francis Younghusband from Ladak to the Pamirs; journeyed alone from Ladak to Pekin, the capital of the Manchu Empire. For the survey work on this tour he received the Macgregor Memorial Medal of the United Service Institution. He made journeys to Abyssinia, Tibet, and Kabul; and also explored the route from Fashoda on the Nile to Lake Rudolf in Central Africa; travels which provided intelligence of inestimable value to the Army. Much of the survey work in parts of the Frontier which had to be carried out in secrecy and attended with grave danger was entrusted to this Indian officer, who subsequently received the Indian Order of Merit and large grants of land.

1914 was a disappointment to both Regiments, they being retained in India. The 12th subsequently went to Mesopotamia, where it distinguished itself before Kut-al-Amara, and in the advance to Baghdad. Meanwhile the 11th had been active on the North-west Frontier. Both supplied many drafts of officers and men to Indian cavalry regiments in France. The 11th later proceeded to the Euphrates valley and saw service in the Kurdish revolt operations of 1919, when isolated detachments behaved with great bravery.

On the return of the 11th to India in January, 1921—the 12th had returned in March, 1919—the old 1st and 2nd Sikh Cavalry became one; which was well, for both had been raised at the same time from the same source, and by the same order.

The uniform, blue *kurta* or long coat with scarlet facings, red cummarband, blue *pagri*, white breeches, is almost unchanged from that adopted when they first came on to the establishment. The *kurta* of the officers is heavily embroidered in gold on shoulders, chest, sleeves, and hem, and is probably the most elaborate of its type as military full-dress. The lance pennons are red and blue.

Major-General The Rt. Hon. Sir Dighton Probyn, V.C., who gave up the command in 1866, became the Honorary Colonel in 1904, and continued to be so until his death, at the age of 92, in 1924. His bearded figure was well known in the Empire's capital, he having become a member of the personal staff of King Edward VII when the latter was Prince of Wales. After the "Peacemaker's" death he was Comptroller to Queen Alexandra. Sir Dighton was a man of very high character who would have nothing to do with anything that was not absolutely right; his influence in the position he held meant an enormous amount, not only to the Throne, but to the Empire and to India. He was succeeded in the Colonelcy by Field-Marshal Sir William Birdwood, first Baronet, an old regimental officer of the 11th, Commander of the Australian and New

Zealand Forces in the Great War, and Commander-in-Chief in India, 1925–30.

For his great services to the Armies of the British Commonwealth, the Field-Marshal was created Baron Birdwood of Anzac in 1938; he thus joined the illustrious company of Clive, Lake, Napier of Magdala, and Roberts, as the fifth General of the Indian forces to be raised to the peerage of Great Britain by the Sovereign. Lord Birdwood is also Colonel of the Royal Horse Guards, and as such, alternately with the Colonel of the Life Guards, Gold Stick-in-Waiting to the King-Emperor, and thus personally responsible for His Majesty's safety; an honour never before conferred upon an officer of the Indian Army.

CHAPTER X

6th Duke of Connaught's Own Lancers (Watson's Horse)

"Afghanistan, 1878-80", "Tel-el-Kebir", "Egypt, 1882", "Punjab Frontier", "China, 1900", "Shaiba", "Kut-al-Amara, 1915-17", "Ctesiphon", "Tigris, 1916", "Baghdad", "Sharqat", "Mesopotamia, 1915-18", "N.W. Frontier, India, 1915", "Afghanistan, 1919".

"*Whoever is willing to take service in a cavalry regiment under the Sirkar, let him come with his horse to the Pass of Haldwani on the 1st September.*"

SUCH was the message disseminated by the Commissioner of Rohilkhand and the Rajah of Kassipur during the autumn of 1857. Within a few days of the appointed date there assembled what was to be the "Rohilkhand Horse". Uniformed in green *kurtas*, yellow pyjamas, red cummarbands, and black turbans, and led by their first Commanding Officer—Captain F. G. Crossman—they for two years joined in suppressing the Sepoy rebellion. They were brought into the Army

THE QUARTER-GUARD, 6TH D.C.O. LANCERS

List as the 16th Bengal Cavalry in 1862, to be disbanded twenty years later upon the reduction of the cavalry. During their short life they took part in the Bhutanese and Hazara expeditions.

No history of the first 16th Bengal Cavalry would be complete without a mention of Risaldar Law Wilson. The Risaldar joined the Regiment in 1857 as a European Sergeant from the British 4th Hussars, he became Sergeant-major, and afterwards Risaldar; he is the only case on record of a European holding what was then Indian Officer's rank.

While the old "Rohilkhand Horse" were on service in Oudh, instructions reached Lahore in mid-1858 which authorized the raising of a fourth regiment of irregular cavalry to be composed of Sikhs. The District Officers were ordered to get busy, and by the October the new Corps was garrisoning Delhi.

They were not in that city long, for early in 1859 orders were received to join the field force endeavouring to intercept Tantia, misreported to be approaching Delhi. Not expecting to be out for more than two weeks, and being ordered to march light, the men took only their picketing gear and a few rupees. After ten days the main force was ordered back to Delhi, except the 4th Sikh Cavalry, who were instructed to join by forced marches the cavalry under General Showers, then 150 miles away pursuing the insurgent troops. By making six marches of forty miles the Corps joined Showers' Brigade, and with him followed on Tantia's heels through Bikaner to Ajmer, and Goona, finally to catch him at Sipree. During this pursuit the greater part of the men were without shelter, food was extremely dear, and at times not to be had; while water and fodder were more than scarce. Men, after walking a sick horse for thirty miles or more, often on arrival in camp had to go on duty, with little hope of a meal until relieved. Yet there were no deserters, and no complaints.

In 1860, Lieut. Watson, V.C.—the Watson of the 1st Punjab

Cavalry—who had originally been appointed to command the Corps, but could not do so, he being on active service before Lucknow, after a period of sick leave took over his command.

Two years later the Corps became the 13th Bengal Cavalry. The Sowars' pay was raised to Rs.30 per mensem, out of which he had to feed, clothe, and equip himself and his horse.

The Regiment were on active service in the Ambeyla Pass, 1863.

Colonel John Watson, V.C., whose name the Regiment bears, vacated the command for that of the Central India Horse in 1871; he subsequently rose to the rank of General and Knighthood. It was Watson who introduced to the mounted arms of the service the rising in the stirrups when at the trot on occasions other than ceremonial; men in those days *bumped* the saddle at all times. Watson's views were hotly contested among others by the great Duke of Cambridge—it is not recorded what that quaint personality said when he saw the Indian Cavalry Brigade which Watson commanded *rising* at Malta in 1878. However, Watson in the end lived long enough to see his comfortable method become established in the service, to the joy, no doubt, of all riders and their horses.

During the Afghan War, 1878, the Regiment took part in both the first and the second Bazar Valley Expeditions. Until 1878 the Bazar Valley, a backwater of the Khyber, had never been explored, nor entered by troops from India. The Afghan fugitives after Ali Masjid escaped into it, and then with the Zakka Khels started harassing our convoys. A punitive expedition was thus necessary for the safety of the army in Afghanistan. It did not, however, punish enough, and when war again broke out in the following year further operations were necessary.

Arabi Pasha having deposed the Khedive of Egypt, a force had to be sent to the Nile. From India went the 13th Bengal Lancers. Before Tel-el-Kebir the Regiment gallantly charged with success a vastly superior force of enemy horse, and by

doing so arrested Arabi's attempt to attack the leading troops of the advancing army. In the actual battle of Tel-el-Kebir—13th September, 1882—after many charges which led to the rout of the enemy, the Indian Cavalry present carried on the long pursuit to Cairo which ended in Arabi's surrender.

It was at Cairo that H.R.H. The Duke of Connaught informed the Regiment that he had applied to Her Majesty the Empress for permission to become their Colonel; which distinction was conferred on their return to India at the end of the year.

Fears of a Russian invasion led to the augmentation of the Army of India, and as a result, the old 16th Bengal Cavalry were re-raised at Ambala in 1885; they, with the 13th, eventually became the present Regiment.

The Frontier outbreaks of 1897 gave two squadrons of the 13th the opportunity to perform one of the most striking achievements of cavalry on the North-west Frontier. On the evening of the 7th August the squadrons were part of a movable column attempting to relieve the Fort at Shabkadr, then surrounded by Mohmands in great strength. The action was not going too well, so the squadrons were ordered to charge the enemy from his right flank. They did so over nearly two miles of stony ground right along the enemy's front. The result was a complete clearance of the tribesmen back into their hills, and victory to a hotly contested force. It was an operation carried out with great dash; one of the Sikh signallers, minus his *pagri*, and armed with nothing but his flag, his hair streaming behind him in the manner of Absalom, rode down several of the enemy.

The 16th, in 1900, joined the allied forces in China about to relieve the International Legations in Pekin, then besieged by the Boxers. At the final assault on the Li Chand Kwang Pass, the Regiment dismounted and supported the German infantry. The 16th, who by then had become "Lancers", were the first to reach the American Legation, and as a token of esteem were presented with the "Stars and Stripes" which

had flown over the building during the siege. The flag now hangs in the Officers' Mess. On their return from China they ceased to be "Lancers" and by 1903 were again styled "Cavalry".

The 16th Cavalry joined I.E.F. (D), as the early campaigners in Mesopotamia were known, in February, 1915. At Shaiba, they, with the rest of the Cavalry Brigade, charged the stubborn Turk and put him to full retreat, the 16th capturing a Turkish Standard, now preserved. Under Townshend, they reached Ctesiphon. Then came the retreat of ninety miles under the very noses of the Turk; a feat only made possible by the sterling work of the hard-pressed cavalry. But when Townshend said on the 4th December "I intend to defend Kut", he sent most of his cavalry away, the 16th being among them; nevertheless sixty men of the Regiment remained with him in the defence, of which only twenty-five returned. The Regiment, much depleted, arrived back in India, October, 1916.

The 13th Duke of Connaught's Lancers (Watson's Horse) their then title, were employed on the frontier in September and October, 1915, one detachment making a charge over the almost identical ground of the Regiment's exploit in 1897. They joined the 7th Cavalry Brigade in Mesopotamia, July, 1916, and in the advance to avenge the Kut disaster, at the second storming of that city captured a standard. They made two outstanding mounted attacks, one against the Tikrat trenches, and the other upon a large convoy near Mosul.

The 16th were in Waziristan for the Third Afghan War, 1919. The Regiments having amalgamated at Kohat in the June, the united Corps were also in the operations in that area.

In the Khajuri Plain affair of 1929–30, the Regiment were specially thanked for their services.

Uniforms in 1914 were for the 13th, blue with scarlet facings, and silver lace; for the 16th, blue with blue facings, and gold lace. The present kit is blue with scarlet facings, and gold lace; the turban being blue, cummarband red.

CHAPTER XI

7th Light Cavalry

" Mysore ", " Seringapatam ", " Maheidpoor ", " Merv ", " Persia, 1915-19 ", " Aghanistan, 1919 ".

THE 28th Light Cavalry, as the present Regiment were known from 1903 to 1922, were the only cavalry regiment of the Armies of the British Commonwealth in the Great War to have fought against the Bolshevik Russians, a duty which fell to their lot in Trans-Caspia in the winter of 1918–19. How an Indian regiment came to be concerned in this backwater of the Great War was the result of the necessity for frustrating the endeavours of the Central Powers to stir Afghanistan into attacking India at a time when the latter's main forces were engaged elsewhere.

The 28th, on the outbreak, were cantoned in Quetta, and being so situated had little expectation of going overseas. When, however, mobilization orders did come in June, 1915, to move to an unknown destination, little thought was ever given to the fact that it could be anywhere but France or Mesopotamia.

Deprived of their horses and mounted on camels, two squadrons were despatched by the Seistan trade route across 500 miles of almost virgin desert into East Persia to form a cordon

OFFICER, 3RD MADRAS CAVALRY, 1840

with the Russians to keep unfriendly emmissaries out of Afghanistan. They were joined by their third squadron and the horses in October, it having been found that contrary to recorded opinion, horses could live in Seistan during the winter. The march along the trade route with the horses was, however, a different matter, and a feat which is deserving of great credit, only five deaths occurring among 400 horses after a march of 500 miles across almost waterless desert.

The first shots fired by the Regiment were at Deh Salem—meaning the Village of Peace—when a patrol captured the German Lieut. Winkelman, attempting to reach the Amir's territory. From then on their work was one of marching, counter-marching, patrolling, with many small fights against marauding banditti, whose depredations the Persian police made no attempt to stop. A squadron was also employed under General Dyer in the operations against the Baluchi Sirhad raiders who had been harassing convoys on the trade route.

The Russians left Persia in consequence of the revolution, and it was necessary to extend the cordon. The Regiment's activities were moved farther north to the Russian border, thus closing the 350 mile gap left by the departing Cossack cavalry.

With the Russians out of the War, further avenues of approach to India were opened to enemy agents, who now could make their way via the Central Asian Railway to the Afghan border at Kushkh; while there was grave danger of Turkish forces following the same route. Troops in consequence were sent into Russia to deny them passage. So onwards went squadrons of the 28th Light Cavalry to join the Menshevik forces and to hearten them in their attempt to throw off the Bolshevik regime.

In the advance to Merv, individual bodies made many charges against the revolutionary troops. Later, a patrol of thirteen men, on being surrounded by one hundred and fifty Bolshevik cavalry, closed their ranks and charged. This deter-

mined band of Punjabis broke through the enemy, cutting down twenty-two of them. Three of the patrol were unfortunately captured, but two escaped individually from their prison in Russia, and after showing extraordinary determination and pluck—one had no knowledge of geography nor in which direction India lay—eventually months later reported back to the Regiment at Lucknow. The story cannot be told here, but both feats could be compared to other famous escapes of history.

The decision to enter into no further commitments in Russia led to the withdrawal of the Regiment back to Meshed in Persia; from where on the outbreak of the Third Afghan War, heavy work fell to their lot in escorting convoys. The lines of communication with Persia run close to the Afghan border, and it would have been an easy matter to have cut them.

On the close of the Afghan War the 28th remained watching the Russian frontier to give warning of any attempt at Bolshevik invasion. Eventually, after over four years of constant hard service in Persia and Turkestan, they returned to India during the winter of 1919–20.

The Regiment were not required to amalgamate as a consequence of the cavalry reorganization of 1921, but their number was changed from 28th to 7th. The abolition of the *sillidar* system also did not affect them, since they were one of the three *regular* cavalry corps. They changed their uniform, however, from the silver-laced buff-faced French grey, to dark blue with French grey facings. The French grey had been worn from the early Madras days, but it has not entirely disappeared, as it is still worn by Indian officers appointed Orderly Officers in London to the King-Emperor.

The 7th Light Cavalry originally were formed from selected details of three regiments of cavalry taken over by the East India Company from the Nawab of Arcot, which were subsequently disbanded for mutiny, and were first styled the 2nd Madras Native Cavalry; they became afterwards the 1st, and then the 3rd, changes made on account of the seniority of their Commandants.

The 3rd Madras Cavalry, the number they were to bear from 1788 to 1903, served in the wars against the French for the supremacy of power in India; in all the Mysore and Mahratta Wars; against the Pindaris; and the operations in the Southern Mahratta territories, 1844–55. In the Great Mutiny, the Corps were split to join various columns operating in the Deccan. As the 7th Light Cavalry, they served in Waziristan from 1921 to 1923, being several times engaged with the Mahsuds.

On the 12th March, 1923, under the scheme of Lord Rawlinson—then Commander-in-Chief—the Corps were selected for the Indianization of the commissioned officer cadre. Almost all the squadron officers are now Indian gentlemen, as will be all those commissioned to the Regiment on first appointment henceforth.

K.G.O.

CHAPTER XII

8th King George V's Own Light Cavalry

" Mysore ", " Seringapatam ", " Ava ", " Central India ", " Afghanistan, 1879-80 ", " Burma, 1885-87 ", " Givenchy, 1914 ", " France and Flanders, 1914-18 ", " Aden ", " Afghanistan, 1919 ".

ALTHOUGH recruited to-day entirely from the sturdy peasantry of the Punjab, this Regiment, one of the oldest of India's Army, originally had its home in the south; most of the men being Madrasi Musalmans, with some Mahrattas and Rajputs; though even in the far off days of the close of the eighteenth century some men from the north found their way into the ranks.

As of almost all the present cavalry, the Regiment is an amalgamation of two, and the senior of the two, which form the 8th King George V's Own Light Cavalry, was raised at Arcot—famous for its siege by Clive—in the year 1787 by Captain Henry Darley, as the 5th Madras Native Cavalry.

British Officer, 26th Light Cavalry, 1914

The Madras Presidency had been engaged in a series of wars with the adjoining and powerful state of Mysore. A sepoy adventurer, Hyder Ali, had seized the throne and had given much trouble, which his son Tippu continued, fighting not only the British, but the states of Travancore to the south, and the Mahrattas to the north. The 5th were raised to assist in coping with this menace.

In 1788 a renumbering of Madras Cavalry took place, and the 5th became the 1st. As the 1st Madras Cavalry they took the field for the first time in 1790, when as a result of Tippu having attacked Travancore, the East India Company declared war upon him. The campaign was concluded some months later in the following year, when Tippu was driven back inside his fortress at Seringapatam. The operations were severe on the horses, and towards the end the 1st, 2nd and 4th Regiments of Madras Cavalry were fighting dismounted. The Cavalry had not then learnt the art of horsemastership, nor was there a proper remount service.

The Mysore Sultan again got up to his tricks and started to parley with Napoleon. He was requested to desist, but refused, and thus war was again made upon him. This time it was decided to capture Seringapatam. The army duly assembled, and on the way, at Malavelly, the British 19th Light Dragoons and the 1st and 2nd Madras Cavalry delivered a skilful charge which utterly defeated the enemy's cavalry. Malavelly Day, March 27th, 1799, is commemorated annually by a Regimental holiday. On the 4th May, Seringapatam was stormed, Tippu killed, and the Mysore menace ended.

The Regiment took part in the pacifying operations under Colonel Wellesley—afterwards Duke of Wellington—rendered necessary by some of Tippu's followers raising an insurgent standard under Dhoondia Waugh. For several months Wellesley was unable to bring him to battle, but the day came; Dhoondia was caught with 5,000 cavalry on the march unprepared. Wellesley formed his two British and two Indian cavalry regiments, of which the 1st Madras Cavalry was one, into one long line and charged. The result was completely successful, Dhoondia and most of his men being killed.

The reorganization of the Nizam's Army in 1826 brought about the establishment of a new cavalry regiment in his forces, viz: the 4th Nizam's Cavalry, raised by Sir John Gordon, Bart, originally of the Coldstream Guards. This regiment, nearly a century later, joined what were then the 1st Madras Light Cavalry to make the present Corps. From the time of its raising until the Great Mutiny, the 4th Cavalry Hyderabad Contingent, as they became in 1854, were constantly on field service against Naiks, Pindaris, Rohillas, Bhils, and others of the marauding fraternity of the Central India of those days.

To return to the senior regiment. The 1st Madras Cavalry had been overseas to Burma in 1826, though what part they played in this campaign is to-day obscure; nevertheless they lost 233 horses in the short campaign of three months, an enormous percentage when it is considered that only two squadrons and Headquarters, having a total horse establishment of 380 horses, took the field.

In the Great Mutiny, the 4th Hyderabadis saw continuous service in the long operations in Central India. The 1st Madras Light Cavalry were unlucky enough to be at Trichinopoly, the station farthest away from the scenes of action; they, however, moved out to Cuddapah and two squadrons chased a body of Rohilla rebels into the Kistna.

The honours for the Second Afghan and the Second Burmese Wars were earned by the 1st Madras Light Cavalry. No serious fighting was seen in the first campaign; while in the second the Regiment were for not quite two years employed by detachments hunting down dacoits, escorting convoys, and doing the multiplicity of jobs cavalry are expected to do in jungle operations of the Burmese kind.

The reorganization of 1903 turned the 4th Hyderabad

Contingent Cavalry into the 30th Lancers (Gordon's Horse), and the senior corps, who had become the 1st Madras Lancers in 1901, into the 26th Light Cavalry; to become later in 1910 the 26th King George's Own Light Cavalry, King George V, as Prince of Wales, having become the Colonel on his visit to India, 1906.

The 30th left Ambala for Bombay and "La Belle France" on the 5th October, 1914, having been standing by mobilized since the 31st August. Arriving on the 11th November, by the end of the month they were occupying trenches around Givenchy where were suffered the first casualties. They were with the Indian Corps throughout the second battle of Ypres. The break up of the Indian Cavalry Corps in France sent them, not as expected to Iraq, but to Peshawar and the Frontier.

The 26th supplied the cavalry for the Aden Field Force from September, 1915, remaining there until 1922. The climate of the Aden Hinterland, where the Turks had an army, was most exhausting, making the work of the mounted troops arduous indeed. Constant patrolling had to be carried out, and numerous skirmishes took place between these patrols and parties of Turks and Arabs. Altogether the 26th took part in sixty-six important engagements, and one hundred and seventy-three minor engagements in the defence of the Aden Protectorate. A special service squadron was also sent to South Persia, June, 1918 to April, 1919.

The 30th again left India for Iraq for two years, December, 1920, returning for the amalgamation with the 26th at Peshawar on the 1st June, 1922.

The present uniform is blue, with French grey facings. At one time the old 26th wore the reverse.

His Highness the Maharajah of Mysore is the Hon. Colonel of the Regiment.

CHAPTER XIII

The Royal Deccan Horse (9th Horse)

"Central India", "Givenchy, 1914", "Somme, 1916", "Bazentin", "Delville Wood", "Flers Courcelette", "Cambrai, 1917", "France and Flanders, 1914-18", "Megiddo", "Sharon", "Damascus", "Palestine, 1918".

THE Nizam of Hyderabad having terminated French influence in his territories, His Exalted Highness undertook in 1798 to maintain, in co-operation with the British Indian Government, sufficient forces for the defence of his state. Such troops were duly raised, but being organized largely on the feudal system, were soon found to be ineffective. The force were ill paid, and when pay was forthcoming, it was often six months or more overdue. The cavalry were *silladar*, that is

to say horse and equipment were provided by the lesser feudatory lights, who also provided the men.

Berar and the northern territories of the Nizam had by this time been so overrun by rebels and marauders from the Mahratta states—plunderers rather than fighters, numbering from 8,000 to 10,000 horsemen, sometimes possessing cannon—who ravaged indiscriminately the whole country around them, with a power so great that on several occasions they compelled the state to come to terms with them. They traversed the Nizam's territory, plundered towns and villages in all directions, and with the indiscipline of the latter's troops, grew bolder and bolder, so that the establishment of a proper disciplined military body to deal with them became an imperative necessity.

Numerous attempts at reform of the Hyderabad Army had been attempted, but His Exalted Highness of the time was not disposed to agree to the proposals. Eventually, in 1811, Mr. Henry Russell, the new Resident, succeeded in creating regiments of regular infantry. Five years afterwards, his cavalry reform scheme also received the approval of the Government of India, and four regiments of the Nizam called "The Reformed Horse" came into being. They were subsequently made liable for service outside their own locality, and the titles changed to "The Nizam's Cavalry". These regiments were the foundations of the afterwards Hyderabad Contingent Cavalry of 1854, the first two regiments of which are the ancestors of the Royal Deccan Horse of to-day.

The Nizam's Cavalry were mounted on locally bred, but apparently very useful small horses, for they were able to make extraordinary marches of sixty miles in a night, seldom or ever leaving a man behind; indeed we are told that the men would have felt disgraced if they had been unable to proceed with their comrades. One of their rules was that no man wounded in the back need ever expect promotion.

An interesting case showing how failures were dealt with occurred in 1831. A detachment had been sent out but had

TYPES OF THE ROYAL DECCAN HORSE, 1937

INDIAN OFFICER

NON-COMMISSIONED OFFICER

not accomplished their object. A *panchayat* consisting of the Indian officers assembled, held a thorough enquiry and decided that with the exception of nine sowars, all the others, officers and men, had failed in their duty. The Resident confirmed the finding and the whole concerned were paid up and dismissed on the spot. The Army Act at this time did not apply to the Nizam's troops.

When the Indian Mutiny broke out, the whole of the Hyderabad Cavalry remained loyal. They fought with great daring in the arduous campaign in Central India, marching many thousands of miles, and distinguishing themselves in many actions; Sir Hugh Rose, the Commander of the forces, describing them as "the wings of the army". In the action at Chichamba, 1859, Captain Clogstoun of the 2nd Regiment acted with such gallantry in charging the rebels with only eight of his men, that the Victoria Cross was afterwards conferred upon him.

The 1st and 2nd Regiments in 1891 sent a combined detachment on active service to Central Africa, six of the sowars receiving Orders of Merit for their conspicuous gallantry in action in the attack on the Chief Makanjira's stronghold.

The Hyderabad Contingent was disbanded in 1903 and the two senior regiments of the cavalry transferred to the Indian Army as the 20th Deccan Horse and the 29th Lancers (Deccan Horse).

In the Great War both Regiments served in France, the 20th Deccan Horse earning the first honours at Givenchy, 1914. After the move of the Indian Cavalry to Palestine, they both served in the Jordan Valley and the battles of the final offensive.

Near Khan-es-Samriyeh in the Jordan Valley, a half squadron of the 29th charged, under a heavy fire, a large force of Turkish infantry and machine guns defending a very strong position. The two troops scored a signal victory, one thousand prisoners—including a Divisional Commander—and many automatic weapons being captured. In this affair, Risaldar Badlu Singh, realizing that the troops were meeting with casualties from a

TRUMPET BANNER, THE ROYAL DECCAN HORSE

small hill occupied by seven machine guns and some two hundred infantry, at once collected the only fighting ranks at his disposal—another Indian officer, a British trooper, and four sowars—and entirely disregarding the danger, with the greatest dash and personal gallantry charged and captured the position; thereby saving very heavy casualties to the men of the two troops concerned. He himself was mortally wounded when in the act of capturing one of the machine guns single handed. Before he died, all the Turkish machine gunners and infantry on the hill had surrendered to him. Such gallantry and initiative could not go unpassed and the King-Emperor awarded him a posthumous Victoria Cross. The Risaldar was a Jat, the first to earn the highest military decoration, and was attached from his own regiment, the 14th Murray's Jat Lancers.

The two Regiments amalgamated at Bolarum in the territory of their birth on the 16th July, 1921. Just prior to this, the 20th had been created a Royal Regiment for their eminent services in the Great War. They made no change in their uniform and they remain the only Indian cavalry to be dressed in rifle green, the facings being white.

CHAPTER XIV

The Guides Cavalry (10th Queen Victoria's Own Frontier Force)

" Mooltan ", " Goojerat ", " Punjaub ", " Delhi, 1857 ", " Ali Masjid ", " Kabul, 1879 ", " Afghanistan, 1878-80 ", " Chitral ", " Malakand ", " Punjab Frontier ", " Khan Baghdadi ", " Sharqat ", " Mesopotamia, 1917-18 ", " N.W. Frontier, India, 1915 ".

" *Daffadar Fatteh Khan volunteered to accompany the Infantry on the heights, and was blown up whilst cutting down one of the enemy in a breastwork. His sword severed a* masak *of powder, which the man happened to be carrying, and a few grains falling on his matchlock ignited the mass, and both the Daffadar and his antagonist were blown into the air. The Daffadar was likewise shot through both arms.*"

SO runs the exciting record of the Corps of Guides for September, 1847, covering the second action in which they were engaged since they had come into being at Peshawar on the 14th December the year previously.

Originally one troop of Cavalry and two companies of

GUIDES CAVALRY—SOWAR, FULL-DRESS, 1911

Infantry just about three hundred strong, under the command of Lieut. H. B. Lumsden, they rapidly made a name for themselves for their fidelity and bravery; so much so, that when the intrigues inside the Lahore Durbar after Mahrajah Ranjit Singh's death necessitated the strengthening of the forces in the Punjab, they were brought down to Lahore in 1848.

They were entrusted with the delicate task of unravelling the plots which aimed at the seduction of the friendly forces; and through their efforts, and the prompt action of Lumsden, a serious mutiny was averted and sufficient evidence obtained to banish the Maharani from the Punjab. The Corps were employed to escort her out; a service then considered of utmost importance, since attempts at rescue were expected.

The unrest in the Sikh territories continued, necessitating further operations by the British. The Cavalry fought before Multan, while the Infantry, by a successful ruse, seized Fort Govindgarh at Amritsar, in addition to occupying other smaller strongholds which declared in favour of the Khalsa insurgents. Both arms fought at Gujrat, and in the ensuing pursuit to Rawalpindi where the rebel Sikhs and their Afghan allies laid down their arms.

Until the Great War the history of both the Cavalry and Infantry of the Corps is so intertwined that the services of the Corps as a whole will be dealt with here up till then, and thereafter the services of the Infantry will be found with those of the 12th Frontier Force Regiment, to which the Infantry were transferred in 1922 as the 5th and 10th Battalions.

After the 2nd Sikh War practically no frontier affair was without the Guides. While Lumsden for a time was Deputy Commissioner of Peshawar, then a post of special and unique importance. He was deputed in March, 1857, to Kandahar on a special mission entailing considerable risk, and so with a small escort of his Corps was absent from Hoti Mardan on the fateful 13th May, 1857, when the Guides were ordered south to suppress the rebel Bengal Army.

By 6 p.m. the same day, the Guides under Captain Daly marched; the Cavalry on their horses, the Infantry two to one camel. They arrived at Delhi, 580 miles away, on the morning of June 9th, having traversed the distance in twenty-six days in addition to fighting an action near Karnal on the way; a feat of movement as yet unparalleled in military history. They journeyed principally between sunset and sunrise, and each man, we are told, seemed to vie with the other in bearing the privations and fatigue that the march involved.

No sooner had they begun to pitch their camp than their services were required, and three hours after arrival they were hotly engaged, all their officers being wounded.

For four months they were constantly in action. The Infantry, with the 60th Rifles—now the King's Royal Rifle Corps—and the now 1st Battalion 2nd King Edward's Own Gurkha Rifles, held the main piquet at Hindu Rao's house, beating off innumerable attacks on that key position, until victory was won. The Cavalry were constantly employed with the mounted brigade in the operations around the city in support of the besiegers.

In the final grand assault the Guides took no part; but they materially aided its success by assailing the enfilading enemy batteries at Kishenganj. The Cavalry supported the 9th Queen's Royal Lancers, protecting the flank of the assaulting columns. Drawn up under an appalling fire from small arms and artillery, the 9th suffered so severely that the Guides Cavalry were ordered to take their place, their steadiness evoking the remark "they stand like the Lancers" from the latter's Commanding Officer.

After the fall of the City of Delhi, beyond a few minor fights in the adjacent countryside, the Corps took no further part in the Mutiny campaign, and were ordered back to their fortress in the Yusafai territory of the north to join the Punjab Frontier Force. They arrived at Peshawar in the beginning of February, 1858, where they were received by the paraded garrison; a royal salute and a *feu-de-joie* being fired in their honour. The

Commanding General—Cotton—greeted them with these words:

> "We respect, we honour you, and we feel proud of being re-associated with men whose deeds of daring have earned our noble profession never dying fame."

Sentiments with which the East India Company expressed their strong and entire concurrence, the Directors remarking that their singular fidelity was shown by the fact that of the 800 men engaged, not one deserted to the enemy, whilst 350 of them were killed or wounded.

Lumsden, who had then reached field rank, returned from his special mission in the June of 1858. A duty which was eminently successful, and which gave the authorities knowledge which proved to be of untold value in subsequent campaigns against Afghanistan. Major Lumsden's services, although different in kind to those which he would have rendered in the field at the head of the Corps he had raised, were no less valued by the Government in London. He subsequently received the Companionship of the Bath and promotion to Lieut.-Colonel.

Lumsden led his Corps against the Kabul-Kel Waziris, 1859, and their brother tribe the Mahsuds during the following year. In the previous affair, some mounted enemy suddenly appeared on the crest of a hill, charging down it as if on a race course—such is their fanatical bravery. We are told that one little fellow of the Guides Infantry, some five feet tall, fixed his bayonet and called upon his officer—a man over six feet in height, and broad in proportion—to get behind him for protection. The offer was not accepted, but the spirit that prompted it and the way the sepoy—by name Alladad—stood by his officer against all comers, is worthy of mention.

Three years later, 1863, the Corps was in the thick of the Ambeyla Campaign against the Hindustani fanatics of Sittana. An enterprise marked by severe hand-to-hand fighting in

precipitous mountain country, the bravery shown by the enemy being equal to our own.

For their great services, Queen Victoria in 1876 conferred upon the Corps the style of "Queen's Own", and the privilege of bearing on their colours and appointments the Royal Cypher within the Garter. The Guides were thus one of the first units of the Indian Army to become a Royal regiment. At the same time, H.R.H. the Prince of Wales, afterwards King Edward VII, became their Colonel.

The ensuing years, 1877–78, saw small expeditions against the Jowaki Afridis, and the Utman Khels; the latter in revenge for the massacre of the coolies constructing the Swat Irrigation Canal.

The British Government now decided that a mission to Kabul was essential if the peace of India was to be protected from the machinations of the Russians at the Afghan Court. And so the Guides were ordered to escort the proposed Ambassador to Kabul. The Mission was refused passage through the Khyber and thus commenced the long operations against the Amir which were carried on until 1880.

In the forcing of the Khyber, the Guides and the 1st Sikhs, now the 1st Battalion 12th Frontier Force Regiment, were principally responsible for the capture of the fortress of Ali Masjid. In the following advance into Afghan territory, the Cavalry fought a striking action at Fattehbad, where Lieut. W. R. P. Hamilton, in leading the charge that led to the disorderly retreat of the enemy, won the Victoria Cross.

The Corps reached Jellalabad and from thence returned to India, where laudatory orders were issued that they had, if such a thing were possible, surpassed their old reputation as models of what Light Horsemen should be: ever ready, ever serviceable, ever soldier-like, and never found wanting.

As history tells, the Amir now agreed to receive a mission at his capital, and Sir Louis Cavagnari was duly installed. For his protection Lieut. W. R. P. Hamilton, V.C., and seventy-six

of the Guides were ordered to Kabul as his escort. They left India on the 26th June, 1879, none but three ever to return; for they died soldiers' deaths in defending the overwhelming attack on the British Embassy. Few regiments can show a brighter record of devoted bravery than that achieved by this small band of "The Queen's Own Corps of Guides" in their efforts to protect the Ambassador of their Queen who fell with them on the 3rd/4th September, 1879. In recognition of their great gallantry double pensions were conferred upon their widows and heirs.

It was not long before the rest of the Corps were on the way to avenge their comrades; the 11th December finding them approaching Kabul, where they learnt that reinforcements were a dire necessity. So without any to do, they traversed the last thirty-six miles in a single march with the briefest of halts, arriving to take part in the defence of the Sherpur Cantonment and the assault on the Asmai Heights. In the Asmai battle, Captain Arthur Hammond displayed such outstanding qualities of leadership and gallantry, that he received the highest honour for bravery in action that the Queen Empress could bestow.

Evacuating Kabul by 1880 they enjoyed for a time comparative peace, save for an expedition against the Bunnerwals (1887), the Hazara Black Mountain tribes (1891), and the Chitral Relief Force (1895). With the last named the Corps were in the assault on the Malakand Pass, actions of the Swat River, and the capture of Mandah.

The general frontier rising of 1897 again brought out the Guides' renowned marching powers, the Corps leaving Mardan in the heat of July to reach the Malakand defences in hill country thirty-six miles distant in sixteen hours. They defended the pass, helped to relieve the fort of Chakdara, afterwards seeing service in nearly all of the other adjacent actions.

For great bravery at Landakai in the Swat valley on August 17th, 1897, Colonel R. Adams and Lieut. H. Maclean of the Guides, together with Lieut. Lord Findcastle of the 16th Queen's

Lancers, received that much prized bronze cross inscribed "For Valour" in attempting to rescue an officer from a superior force of fanatical tribesmen; Lieut. Maclean being mortally wounded in the attempt.

Another forced march across rocky ground from Mardan to Shabkadr, thirty-five miles in sixteen hours, of which four were spent in fording the Khaili River, brought the Guides into the Mohmand small war of 1908. They suffered so severely from cholera in the operations of the time against the Khoda Khel that for a while they had to be segregated.

So we come to 1914, when the story will be that of "Queen Victoria's Own Corps of Guides (Frontier Force) (Lumsden's) Cavalry" under which title the mounted branch of the Corps took the field against the enemies led by the Prussian Eagle in that fateful year.

Although the border services of the Cavalry of the Corps are only recognized by the honour "N.W. Frontier, India, 1915" they were constantly at it, keeping watch and ward until November, 1917, when proceeding overseas they joined the 11th Cavalry Brigade in Mesopotamia to take part in the battles of Sharqât and Khan Baghdadi. They were with the force occupying Mosul when the Armistice called a halt; but India was not for them, they continuing further service with the 3rd Cavalry Brigade, in Khurdistan, putting down the Arab revolt around the Beziyan Pass in the June of 1919.

They again went further afield, one squadron marching the 378 miles from Qaraitu to Kasvin in Persia, where the Bolsheviks were threatening invasion in October, 1919; a second squadron followed shortly after, the whole Regiment joining by May, 1920. The Guides Cavalry stayed in Persia for over a year, and all ranks present received the newly instituted General Service Medal of King George V with the clasp "North-West Persia" in 1923.

The name of the founder, borne as a subsidiary title from 1904 to 1922, appears to have been dropped on the separation of the Corps in the latter year. Lieut.-General Sir Harry Lumsden, who had for a time commanded the famous Hyderabad Contingent, 1862-69, died 1896, aged seventy-five years. Sir Harry may claim to be the inventor of the khaki uniform, and the Guides the first corps to wear it. They have continued to do so, both for service and full dress, the latter officially known as drab, the facings being red.

A note on the history of khaki as a uniform appears at the end of this book.

(*India Office photo*)

MECHANIZED INDIAN CAVALRY—A HEAVY TANK ON ROAD PATROL

CHAPTER XV

Prince Albert Victor's Own Cavalry (11th Frontier Force)

" Delhi, 1857," " Lucknow," " Ahmad Khel," " Kandahar, 1880," " Afghanistan, 1878–80," " Kut al Amara, 1917," " Baghdad," " Khan Baghdadi," " Sharqat," " Mesopotamia, 1915–18," " Afghanistan, 1919."

IMMEDIATELY after the annexation of the Punjab, five regiments of irregular cavalry were formed for the protection of the new frontier from the Black Mountain of Hazara to the limits of Sind. The P.A.V.O. Cavalry is the descendant of the first and third of those regiments of the Punjab Cavalry.

Henry Daly of the 1st Bombay Fusiliers raised the 1st at Lahore, and W. G. Prendergast of the 8th Light Cavalry the 3rd at Peshawar, in the spring of 1849.

The 1st received their baptism of fire under Sir Colin Campbell in the warlike enterprise against the Kohat Afridis (1850), the Ranizais and the Utman Khel (1852), while 1856 saw them with Neville Chamberlain against the Miranzai.

The 3rd, after formation, came south to Amritsar where, in February, 1850, they suppressed the mutiny of the 66th Bengal Native Infantry at Fort Govindgarh.

The junior Regiment suffered severely in 1852 in the loss of 300 horses from "Surra" a deadly disease which, curiously, was quaintly described by the Commanding Officer of the time as "Necromancy". Whether it was magic or not, it crippled the Regimental Chunda or Horse Fund for years, and a special loan had to be made by the Government. In those days, the Indian cavalryman, save in certain regiments, provided his own horse, and no arrangements existed for compensation other than injury, etc., on active service; which so far the 3rd had not yet seen.

In 1857, however, a small body of fifty sabres, under Lieut. Watson, were sent against the Bozdars, a truculent tribe on the Baluch border; while the Regiment, although they remained guarding the frontier all through the Great Mutiny, saw plenty of minor fighting with would-be insurgents.

The 1st Punjab Cavalry were more fortunate, and sent down to Delhi a squadron. This squadron under Watson, who had just been transferred from the 3rd, took part in the operations of the siege; and afterwards in the battle for Agra, where they captured three guns and five standards. They were in the force relieving Lucknow, and in the contingent Rohilkhand operations, rejoining the regiment at Bijnor in June, 1858.

Watson, the intrepid squadron commander, earned a well-merited Victoria Cross for leading their charge against a body of rebel cavalry on the 14th November, 1857.

The remainder of the 1st had been employed suppressing mutiny at Multan, Ambala, and Karnal, but they arrived in the principal disaffected areas in time to serve in the Oudh campaign under Sir Colin Campbell, or rather Lord Clyde as he had by then become. These operations were long, not finally ceasing until 1859. Many detachments had to be furnished to assist in restoring the civil government. Upon one occasion, near Muzaffarnagar, when things were going awry, one detachment was led in the charge which won the day by a member of what we should now call the Indian Civil Service.

The Regiment remained in Oudh for some time, and on Lord Canning—the first Viceroy—proceeding on his grand tour of pacification from Cawnpore to Lahore, had the honour of assisting his Body-Guard to escort him. A march which lasted four months.

The 1st subsequently re-joined the Punjab Irregular Force whose work Kipling correctly describes in *The Lost Legion :*

> "You must know that all along the North-west frontier of India there is spread a force whose duty is quietly and unostentatiously to shepherd the tribesmen in front of them. They move up and down from one desolate little post to another; they are ready to take the field at ten minutes' notice; they are always half in and half out of a difficulty somewhere along the monotonous line; their lives are as hard as their own muscles, and the papers never say anything about them."

In 1860, the 3rd were employed on the trying and arduous work of the Mahsud Blockade.

During the Afghan campaigns, 1878–80, the 1st occupied Kandahar, and subsequently fought with extreme bravery at the battle of Ahmad Khel, April, 1880. The 3rd were with Roberts at Kabul, where they arrived the same year in time to be detailed for the Kandahar Relief Force and participation in the following victory.

Both Regiments were in the Mahsud campaign of 1894–95, the 3rd carrying out a remarkable march from Kohat to Bannu—eighty miles—in twenty-eight hours.

In the large border insurrections of 1897–98, the 1st were in Tochi; the 3rd in the Kurram Valley and on the Samana Ridge. Despite the fact that every single man of both Regiments had been engaged in the long operations, they had been so split up that on no occasion had two squadrons been employed together, so that the battle honour "Tirah" did not fall to their lot.

Lord Kitchener in 1902 broke up the Punjab Frontier Force and incorporated them in the Indian Army. The 1st and 3rd Punjab Cavalry thus became the 21st P.A.V.O. Cavalry (F.F.) (Daly's Horse), and the 23rd Cavalry (F.F.) respectively.

Both Corps were detailed for internal security in 1914, and thus prepared drafts for other regiments. The 23rd were unfortunate to be in Lahore, then the centre of the "Ghaddar" conspiracy. The Sikhs of the 23rd were subjected to constant propaganda from the conspirators, a party of Sikhs became infected, and in 1915, several men were tried by Court Martial and sentenced to death. The subsequent loyalty, gallantry, and good behaviour by this class when the Regiment eventually went to Mesopotamia, entirely obliterated this "bar sinister" upon what had long been a perfect shield of honour.

The 23rd proceeded in May, 1915, to the Persian Gulf, and were responsible for the patrol of part of the Anglo-Persian Oil Company's pipe line, a vital source of fuel supply for the Royal Navy. They were in this area until October, 1917; and so successful were the Regiment's methods, that the district—previously an unruly one—was brought into complete order and control. One squadron fought in the defence of Kut-al-Amara.

The 21st went out to the Tigris in September, 1916, and after the recapture of Kut in February, 1917, took part in the ensuing pursuit to Baghdad. In the April, with Cassells' column, they, at the action on the Shatt-al-Adhaim, in the course of a long and exhausting pursuit, captured 800 prisoners. They came back from detached duties to take part in the tactical move which put the cavalry astride the Aleppo Road and closed the Turkish line of retreat which resulted in the surrender of the entire enemy force.

The 21st having been relieved from the oil fields, joined Cassells' force at Baghdad, and later, under him, played their part in the successful pursuit to Khan Baghdadi which put an end to Von Falkenheyn's plans for an advance on Baghdad

via Aleppo and the Euphrates; a threat that had indeed been a very real one. Shortly after this the 21st were in the operations around Kirkuk, taking part in a charge against a Turkish battalion on the march, the Regiment capturing 350 and sabreing another 100. They went on in the advance to Mosul, seeing very serious fighting. With the fall of Mosul, the 21st returned to India.

Both Regiments were ordered to the Frontier in 1919, but beyond minor affairs against unfriendly tribesmen, met no engagement with Afghan troops.

On the 6th June, 1921, the amalgamation of the 21st and 23rd took place, and in 1927 the present title was conferred.

The bronze badge of the amalgamated Regiment is a combination of that of the 21st and the 23rd. That of the latter being crossed swords, and the former's the Kandahar Star granted for the famous march. The two have been superimposed, and the Imperial Cypher substituted for the letters P.A.V.O. in the centre.

The uniform of both Regiments, which was practically identical, being blue with red facings, has been retained and follows the general Indian cavalry pattern of to-day. The 23rd in full dress wore gold-laced accoutrements of black leather—usually gold lace is superimposed upon red or blue leather in the Indian Army—they prized this very much, and the feature has been adopted for the present Corps.

Prince Albert Victor, better known as the Duke of Clarence and Avondale, the eldest son of King Edward VII and Queen Alexandra, and consequently heir to the British Throne, became the Honorary Colonel of the 1st Punjab Cavalry in 1890. He died in 1892. Hence the subsidiary title, which alone is shared with the 2nd Battalion 7th Rajput Regiment

CHAPTER XVI

Sam Browne's Cavalry (12th Frontier Force)

"Delhi, 1857," "Lucknow," "Charasiah," "Kabul, 1879," "Ahmad Khel," "Afghanistan, 1878-80," "Kut-al-Amara, 1917," "Baghdad," "Mesopotamia, 1916–18," "N.W. Frontier, India, 1914–15," "E. Africa, 1917," "Afghanistan, 1919."

WHEN Lieut. Samuel James Browne of the 36th Bengal Native Infantry was ordered in 1849 to recruit at Lahore a cavalry regiment for service on the Punjab Frontier, little could he have thought that nearly a century later that Regiment would bear his name; a name which is a household word in most armies of the world from the article of military accoutrement he was later to devise. The original title of the Regiment was the 2nd Punjab Cavalry of the Punjab Frontier Force—the name of Sam Browne not being officially bestowed in the title until they were forty-five years old.

While Lieut. Browne was recruiting at Lahore, one Captain Robert Fitz-Gerald of the 12th Bombay Native Infantry was

General Sir Samuel J. Browne, V.C., G.C.B., K.C.S.I.

busy on similar duties at Multan where he founded the 5th Regiment of Punjab Cavalry, also for the Frontier Force.

These two regiments in 1921 became one, the present title being conferred in 1927. It is as the 2nd and the 5th Punjab Cavalry that their early glories were earned and as such their history will be traced until 1904 when further changes of title occurred.

Both Regiments shortly after their foundation were policing the Frontier—though one reads with dismay that on the arrival of the 2nd, the General commanding the force expressed himself not pleased with their appearance. This, however, was soon rectified; Waziristan, then just as troublesome as now, soon made them into shape. Much of the Frontier bother in those days, however, appears to have been due to horse and cattle thieving, and we are told that the *sowars* of the 5th had by 1856 become as bold and successful in recapturing cattle, as they used to be formerly in plundering them.

In the great revolt of the regular Bengal Native Army, squadrons of both Regiments were sent south to the relief of Delhi. These took part in the besieging operations, and after the fall in the battle for Agra, where Captain D. M. Probyn of the 2nd (afterwards of Probyn's Horse) won the Victoria Cross for a series of gallant exploits culminating in his capturing a rebel standard. The squadrons marched to join Sir Colin Campbell at Cawnpore—where the remaining squadrons of the 2nd joined up—took part in the relief of Lucknow and in the succeeding campaign in Oudh and Rohilkhand. At Seerporah on August 31st, 1858, Brevet-Major Sam Browne of the 2nd accompanied by a single trooper charged and captured a rebel gun. He was severely wounded and lost his left arm, but he won that most coveted of decorations, the Victoria Cross.

In one of the actions around Bareilly, a company of the Black Watch, after losing heavily in officers and n.c.o.'s, were rallied by Risaldar Hakdad Khan of the 2nd, and as a compli-

ment to the Regiment when they left the forces in the field for their home in the north, they were played out of Bareilly by the Band and Pipers of that famous Scottish Regiment; which goes to show that no question of race ever arises when natural leadership and bravery assert themselves before the common enemy.

The single squadron of the 5th also distinguished themselves, the two Indian Officers receiving the Order of British India and nine other ranks Orders of Merit.

After the Mutiny both Corps were constantly on service holding the Frontier outposts, the following being two typical incidents of the time. On March 18th, 1860, Risaldar Saadat Khan with 150 men of the 5th and 37 Mounted Police attacked 3,000 Mahsud Waziris at Tank, killing 300 and putting the rest to headlong flight: and on January 6th seven years later Jemadar Imam Khan, also of the 5th, with 27 *sowars* charged a body of raiders 1,000 strong, killing 150 and taking prisoner almost the whole of the remainder.

It was during this period that the famous Belt was born. It is said that Sam Browne, having lost his arm, now found considerable difficulty in carrying his sword when dismounted. He therefore turned his attention to an idea with which he had been toying some years before, the designing of a belt which would not only carry the sword comfortably when mounted and dismounted and leave the arms free, but which would also securely carry the pistol so that if it did go off accidentally, as they often did, no harm would come to the wearer. Sam Browne perfected his own invention, for the original belt, which is now preserved in the Officers' Mess; differs very little from the regulation pattern of to-day.

In the second Afghan War, 1878–80, the 2nd distinguished themselves at Ahmad Khel while serving with the Khandahar Field Force, the 5th being in the Kurram Valley and later with Sir Frederick Roberts at Charasiah. The 5th Punjab Cavalry and the 9th Lancers of the British Army for their gallantry at

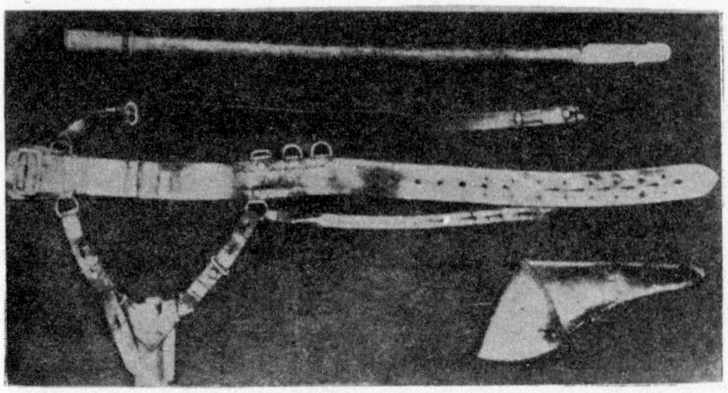

"The Original Sam Browne Belt"

Charasiah were accorded the honour of escorting Sir Frederick when he made his triumphal entry into Kabul.

There was still fighting to be done by the Kabul Field Force, and at the storming of the Asmai Heights, December, 1879, one of the sortie actions around Kabul, Captain William Vousden received the Victoria Cross for exceptional gallantry in making repeated charges with a small body of the 5th against overwhelming numbers of the enemy, passing through their ranks again and again until they were completely put to rout. The ten survivors of the men received Orders of Merit.

Until 1914, neither Corps took part in any campaign for which honours were subsequently granted. The 2nd Punjab Cavalry took part in the Zhob Valley affair of 1884, and Miranzai, 1891, all ranks receiving the medal with the clasp "Samana, 1891": they also had a share in the Kabul Khel Wazir Expedition, 1902, under General Egerton.

1904, as stated, brought a change in title and the two now became the 22nd Sam Browne's Cavalry F.F. and the 25th Cavalry F.F.; in the case of the latter the change had actually been made a year earlier. It was under these titles that they fought in the World War.

The 25th were active on the North-west Frontier until 1915, taking part in the action at Dardoni—the first in which the Pathans are believed to have ever left out their dead and wounded all night. They then proceeded to East Africa seeing fighting at Nahungo and Chingwea, and in the pursuit of Von Lettow-Vorbeck to the Portuguese African Border until prevalence of *tsetse* fly prevented further cavalry operations.

The 22nd in 1916 proceeded to Mesopotamia, and working up the Tigris after Kut-al-Amara participated in the actions and outpost affairs of Duqma, Ranidasi, and around the Khalis Canal, going on to the Khurdistan country in the last year of the war. They eventually returned to India in 1920.

The junior Corps were back on the Frontier in time to resist the invasion by the Afghan forces in 1919. Operating from

Quetta they were present at the investment and capture of Spin-Baldak Fort.

Original uniforms of the 25th had been a dark Prussian green *Alkaluk* with scarlet facings and a scarlet *pagri;* upon the amalgamation the scarlet coat with blue facings and the blue turbans of "Sam Browne's" were adopted for the whole.

Sam Browne, destined to survive the vicissitude of war and hardship and to earn for himself the highest military honours and an imperishable name, died General Sir Sam Browne, V.C., K.C.B., K.C.S.I., in 1901. Robert Fitz-Gerald was denied such opportunities and soon after raising the 5th Punjab Cavalry was invalided to England where he died in 1853. The beautiful marble pulpit in St. Paul's Cathedral is his memorial.

The active life of Sam Browne's Cavalry has ended, for in 1936, after nearly a hundred years of service as a field unit, the Regiment became the Training Regiment of the 2nd Indian Cavalry Group with a permanent station at Ferozepore.

CHAPTER XVII

13th Duke of Connaught's Own Lancers

"Ghuznee," "Afghanistan, 1839," "Mooltan," "Punjaub," "Central India," "Afghanistan, 1879–80," "Burma, 1885–87," "Kut-al-Amara, 1917," "Baghdad," "Sharqat," "Mesopotamia, 1916–18," "North-west Frontier, India, 1917," "Afghanistan, 1919."

THIS Regiment was raised originally from the old Bombay Squadron of Cavalry formed for service under Lord Lake which, in 1805 and subsequent years, took part in many engagements including the seige of Bhurtpore.

In 1817 it was decided to split the Squadron, and from it were raised, with two troops each as a nucleus, the 1st and 2nd Bombay Light Cavalry; the two senior cavalry regiments of the Bombay Presidency Army.

From the time of their raising until 1838, both units saw almost constant service in Gujerat, Sind, and Kathiawar.

The whole of the 1st, with a detachment of the 2nd, were in the grand army under H. E. Sir John Keane, which invaded Afghanistan in 1839. They were present at the capture of

Ghuznee, taking a prominent part in the pursuit of the enemy' marched on to Kabul, eventually returning to India in 1840.

Eight years later the 1st joined the Bombay Division for service against the Khalsa. They assisted in the storming of Multan, and after the fall of the fortress remained as part of the garrison.

May, 1857, found the 1st at Nasirabad. The men, although appealed to by the disloyal *pandies* of the Artillery and Infantry in the Cantonment, refused to mutiny and, falling in under their officers, made a most gallant attack upon the mutineers. Their charge, though scattering the rebel foot, did not, unfortunately, succeed in capturing the guns, and after suffering losses, the Regiment was ordered to withdraw and escort the refugees to safety at Beawar.

The 2nd in that fateful May were at Neemuch. They manned and horsed two nine-pounder guns and used them with great effect in a successful attack on the fort of Nimbhara.

Both Regiments between them, were in nearly all the operations of the Central Indian campaign, 1858–60, the 1st being specially mentioned for the part played in the actions before Gwalior. The 2nd made a charge during the capture of the Fort of Jeerun, and another near Neemuch, when 4,000 rebels were put to flight; they also captured the Nawab of Kalaree with all his headmen. During these days some troops of both Regiments marched 4,319 miles in 632 days. One of the 2nd's subalterns—afterwards General Sir James Blair, K.C.B.—earned the Victoria Cross.

Disraeli's move in 1878 to show Russia that the British Empire had military resources other than those in Europe, sent the 1st to Malta and Cyprus for that summer—a troop of eighty-three from the 2nd being attached to them.

The 2nd went to Afghanistan 1878–89, serving under General Phayre in the operations leading to the occupation of Kandahar.

The 1st had the good fortune to be selected for service in Burma, 1885. Split up into small detachments, all of which

TYPES OF MEN ENLISTED INTO THE MECHANIZED INDIAN CAVALRY. LEFT TO RIGHT—MUSALMAN RAJPUT (RANGHAR), SIKH, PATHAN.

were continuously engaged for eighteen months, they were away from India for nearly two years.

In the mixed brigade of Indian troops under Brigadier-General Egerton sent to Suakin on the Red Sea in May, 1896, were the 1st Bombay Lancers, as they had then become. The object of the force was to suppress one Osman Dinga, but as he failed to put in an appearance, the Regiment returned to India, having thus seen three expeditions overseas.

With the re-numbering of the Indian Army, 1903, the Regiments became the 31st Duke of Connaught's Own Lancers, and the 32nd Lancers. The Duke of Connaught, who at one time commanded the Bombay Presidency Army, became the Colonel-in-Chief of the 1st Bombay Lancers in 1890, and the veteran Field-Marshal still so remains.

The North-West Frontier claimed the services of the 31st for the Great War.

The 32nd mobilized in the autumn of 1916 for Mesopotamia, becoming the Corps Cavalry to the Force. They did good work in the numerous engagements leading to the capture of Baghdad, and the wing attached to the 3rd Corps are believed to have been the first troops to enter that city. They continued the pursuit of the Turk, one wing, together with the then 21st Cavalry, charging a Turkish Column, capturing about 1,000 of the enemy.

On the 22nd April, 1917, was fought the battle of Istabulat. In this action a detachment of the 32nd achieved great fame by a mounted charge led personally by the Commanding Officer—Lieut.-Col. W. W. G. Griffith—against the Turkish entrenched position. A charge which has been described as the Balaclava of the Indian Cavalry, all the officers, and almost all of the men, becoming casualties.

Before returning to India the Regiment saw heavy fighting in the Kurdish rebellion of 1919. In one action, when without water and with only very little ammunition, they resisted for three days an enveloping force of Kurdish tribesmen several

thousands strong; whilst in 1920 they captured a Turkish standard from the insurgents in the operations around Kirkuk.

The 31st earned the honour for the 3rd Afghan War, in which they played a conspicuous part in the operations in and around Dardoni, and the Tochi Valley, their services being specially brought to notice.

Between the close of the operations of 1919 and the amalgamation, the 31st did a spell of garrison duty in Palestine, being the last Indian Cavalry Regiment to serve there. Owing to this the two Regiments were the last to amalgamate under the post-War re-organization, and thus the last to lose their old identity, the re-union in September, 1923, ending a separation which had begun over a century before.

The 13th Lancers were the only Regiment of Indian Cavalry to maintain a mounted band after 1914; the 31st had done so since 1846. The Drum Banners are blue, with lace and embroidery in silver. The bandsmen are Pathans, Sikhs and Moslem Rajputs—the classes enlisted into the Regiment.

The present uniform is a blue kurta with scarlet facings and cummarband, blue pagri, white breeches, and blue putties. Pre-War uniforms were green with white facings; in the early days French grey was worn for a time.

For many years the 31st were the only Indian Cavalry regiment to enlist Mahrattas, a full squadron of whom existed prior to the amalgamation with the 32nd. Mahrattas are not now included in the organization of Indian Cavalry, and thus the last trace of the old Mahratta Horse has disappeared. These men first came into the Regiment in 1862, as drafts from the Mahratta Horse. Other regiments enlisting them had great difficulty in recruiting because the *sillidar* system did not suit Mahratta characteristics. Mahrattas are very much attached to their homes and do not readily enlist in the Army unless obliged to do so for lack of funds. They were thus obviously unable to produce the *assami*, necessary under the *silladar* system, and consequently usually joined the Infantry. The

THE MOUNTED

THE LAST MOUNTED

BAND IN 1936

PARADE—9TH APRIL, 1938

31st Lancers, however, continued to enlist them when other regiments ceased to do so, by means of a Mahratta Fund which produced *assami* money for recruits on loan, when necessary.

The first standards presented to both the 1st and 2nd Regiments of Bombay Light Cavalry are preserved in the Officers' Mess. Amongst these is the *Kala Baba*, the blue or royal standard of the 1st Bombay Light Cavalry, possessing a superstition in the old days that its presence in battle brought success to the regimental arms. Until quite recently, recruits were sworn in on this, "1st Squadron or Royal Standard", whilst holding the fringe in their right hand. They are now sworn in on a regimental flag, which was made to take the place of the old standard, pending its replacement by a new one. This standard was also carried on certain parades until it became too fragile some years after amalgamation. Another standard still preserved is of special interest, since it is heraldically incorrect, as from whichever edge it is flown the devices are either reversed or upside down: the reason for this curious emblazonment is unknown.

In the spring of 1938 it was announced that mechanization of a part of the Indian cavalry was inevitable if the Army of India was to keep pace with those other armies of the world from which India might expect attack should her defences become obsolete. Of the two regiments selected to be the pioneers in the modernizing scheme, one was the 13th D.C.O. Lancers.

CHAPTER XVIII

The Scinde Horse (14th Prince of Wales's Own Cavalry)

"Meanee," "Hyderabad," "Cutchee," "Mooltan," "Goojerat," "Punjaub," "Persia," "Central India," "Afghanistan, 1878-80," "Somme, 1916," "Morval," "Cambrai, 1917," "France and Flanders, 1914-18," "Megiddo," "Sharon," "Damascus," "Palestine, 1918," "N.W. Frontier, India, 1914-15, '18."

ONE hundred years ago the modern province of Sind and its hinterland of Baluchistan were not the peaceful territories they are to-day. Sind had not been annexed, and the communications with the British Indian army in Southern Afghanistan were constantly harried by the Baluch tribes.

Baluchis are much akin to Arabs in their customs and possess a curiously high sense of honour. Those of the plains in those days were mounted on hardy ponies which could make long marches without water; while they were great adepts at mounted combat with the sword. The hill Baluch although mounted, preferred to fight on foot. Both classes were expert shots, possessing matchlocks of superior range to the weapon then in use in our forces.

THE STANDARD OF THE SCINDE HORSE (RIGHT SIDE)

THE STANDARD OF THE SCINDE HORSE (LEFT SIDE)

The country was a wilderness, except in the very few places where there was water. The climate, except for the three winter months, one of intense heat ranging up to 135 degrees in the shade without any relief by cool nights. Thus, Scinde, as it was then spelt, was the most unpopular place in all Hindustan with the soldier.

By 1838 molestation by marauders had become so bad that the Government decided to raise a special body of horse for service on the Scinde border; and so the Scinde Irregular Horse came into being at Hyderabad in the following year. The nucleus was the Cutch Levy of the Poona Horse detached for duty in Scinde.

The reverses of the British Indian forces in Afghanistan in the late months of 1841 caused the whole border to become disturbed. To check this in the west-north-west, Lieut. John Jacob of the Bombay Artillery was given command of the Scinde Horse. Jacob's aggressive policy of striking first soon had the desired effect, and in 1842, for the first time, the army on its way back to India from Southern Afghanistan was not troubled by marauders. All this was due to Jacob and the Scinde Horse he commanded.

In the cold weather of 1842 the conquest of Scinde was decided upon; and the Scinde Horse joined the army of Sir Charles Napier at Sukkur. Until then it had never paraded as a regiment, nor had the men drilled or trained together in formations larger than a troop. The active service and hard life in the Scinde Desert had made the unit a tough one, and Sir Charles described the troopers as "wild picturesque fellows, very like stage banditti". That was, however, before they were put into uniform.

The first battle of the War, fought at Meanee near Hyderabad, was a desperate one, and it was only through the great dash displayed by the Bengal Light Cavalry and the Scinde Horse that the action was decisive. The latter captured the enemy's principal standard, one surmounted by a

silver open hand, and still carries it to-day on ceremonial duties.

The Mirs of Scinde made their submission, but their followers were not prepared to accept defeat. Napier was unable to move out of Hyderabad before the March, and the delay consequently gave the leaders encouragement. Eventually the armies came to battle, and a brisk combat followed. The Cavalry, thinking the enemy to be in retreat, when in fact they were re-inforcing their flank, charged, technically at anything but the correct moment, with astounding results: the enemy fled and was pursued for miles. The leader Sher Mohamed, however, got away. Thus was the battle of Hyderabad.

Sher Mohamed was pursued constantly until the June when he surrendered to the attack of the Scinde Horse at the village of Shahdadpore. With this fight ended the campaign, for their services in which the Regiment was granted an Honorary Standard.

By 1846 Jacob's methods, and the services of the Regiment he commanded, had so impressed the authorities that another corps was ordered to be raised under his command. This second regiment was formed by splitting the first into two, and completing the establishment by recruits. The men were all from Hindustan proper, Mohammadans predominating. Jacob held strong views against the recruitment of Pathans or local tribesmen; while he considered the Sikhs inferior soldiers—this was doubtless due to the fact that service in Scinde was not attractive to the best classes from the Punjab.

By the time the two Regiments were ready for service, Upper Scinde was again in a state of chaos. The small garrisons were shut up in mud forts and were completely inactive. Raiding again began, and the climax came when fifteen hundred Baluch penetrated the British posts near Shikarpur, and the cavalry regiment sent in pursuit shirked engaging them.

The Scinde Horse were ordered up from Hyderabad, and John Jacob placed in charge of the frontier. A different

"order" started; the troops were kept on the move, the forts abolished, an effective intelligence scheme introduced, and the Baluchistan border once more quietened down.

The Corps took up their permanent quarters at what is now Jacobabad, one of the hottest stations in India, and although service in Scinde was intensely unpopular, service in the Scinde Horse was not—thanks to the admirable administration of Jacob, and the constant active service.

The men at this time wore olive green coats and red *pagris*; the officers silver helmets; and curiously they do not seem to have suffered from the sun, for they continued to wear these until 1882. The uniform to-day is dark blue with scarlet facings. Each man carried a leather water bag under his horse's belly—a novelty. The Regiments had no wheeled transport, two men sharing a camel or a pony. They could get on the move for a long campaign within twelve hours.

Herbert Edwardes from Multan sent urgent demands to Jacob for assistance in 1848; as a result a composite detachment of five hundred men were promptly despatched to the Punjab. But beyond, however, charging a body of the enemy on the road outside Multan City, they saw no service in that siege other than escort duty.

Joining Lord Gough's force after the fall of Multan the Detachment with the 9th Queen's Royal Lancers charged the Afghan Horse at Gujrat, two standards being captured and numbers of the enemy slain. With the other cavalry they carried on the pursuit to Rawalpindi, and it was an advanced piquet of the Regiment that received the Sikh Chiefs when they came in to surrender. The chase of the Afghan troops was continued to the Khyber, the dash and organization of the Scinde Horse exciting great admiration in the Punjab.

The Detachment remained at Peshawar some time, and the Afridi raiders who tried their measure with them between Peshawar and the Pass so got the worst of it, that they left the Corps severely alone.

In 1857 an expeditionary force under Outram was sent to Persia, Jacob commanding the irregular cavalry of which the 1st Regiment formed part.

They returned in time to take part in the Central India campaign suppressing the Mutiny, work which they were engaged upon until 1858. The 2nd Regiment remained to patrol the Baluch frontier where raiding, encouraged by tales of the Mutiny, had again broken out. Neither Regiment had any disaffected elements, although they were recruited from Hindustan districts where mutiny was at its worst.

Late in 1857 a third regiment was raised which did very well in the Mutiny operations, as well as in Abyssinia ten years later, earning therefore special thanks in despatches. The 3rd Scinde Horse also saw considerable service in Afghanistan from 1878. At Khushk-i-Nakhud two squadrons were surprised whilst off-saddled and under inspection. Saddling-up under fire they charged the 1,500 enemy tribesmen and put them to complete rout. The 3rd Regiment were also involved in the disaster at Maiwand. By orders they remained motionless under a very heavy fire for over five hours. Eventually, when ordered to advance and relieve the pressure on the almost overwhelmed infantry, orders and counter-orders followed in such quick succession from the Brigadier that hopeless muddle prevailed. The Brigadier, although personally a very brave man —as was the force commander General Burrows—was an infantryman of long service in the Police, and the result from the mismanagement of the cavalry and of the entire force was not due to lack of courage by the men; for among the 220 *sowars* of the 3rd Scinde Horse engaged, nine Indian Orders of Merit were earned.

The 3rd Regiment were disbanded in 1882 upon general reductions of the cavalry; and by 1885 the titles of Scinde Horse ceased to be borne by the remaining Regiments, they becoming the 5th and 6th Bombay Cavalry (Jacob-ka-Rissallah). Three years later the subsidiary titles were again changed to

Scinde Horse for the 5th Bombay Cavalry, and Jacob's Horse for the 6th. The 1903 changes made them the 35th Scinde Horse, and the 36th Jacob's Horse.

The 35th Scinde Horse remained in India throughout the Great War training drafts and seeing to internal security, as well as being in the Frontier operations, 1914 to 1916.

The 36th Jacob's Horse proceeded to France in the Lucknow Brigade of the 1st Indian Cavalry Division, to first see action at Festubert in January, 1915, in trenches waist deep in water. They took part in the second battle of Ypres, but had no serious employment until the Somme, their first Great War honour, and where they were engaged in both dismounted and mounted actions; though the expected cavalry chance did not come. The Regimental title was a source of confusion to their British comrades-in-arms, and they were at first humorously mistaken for one of the Jewish units of the King's—mistakenly called Kitchener's—New Army. All 1917 Jacob's Horse were in the line, and in the great action at Cambrai they particularly distinguished themselves by making many attacks in an endeavour to hold up the advancing enemy.

Moving to Palestine, where they joined the 4th Cavalry Division under the command of an old Scinde Horseman in the shape of Major-General Barrow, now General Sir George Barrow, G.C.B., K.C.M.G., and Colonel of the Regiment. This gallant officer risked his career in September, 1917, by disputing orders which he felt would mar the start of the great cavalry drive that began over the Sharon plain on the 19th of that month.

General Barrow was right; the cavalry got under way without impeding the attacking infantry, and the race for the Musmus Pass began. Had the original orders to wait until the infantry had cleared the front before the cavalry were to move been adhered to, the Turks would, in all probability, have had time to man the Pass, and history might have been very different: as it was we only won the race by a short hour.

On the successful cavalry went, and by the 6th October, Jacob's Horse were on the outskirts of Damascus. Since the 11th September they had travelled three hundred miles in nineteen days, with only one day's complete rest. The average weight carried by the horses over the stony and barren country was seventeen stone, while the loss in horses to Jacob's was not forty out of their five hundred. In these days of the automobile many probably do not realize what these figures mean; but to those that do, it indicates that the standard of horsemastership, the first essential of every cavalry soldier, was indeed a very high one.

The greatest known cavalry pursuit before that of Allenby had been by the French General Murat after the battle of Jena (1806). Murat marched four-hundred and twenty-five miles in twenty-three days. Due to poor horsemanship his losses in horses were enormous, and had his march not been through rich country, where remounts were readily available, his pursuit would have failed.

For Jacob's Horse the Great War was now over, but not until 1921 did they return to India, they being required to garrison Syria and Palestine.

The turn now came of the Scinde Horse; and on the 17th March, 1920, they marched out of barracks for Mesopotamia under an archway containing the Koran, the Sikh *sowars* avering that there was only one God for all, and as long as a man was a good man his creed was of no importance. Thus does the tone of the good regiment rise above differences of religion and race.

During the eighteen months' stay of the 35th Scinde Horse with Sir Aylmer Haldane's force they had six months of strenuous work in putting down the serious Arab rising, squadrons being constantly engaged with the columns relieving the scattered besieged posts. During the retreat of the "Manchester Column" from Hillah, two squadrons did their utmost, under trying conditions, to cover the retreat of the column, fighting

THE LAST MOUNTED PARADE OF THE SCINDE HORSE

RAWALPINDI—APRIL, 1939

with great gallantry to ensure the safe withdrawal of the guns. Later they charged the insurgents before Kufah, inflicting upon them heavy casualties.

The Scinde Horse left Mesopotamian soil to re-unite at Jubbulpore with Jacob's Horse on the 1st November, 1921, and on the 1st January, 1922, became the 14th Prince of Wales' Own Scinde Horse.

The two Regiments never forgot the traditions laid down by John Jacob. He lived in Sind practically continuously, and without leave, from 1839 under conditions of heat and discomfort until overwork killed him at the age of forty-seven in 1858. His name was held in respect in many out-of-the-way parts of Baluchistan for more than half a century after his death.

The Regimental Trumpet call is the old "Advanced Guard" of 1840, derived from the fact that Sir Charles Napier described the Corps to Jacob as "my old advanced guard".

In 1938 the Scinde Horse became part of the "advanced guard" of the mechanized Indian mounted services. One of the two cavalry regiments first selected to become equipped with armoured fighting vehicles, they continue their glorious traditions happy in the thought that as the mind of General John Jacob was considerably in advance of the age in which he lived, he would have welcomed the orders which made his old corps the precursors of motorization.

CHAPTER XIX

15th Lancers

"Afghanistan, 1879–80," "Afghanistan, 1919."

TO those bred in superstition the number of this regiment should be 13, for neither the 17th Cavalry, nor the 37th Lancers (Baluch Horse), who amalgamated at Lucknow on the 15th February, 1922, had the good fortune to be employed as regiments in actual battle, and thus the 15th Lancers—the title of the amalgamated corps—have not as many emblazonments as their more fortunate sisters.

The original 17th were raised in 1857 from several sources, principally the Muttra Horse and the Rohilkhand Police, and became "Robart's Horse" in the following year.

Colonel Robart, the first Commandant, was one of those remarkable men to whom the Indian Cavalry owe so much. Possessed of large private means—the Colonel was connected with the famous London banking firm of Robarts Lubbock & Co.—he was lavish with his money in his endeavour to make the Corps the crack corps of the new Bengal Army. He lived as a *Nawab*, adopted an Afghan family, and made India his home. He never kept accounts, and when asked by Government for a statement in order to settle financial matters could

THE MOUNTED PIPE BAND, 17TH BENGAL CAVALRY

not, or would not, do so. When he died, Government benefited to the extent of over a lac of rupees as a result of his generosity; in fact no one knew which horses in the ranks were his, and which were the Regiment's.

Through his adopted family he recruited only wild Afghan tribesmen, and kept on curiously intimate terms with them. Discipline as we know it was but scant, nevertheless Robart's word was law, the men knew it, while the rest of the Cantonment looked on with some sort of fear of the barbarous-like troopers.

The Regiment in 1861 became the 17th Bengal Cavalry, the men receiving their baptism of fire in the Bhutan Expedition of 1866.

They were employed guarding the communications from Jamrud to Dakka in the Second Afghan War, eventually reaching Kabul where Sir Frederick Roberts inspected them. The records do not show whether any actions were fought by the Regiment at that time, but the presumption must be negative, as no battle honours are borne for this campaign other than "Afghanistan, 1879–80".

The immense expenditure over the war in Afghanistan led to the reduction of several units of the Indian Army. Of the four cavalry regiments to be disbanded, "Robart's" was one. For what reason, none was stated, but it is probable that the Commandants who succeeded Robart found difficulty in instilling the discipline so essential to a regular regiment; and so after a short life of twenty-four years the 17th Bengal Cavalry was disembodied on the 13th June, 1881.

In 1885 the incident at Penjdeh, when the Russians attacked an Afghan force at a time when the two countries were supposed to be amicably settling their boundaries with the aid of a joint commission, led to fears of war with Russia. Many of the disbanded corps were restored to the establishment, but with an improved class composition, and so in the September of that year the 17th Bengal Cavalry were re-raised at Mian Mir from Punjabi Musalmans and Pathans—a composition

retained until the final amalgamation. They were thus one of the few entirely Moslem regiments of the Indian Army, and this accounted for their badge, the Star and Crescent upon crossed lances.

Bad luck seems to have been theirs from the first; for they lost their chance of active service against the Boxers in 1900 through outbreaks of cholera, their place in the China Expeditionary Force being taken by Watson's Horse, now the 6th D.C.O. Lancers.

The dice again spun against them in 1914. Put under orders for France at Allahabad in the August, the detached squadron at Calcutta contracted horse sickness and could not move. A squadron in replacement was supplied from the 27th Light Cavalry, and the Regiment in high hopes entrained for Bombay to discover on arrival that the journey had brought out cases of latent glanders in the horses: how contracted to this day remains a mystery. Despite the endeavours of the veterinary officers the Regiment could not go and their chance went to the 29th Lancers (Deccan Horse). The Regiment with sad hearts returned to Allahabad once more denied the opportunity of proving themselves on the field of war. Later, all of the original officers, and almost all of the men, were drafted to units overseas, where they rendered distinguished and meritorious services.

A complete Pathan squadron went overseas to East Africa in January, 1915, to reinforce the small garrison protecting the Uganda Railway until such time as the hosts from South and West Africa, and even the West Indies, could arrive to take offensive action.

A patrol of the Squadron had the misfortune to be surprised by two hundred Germans when off-saddled and at rest near Kilamanjaro. Despite the breast high grass, which rendered it difficult to observe or control, great bravery was shown, and after the two British Officers had been killed, and the Patrol almost surrounded, Jemadar Wazir Khan with great skill suc-

ceeded in extricating the survivors. The Jemadar afterwards received the Indian Order of Merit; as did Lance Daffadar Khan Sahib who, being wounded in the leg, hid in the bush until the enemy had departed; using a lance as a crutch he managed to reach the outposts thirty miles away after a struggle of six days without food, and no water to speak of, bringing with him his rifle and 100 rounds of ammunition. Stout work!

The Squadron were in many fights in the ensuing advance, but after the taking of Kissaki took no further part in the pursuit of the German forces.

Tsetse fly, horse sickness, and fever had taken their toll of the men and horses; and so just thirty strong the men of the 17th Cavalry—who were for a time the only cavalry on the East African front—returned to their native India in the January of 1917. They had marched many thousands of miles, not only to combat the enemy, but also the forces of nature, and they fully showed that their Regiment could and would give a glorious account whenever and wherever the opportunity arose. Being only a squadron the honours they earned in the Great War were denied to the Regiment.

The internal disturbances of 1919, and the outbreak of war with the Afghans, gave the 17th Cavalry plenty to do. One squadron was at Lahore, another on the Baluch Frontier, and a third in the Peshawar District. The splendid tone of the Regiment is evidenced by the fact that one patrol caused the arrest of four fully armed raiders which was only affected by the intimate knowledge of the people and the marauders' district, from which the patrol themselves came. These operations were against the Regiment's co-religionists, and through largely the countryside of their homes, and thus no one in the ranks could remain nonchalant to a war which affected them so closely. The credit for such a fine spirit as was displayed was due to the Regimental Priest, Khan Bahadur Moulvi Qazi Abdul Hakim Khan, who joined the Corps upon the re-raising, and remained with them until the amalgamation. He was, so

(*India Office photo*)

MECHANIZED INDIAN CAVALRY—A LIGHT TANK IN A STEEP CLIMB

we read, a man who succeeded in keeping friends with every one and yet never lost hold of the Regiment as their religious leader. His influence was unbounded and through him no word was ever said against the loyalty of the 17th Cavalry. His life, it is said, was a shining example of true religion.

The 37th Lancers (Baluch Horse) were also raised in 1885 at Shikarpur, Sind, as the 7th Bombay Cavalry with the subsidiary title of Jacob-ka-Risallah, the latter probably arising from the fact that many of the men were ex-soldiers of the 3rd Scinde Horse disbanded a short while previously. They became the 7th Bombay Cavalry (Baluch Horse) 1886, "Cavalry" being substituted for "Lancers" in 1890, "Bombay" being dropped and the number changed to 37 in 1903. Recruited entirely from Baluchis and Pathans of the Derajat District they were also an entirely Moslem regiment.

The 37th saw no service until 1919, when a squadron was in the defence of Thal then invested by the Afghan General, Nadir Shah, another squadron being in the relieving column. The operations, unfortunately, gave the Regiment as a whole no opportunity to show their mettle. In fact in the 1919 Afghan operations the *arme blanche* of the cavalry was only used once and then by British Dragoon Guards near Dakka.

The present uniform is dark blue with buff facings; those of the 17th had originally been white It is of interest to record that the 17th Cavalry had for a time a mounted pipe band headed by a kettle-drummer, and thus stole a march on "The Scots Greys", the only Scottish Regiment that have no official players of Scotland's national instrument.

The 15th Lancers became in 1937 the training regiment for the 1st Indian Cavalry Group, and thus no longer will have opportunity for battle; but in the permanent home at Jhansi recruits will be fully trained to take their places in the active regiments to which they will be posted—a responsibility no less important to the success of the army in war than that assigned to the troops in the field.

2ND MADRAS CAVALRY
OFFICER'S UNDRESS UNIFORM, 1842

CHAPTER XX

16th Light Cavalry

"Sholinghur", "Carnatic", "Mysore", "Seringapatam", "Burma, 1885-87", "Afghanistan, 1919".

THE 16th Light Cavalry are the oldest cavalry regiment in the Indian service. The exact date of raising is unknown, but is believed to be about 1776, when they were formed as the 3rd Regiment of Cavalry in the service of the Nawab of Arcot in the Carnatic.

The Nawab's regiments, of which there were four, were commanded by officers of the East India Company's service; the men, however, were almost always more or less in a state of mutiny due to the late arrival of their pay. Matters going from bad to worse, the Company decided to take the regiments into their own service temporarily, and in 1780 as temporary forces, they were employed under Sir Eyre Coote against Hyder Ali. They fought at Sholinghur, bringing about Hyder Ali's defeat. The 16th Light Cavalry being the only surviving regiment, are to-day the only cavalry to bear honours for that victory.

In 1784 the Nawab's regiments, who had been paid and mounted by the Company in the Mysore War, were taken over

permanently, and the 3rd Regiment entered the Company's service as the 3rd Regiment of Madras Native Cavalry. This transfer apparently was not to the liking of the majority of the men of the four regiments, and arrears of pay not being forthcoming, they mutinied, confining their officers. At the subsequent enquiry, it transpired that the 3rd Regiment, commanded by Captain Stevenson, had not heartily joined in the mutiny, and thus, while the other three units were disbanded, the 3rd were retained, and became the 1st Regiment of Madras Native Cavalry.

They did not, however, remain the 1st for long, as the Company desired that the senior regiment should be that commanded by the senior officer; an absurd ruling which led to many changes and much confusion. Between 1784 and 1788, the Regiment became the 4th and the 2nd, which latter number as Madras Cavalry or Lancers they succeeded in retaining until 1903.

The Regiment took part in the wars against Tippu of Mysore, in 1790, but after Bangalore, owing to the heavy horse casualties and the lack of remounts, were split up on garrison duty. The records show that three of the regiments had only seventeen horses between them.

In July, 1792, we learn that no man was to be enlisted in future unless he was willing to clean his own horse until he became a non-commissioned officer.

Operations against Tippu again starting in 1799, the Regiment were before Seringapatam, and in the action at Shikarpore which drove the bandit Doondiah, into the Mahratta country. Doondiah, however, rallied his followers and the now 16th Light Cavalry were with the force under Wellesley that harried him until his final defeat and destruction by the cavalry charge at Conahul. The subsequent years saw much activity, putting down free-booting and restoring law and order in the Carnatic plains.

They joined the Army of the Deccan for service against the

Mahrattas, and while participating in no actions for which honours were granted, were actively engaged in pursuing the Peshwa, the Rajah of Satara, a friendly chief who had been held captive by the former, being rescued by a squadron after the cavalry action at Ashtee.

After the Mahratta Wars, a long period of inactivity ensued, and no active service of any kind came until 1880, when a squadron was sent from the south to join the 1st Light Cavalry—now the 8th K.G.V.O. Light Cavalry—then operating in the Bolan Pass. This squadron joined Wilkinson's Cavalry Brigade marching upon Kandahar, and finally returned from Afghanistan in 1881. Only a squadron being employed no honours were granted to the Regiment.

The 2nd Madras Cavalry, as the present 16th were then known, became "Lancers" in 1886, just prior to their joining in the conquest of Upper Burma. Split up into innumerable detachments, who fought many hand-to-hand encounters, either with the enemy proper or with dacoits, who then infested the Burmese jungles, the Regiment rendered a good account of itself, an Indian officer, Subedar Mohomed Dawood, being awarded the Order of Merit for saving the life in action of a soldier of the Hampshire Regiment.

The abolition of separate armies in 1903 changed the title to the 27th Light Cavalry, which was retained until 1922.

The Great War gave the 27th Light Cavalry little opportunity; they sent, however, their Machine-Gun Section for service with the Hariana Lancers with Townshend's force on the ill-fated road to Baghdad. One squadron proceeded to Persia in May, 1918, and were specially mentioned for their services in the actions of Dasht-i-Arjan and Mian Kotal in December, 1918, and January, 1919. Another squadron formed part of the punitive expedition against the Marris on the North-west Frontier, March, 1918.

In May, 1919, when wholesale defections of the Militia and Constabulary on the Indian side of the Border occurred, and

The only Double Set of Cavalry Standards carried

in the Armies of the British Commonwealth.

the Afghan government decided that the time was ripe for an invasion, the Regiment was at Dera Ismail Khan. A squadron was at once despatched to hold the Gomal Pass. This squadron of only three troops charged a tribal *Laskkar* 400 strong, inflicting severe casualties, and thereafter, moving from place to place with such rapidity, and attacking every enemy body it could find, gave the impression to the foe that there was a much larger force of cavalry in the area than was actually the case. Another squadron charged an enemy *Laskkar* 500 strong and drove them into the hills. While the whole Regiment destroyed the capital of the Sherranis by a brilliant night manœuvre, 24th–25th June, 1919.

The first uniforms were red, following the pattern of the British, these were changed to French grey in 1818, and remained this colour with buff facings right up to 1914. Upon the reorganization into the group system, dark blue with French grey facings was adopted.

The *sillidar* system was never introduced into the Regiment, although such was apparently contemplated in 1865 and again in 1902. Men of the south are now no longer enlisted, the ranks to-day comprising Jats, Kaimkhanis and Rajputana Rajputs.

The Regiment were one of the first to be selected for Indianization, and the majority of the officers are now Indian gentlemen.

The 16th Light Cavalry are one of the few regiments of Indian cavalry that carry standards. Standards were first presented in 1788, but were abolished in 1869. On May 2nd, 1920, however, special sanction was accorded by the Government of India to their replacement, and the Regiment, unlike their contemporaries of the British service, carry two standards. In the British Army standards and guidons to the extent of one per regiment are only carried by heavy cavalry, i.e. Dragoons and Dragoon Guards: light cavalry, i.e. Hussars and Lancers carry no standard.

Although officially described as standards, the banners of the Regiment are in effect guidons. The guidon was the swallow-tailed flag carried originally by a knight: when the knight was promoted he cut off the swallow-tails and his guidon became a square flag known as a standard or banneret—hence knights-banneret (baronet). Thus senior regiments carry standards, junior regiments guidons. The new standards also do not follow the usual practice of being the colour of the regimental facings, they being buff, the base for the lettering red with gold embroidery—the lace of the old Madras cavalry was silver—and whilst carrying the modern battle-honours, bear the old Madras number of the Regiment.

The Regiment to-day is numbered 16 in the line, but yet on parade has precedence over all the other Indian cavalry regiments. They are thus unique in the armies of His Majesty King George VI, who honoured them by becoming their Colonel-in-Chief on his coronation day, 1937.

THE STANDARD OF THE POONA HORSE

(The drawing represents an Indian officer in present-day full-dress uniform carrying the standard, the latter, however, on parade is carried by a non-commissioned officer.)

CHAPTER XXI

The Poona Horse (17th Queen Victoria's Own Cavalry)

"Corygaum," "Ghuznee, 1839," "Afghanistan, 1839," "Candahar, 1842," "Ghuznee, 1842," "Cabool, 1842," "Meanee," "Hyderabad," "Reshire," "Bushire," "Kooshab," "Persia," "Central India," "Abyssinia," "Kandahar, 1880," "Afghanistan, 1879–80," "China, 1900," "La Bassee, 1914," "Armentieres, 1914," "Somme, 1916," "Bazentin," "Flers-Courselette," "Cambrai, 1917," "France and Flanders, 1914–18," "Megiddo," "Sharon," "Damascus," "Palestine, 1918," "Shaiba," "Ctesiphon," "Tigris, 1916," "Mesopotamia, 1914–16," "Afghanistan, 1919."

THE first forebears of this Corps were raised in 1817 as part of an auxiliary force for service in the territory of the Peshwa of Poona. Bajee Rao, the Peshwa, had been unable to preserve order, and as the situation was far from satisfactory the Peshwa agreed by treaty that a force of all arms, under the control of the Governor-General and officered by the East India Company's officers, should be established to the cost of which the Poona

revenues would contribute. The cavalry of the force was to consist of ten *risallahs* of 500 men each—one was transferred from the already famous Skinner's Horse of the Bengal Army.

Organization was in progress when the services of the force were first requisitioned. The Peshwa, whilst ostensibly preparing to make common cause with the Company's forces in the Pindari War, was actually preparing to lead a Mahratta confederacy against them. The Resident at Poona—Mountstuart Elphinstone—suspected this, and ordered the Government troops up to Kirkee. One thousand ranks of the new Poona Auxiliary Horse were ordered to join them. This reinforcement arrived too late for the battle of Kirkee in which the Peshwa unsuccessfully attacked the British Indian forces, but onwards from then many detachments of the Horse were employed in various fights and skirmishes until the end of the war.

The most famous exploit in which the Corps were engaged was at Corygaum, where a detachment of 300 Horse, together with what is now the 2nd Battalion 4th Bombay Grenadiers, and a section of Madras European Artillery were surrounded by the whole of the Peshwa's army of 28,000 horse and foot, as well as guns, under Bajee Rao's personal command. This remarkable and desperate action was fought on New Year's Day, 1817, and after a day's hand to hand fighting, in which the Medical Officer led the sepoys in a bayonet charge, the small force so successfully resisted the enemy's attack that the Peshwa withdrew his forces in disgust. The battle of Corygaum was definitely one of the proudest triumphs of the old Bombay Army.

By 1819, on the dissolution of the Mahratta Confederate States, the Corps was reduced gradually, until by 1838 only two *risallahs* remained. These were principally employed in armed police duties in Deccan districts.

In 1820 were formed the 3rd Bombay Light Cavalry, a detachment of which first saw service at the reduction of the

Fortress of Kithoor, a detachment of the Poona Horse being present also. These two corps, the Poona Horse and the 3rd Bombay Light Cavalry, not quite a century later became one: as such an amalgamation had to be, it was admirable in many ways as both Regiments were raised in the same Cantonment, Sirrur, and both had fought side by side in many campaigns.

The Poona Horse alone went to the first Afghan War in 1839, being at Kandahar and the storming of Ghuzni. They were also with the force sent to punish the treachery of the Khan of Kelat, fighting in which no quarter was given nor expected. The Khan even murdered his favourite women before taking his own life, rather than surrender.

Both Regiments were together in the 2nd Afghan War with the Bombay Cavalry Brigade. Before Kandahar in 1842 the Poona Horse captured a standard; and at Oba, the 3rd Bombay Cavalry became heavily engaged—an action incidentally which found no mention in official despatches. Both were at the battles for Ghuzni and Kabul, which released the prisoners taken in the disasters of 1840, who had been reduced to slavery by the Afghans.

The Scinde War of 1842–43 brought both Regiments on the scene; the Poona Horse fighting hard at Meanee, the 3rd arriving in time for the victory of Hyderabad where they made the final charge with the Scinde Horse which ensured the victory. The Poona Horse were also hotly engaged in parallel mounted action at the same time.

The Shah of Persia invaded Afghanistan in 1856 and captured Herat. This was contrary to his treaties and he was called upon to retire. He refused to do so, and so war was declared upon him. An expeditionary force was sent from India and with it went the Poona Horse and the 3rd Bombay Light Cavalry. The force aided by the Indian Navy landed near Reshire in December, 1856, captured that old Dutch port, reduced the fortified town of Bushire, and then rested to await the rest of the army sent from India. Early in January, Sir

State Trumpet and Banner carried by the State Trumpeters from the Indian Cavalry at the Delhi Durbar in 1902

The Trumpet is a replica of the State Clarions of the Sovereign carried by Trumpeters of the His Majesty's Household Cavalry since the reign of King Charles II, while the Banner follows the English design with the cherubs of the Stuart monarchs.

James Outram, the appointed commander arrived with the remainder of his forces. Hearing that the Persians expected reinforcements he decided to attack their army 7,000 strong assembled near Khushab—the British Indian forces numbering 4,500. Attack the enemy he did, the Poona Horse and its sister regiment making two of the most brilliant charges ever known in the annals of cavalry. The Poona Horse practically annihilated the Persian 1st Khusgai Regiment of Fars; spiked the guns and captured the Standard. The 3rd charged another square, and in the end, the enemy, completely demoralized, fled, and only escaped total disaster owing to there being no fresh troops available to pursue.

Two officers of the 3rd Bombay Light Cavalry, Lieuts. A. T. Moore and J. G. Malcolmson, earned the Victoria Cross for their bravery in the battle.

The Standard captured was surmounted by a silver hand, the palm of which is inscribed in Persian "The hand of God above all things", and dates from about the time A.D. 1066. The Regiment were permitted as a reward to carry it on parade; the standard has long since been replaced, but the hand is still carried on the standard staff.

Both Regiments were in the Central India campaign under Sir Hugh Rose, 1858. Lieut. H. E. Wood, afterwards Field Marshal Sir Evelyn Wood, of Her Majesty's 17th Lancers, winning the Victoria Cross when in charge of a troop of the 3rd Bombay Light Cavalry at Sindwaha on the 19th October, 1858. He attacked almost single handed a band of mutineers making a stand, and routed them completely. Neither Regiment lost a single man by desertion or dis-affection, although all came from mutinous areas. The 3rd were also in the subsequent operations in Rajputana and Malwa.

Ten years after the Mutiny, King Theodore of Abyssinia, annoyed with the failure to receive an answer from a letter addressed to Queen Victoria, imprisoned many of the Queen's subjects, as well as those of other countries. Sir Robert Napier

and a British Indian force were sent to release them. The 3rd Bombay Cavalry went overseas, but on landing their horse casualties from "African glanders" was so severe, that it looked as if they would be deprived of their proper rôle. Remounts, however, were forthcoming and they took part in the advance to Malaga, which fell to the British Indian arms, Theodore taking his life. Little opportunity for action was afforded to the cavalry owing to the country, but the continual piquetting, outpost, and escort duty was no easy work.

In 1876 the Prince of Wales, later King Edward VII, became Colonel of the 3rd, who were also accorded the style of "The Queen's Own".

Both Regiments were besieged in Kandahar in 1880 until relieved by the force under Roberts marching from Kabul. The 3rd were in the Helmund, as well as at Maiwand, both at Deh Khoja, the operations of the Kandahar garrison before their investment. Both joined in the final battle at Kandahar.

Prince Albert Victor of Wales, the eldest son of the Prince of Wales, became Colonel of the Poona Horse, 1891. He died in 1892, and that Regiment carried the subsidiary title "Prince Albert Victor's Own" until 1921.

The 3rd (Queen's Own) Bombay Light Cavalry went east to China to join the International force, suppressing the Boxer rising in 1900, though nothing spectacular fell to their lot.

The 34th Prince Albert Victor's Own Poona Horse, their title from 1903, sailed for France in 1914, arriving to reinforce the hard-pressed troops of Britain before La Bassee; in addition to that honour they were the only Indian Cavalry to earn Armentières, 1914. They were in mounted and dismounted combat from the Somme until Cambrai: thereafter in the 14th Cavalry Brigade of the Desert Mounted Corps in Palestine.

At Festubert, on the 24th November, 1914, Lieut. F. A. DePass of the 34th Poona Horse won the Victoria Cross in a most gallant and intrepid way. Accompanied by sowars

Abdulla Khan, Fateh Khan, and Firman Shah he entered a German sap, the fire from which was causing numerous casualties. In the face of the enemy's bombs he crawled up to the loop-hole of the traverse, coolly inserted a charge of gun-cotton and fired it, with the result that the sap was completely demolished. The following day he also rescued under heavy fire a wounded sepoy who was lying exposed in the open. Lieut. DePass lost his life on this day in a second attempt to capture the sap which had been repaired and re-occupied during the night by the enemy. The three sowars who so bravely supported their officer in this affair all received the Indian Distinguished Service Medal.

The 33rd Queen Victoria's Own Cavalry were the first cavalry to go to the Persian Gulf and the only regiment whose Mesopotamian honour starts from 1914. When after a long and victorious advance towards Baghdad, an unsupported British Indian force was compelled to retire from the battle-field of Ctisephon, it was only through the work of the cavalry of the force among whom were the 33rd, that the infantry were able to extricate themselves from disaster, and able to retire to make their determined stand at Kut. When investment was certain, most of the cavalry was sent to join the hoped for reinforcements; but in vain did the 33rd join in the many attempts to relieve their late comrades, and with their ranks much depleted they were relieved and sent to their homeland to recruit and refit.

The 33rd did not return overseas but to the Frontier, and with the rest of the 1st Indian Cavalry Brigade which they had joined at Risalpur, were with the striking force invading Afghanistan to Dakka in 1919.

The 33rd and the 34th amalgamated in 1921 as the 17th Queen Victoria's Own Poona Horse, but five years later discarded that title for their present simple and almost baptismal name.

The original uniform of the 3rd Bombay Light Cavalry was a French grey jacket with orange facings and silver lace, but

in 1883 this became dark green, and in 1901 blue. The Poona Horse also wore green, but changed to blue about the same time. The present dress is the almost general blue of the Indian

THE OFFICERS OF THE 6TH BENGAL CAVALRY IN 1873

Cavalry with French grey facings. The Cypher of Queen Victoria within the Garter, an honour granted to the old 3rd Bombay Cavalry, is still worn on the appointments.

The Poona Horse has more battle honours than any other cavalry regiment in the King's service.

CHAPTER XXII

18th King Edward VII'S Own Cavalry

"Punniar", "Moodkee", "Ferozeshah", "Punjaub", "Tel-el-Kebir", "Egypt, 1882", "Burma, 1885-87", "Punjab Frontier", "Somme, 1916", "Morval", "Cambrai, 1917", "France and Flanders, 1914-18", "Megiddo", "Sharon", "Damascus", "Palestine, 1918", "Shaiba", "Kut-al-Amara, 1915", "Ctesiphon", "Tigris, 1916", "Mesopotamia, 1915-16".

THE necessity for creating a reserve army as a result of the reverses suffered in Afghanistan, and the seriousness of the internal situation in India, caused the formation of the 8th Bengal Irregular Cavalry at Fatehgarh, 1842, the senior ancestor of the present Regiment. They were first "blooded" at Punniar, where the state troops of Gwalior disputed the right of the Indian Government to support the Gwalior Durbar then wishing to be on friendly terms. After Gwalior came the first Sikh War in which the 8th B.I.C. were specially thanked by the Governor-General for their services at Sobraon.

The occupation of the Punjab required still more troops, and so at Meerut and Cawnpore in 1846 the 16th Bengal Irregular Cavalry—a number changed to 17th a year later—were raised, and thus was born the junior ancestor.

The scene now changes to Bareilly, 1857, where the 8th B.I.C. were stationed, and where events occurred which almost led to their extinction. They were believed to be loyal and so two additional troops were raised to suppress the revolt. The old ranks may have intended to remain staunch, but the new recruits were apparently such apt propagandists that before the end of the May the greater part of the troopers had deserted. All the Indian officers, and most of the non-commissioned officers, however, stood firm, and by the following February, the Corps was recruited up to full strength; while the Regiment's subsequent fine behaviour in the Oudh campaign definitely set aside any ideas of their disbandment.

The other Regiment, the 17th, were on the Frontier; they remained true, although recruited from rebel districts, and no less than ten special awards were earned by Indian ranks for outstanding gallantry displayed in the tainted areas of the Punjab where, under John Nicholson, they had been employed preserving order.

Changing their titles in the post-Mutiny reorganization, the 8th became the 6th, and the 17th the 7th, Bengal Cavalry, in 1861.

From 1861 for twenty-one years, the 6th were without active employment, but an officer—Captain R. W. Sartorius—won the Victoria Cross on the 17th January, 1874, at Aboogoo, Ashantee, West Africa, for having rescued, under heavy fire, the badly wounded African Sergeant-Major of the Hausa Corps to which Captain Sartorius was attached.

Wearing khaki for the first time, the 6th were present at Tel-el-Kebir, 1882, and afterwards in the Cairo garrison; where the Mohammedan squadron had the great honour of escorting the Holy Carpet for the Prophet's shrine at Mecca from the

Silver Trumpet and Banner of the Trumpet-Major of The 18th King Edward VII's Own Cavalry. The Colour of the Banner is Red, Purple and Gold.

Citadel to the station. For services rendered in Egypt, the Prince of Wales, afterwards King Edward VII, became their Colonel, and the style of "Prince of Wales' Own" was conferred, as well as the honour of wearing the Prince of Wales' Plume on their appointments.

The 7th Bengal Cavalry were twenty months in Burma, 1886–87. The country being unsuitable for cavalry, most actions against the enemy stockades had to be made on foot through dense jungle.

In the Tirah campaign, the 6th were in several actions in the Kurram Valley. At Thabi on the 29th November, 1897, two squadrons, together with two companies of the 5th Gurkhas, were heavily engaged extricating the 12th Bengal Infantry who had had the misfortune to get into a difficulty.

Neither Regiment was affected by the re-numbering of 1903, except that "Bengal" dropped out of the title. The 6th became "King Edward's Own" and received the honour of the Royal and Imperial Cypher, 1906. At King Edward's Coronation Durbar, 1903, Major A. G. Maxwell of the Regiment was the Chief Herald and read the Royal Proclamation.

As most of the men of the 7th came from the Hariana Doab—the terrain between the Jumna and Sutlej rivers—they became, in the year 1903, the 7th Hariana Lancers.

The closing days of 1914 found the 6th K.E.O. Cavalry in France at La Bassee as part of the 2nd Indian Cavalry Brigade, and from then on they participated in the trials and glories of the Indian Cavalry Corps in France and Palestine. After the Armistice they remained on in Asia Minor and Egypt, arriving back in India, October, 1920.

The 7th Hariana Lancers with the Machine Gun Detachment from the 27th Light Cavalry arrived in Mesopotamia on the 9th March, 1915. They fought in the three-day battle of Shaiba, where on the 13th April, Major G. C. M. Wheeler posthumously received the Victoria Cross for gallantry in leading the attack on the enemy position. They were the only

cavalry present at the first capture of Kut-el-Amara; and proceeded on with Townshend's force to Ctesiphon. On his subsequent retirement to Kut, a squadron became his Divisional Cavalry and remained with him until the capitulation on the 29th April, 1916. The men of this squadron, all Hindustani Musalmans, suffered severely during the siege, and their afterwards long period of honourable captivity. The remainder of the 7th were in action more or less continually after the fall of Kut until they returned to India in October, 1916.

The two Regiments amalgamated, together with one squadron of Rajputs from the 32nd Lancers, at Risalpur, 1921. His Majesty King George V assumed the Colonel-in-Chiefship the same year.

The first uniform of the 6th was green with red facings, red turbans, and later, in full dress, green cloth hussar busbies for all ranks. The coat subsequently became blue with red facings. The 7th wore scarlet with blue facings. On the reorganization of 1921, blue with French grey facings and gold lace was adopted. The Regimental colours are purple, gold and red, the racing colours of King Edward VII; a feature which they share with the Honourable Artillery Company, the ribbon of whose long service medal is devised from the same source.

Major-General, Nawab Malik Sir Umar Hyat Khan, G.B.E., K.C.I.E., M.V.O., A.D.C., Tiwana, Honorary-Colonel of the 19th King George V's Own Lancers

CHAPTER XXIII

19th King George V's Own Lancers

"Taku Forts", "Pekin, 1860", "Ahmad Khel", "Afghanistan, 1878-80", "Tirah", "Punjab Frontier", "Somme, 1916", "Bazentin", "Flers-Courcelette", "Morval", "Cambrai, 1917", "France and Flanders, 1914-18", "Megiddo", "Sharon", "Damascus", "Palestine, 1918".

OF all the regiments that bear the title, King George V's Own, none have such an intimate connection with that beloved monarch as the 19th Regiment of the Cavalry of India. The Colonel, The Lord Wigram, was from 1905 an Equerry, then Assistant Private Secretary, and subsequently Private Secretary to His late Majesty until the end of his reign. The Honorary Colonel, Major-General, Nawab Malik Sir Umar Hyat Khan, Tiwana, an Aide-de-Camp—the first Indian a non-ruling prince to be made so. While the then 18th King George's Own Lancers, the senior partner of the present regiment, had the honour of forming the Indian Cavalry Escort to His Imperial Majesty on his state entry into Delhi in 1912 The Honorary Colonel, as Captain Malik Umar Hyat Khan, Tiwana, was the Assistant Delhi Herald at the Imperial Durbar, and proclaimed in Urdu, King George V, Emperor of India.

Another British officer of the 18th Lancers—Major Corbyn—and an Indian officer—Risaldar Gul Mahwaz Khan—were members of His Majesty's suite.

The Regiment's connection with the Royal house goes back to 1906, when King George V, as Prince of Wales, during his visit to India became Colonel-in-Chief of the 18th, their title in consequence becoming the 18th (Prince of Wales' Own) Tiwana Lancers. Tiwanas are a tribe of Musalman Rajputs who come from the Jhelum Valley and the Shahpur District of the Punjab and are renowned for their sporting and soldier-like qualities.

This regiment, the 18th Lancers, were originally raised as the 2nd Mahratta Horse at Gwalior in 1858 for duty in Central India. A detachment of Tiwana Horse was subsequently co-joined and on the reorganization after the Mutiny, the unit became the 18th Bengal Cavalry.

The 19th Lancers, who amalgamated in 1921 with the 18th to form the present Corps, were founded by Lieut. Fane of the Madras Native Infantry—curious that most of the Mutiny period cavalry should have been raised by infantry officers, and still more curious that the Corps came into being when other cavalry units were being disbanded as surplus. The fact was that a regiment was urgently wanted for service overseas in the cavalry brigade being sent from India to enforce the treaty which allowed Queen Victoria to have an envoy in Pekin, the Chinese capital. As the regiment would have to cross the sea, it was necessary that the men should be volunteers, an old custom in the Bengal Army dictated by rules of caste. From the disbanded regiments Fane called for men, and the ranks were soon filled, principally from the 3rd Regiment of Hodson's Horse, those taken being Sikhs, Pathans, and Punjab Musalmans.

The new regiment reached Calcutta, and together with Probyn's Horse, who had also volunteered for service, were soon on the way: Fane's Horse, as they were known, embarking

in flats were towed to Hong Kong, where the expeditionary force, with which French forces finally joined, was assembling.

The route to Pekin lay up the Peiho river, the entrance to which is guarded by the Taku Forts, the garrison of which had repulsed with heavy loss a British Naval landing party two years before. The subsequent expedition landed, and on the 13th August, 1859, proceeded to storm the Chinese stronghold. The Tartar Cavalry, with great gallantry, charged the besieging artillery; but they had not reckoned with the Indian Cavalry Brigade, whose two regiments of horse—Fane's and Probyn's—supported by half of the King's Dragoon Guards, counter-charged with such intrepidity that the Tartars fled. The pursuit continued for miles until exhaustion called a halt. Through this excellent work, the capture of the Taku became a very simple matter. The force advanced to Pekin, and in the final battle for the city, the cavalry, with Fane's in the van, so routed the Manchu-Mongolian-Tartar horsemen that never again did they wait to cross swords with the gallant and dashing Punjabis. The campaign, during which five Indian members of Fane's Horse earned Orders of Merit for outstanding deeds of bravery, ended in the November, and Fane's Horse on their return to India became the 19th Bengal Cavalry (Lancers).

The 18th and 19th were next on active service in the wars with the Kabul ruler over the foreign policy of Afghanistan. The 18th were employed keeping open the communications; very strenuous work when such ran through Waziri and Mahsud territory, the inhabitants of which missed no chance of causing embarrassment to the field army. The 19th were in the garrison of Kandahar, and when it became necessary for a force under Sir Donald Stewart to march northwards to Kabul in an attempt to pacify the country and to join hands with Sir Frederick Roberts, the 19th went therewith to fight most gallantly in the fanatical *ghazi* action at Ahmad Khel.

It was now the turn of the 18th to add to the laurels; this they did through the hard work put in in the Kurrum Valley

(India Office photo)

THE NEW INDIAN CAVALRY—ARMOURED CAR CREWS PARADING IN FRONT OF THEIR VEHICLES BEFORE MOVING OFF.

and subsequently as Divisional Cavalry to the Tirah expedition during the general Frontier "blaze" of 1897–98.

Except for the Zakha Khel and Mohmand operations of 1908, in which the 19th were participators, the years following the Tirah were ones of peace for both Regiments. But in the South African War, 1899–02, Lieut. F. S. Maxwell of the 18th, then attached to Roberts' Horse, won the Victoria Cross for conspicuous gallantry when U and Q Batteries, R.H.A., were ambushed at Sanna's post on the 30th March, 1900. And Lieut. Malik Umar Hyat Khan, Tiwana, went to Somaliland in command of the 54th Camel Cadre in the expedition against the Mad Mullah two years later.

When the Great War opened, a detachment of the 18th were escorting a boundary commission in North Persia. Their British officer was ordered to return alone via Turkey, and so he left his dog and two tail-less Persian chickens with the Detachment, who were ordered to march back to their homeland. The story of how this little band of twenty men—one died on the way—under an Indian officer named Tiwakli Khan, eventually reached India after traversing some 1,600 miles of turbulent and hostile territory, and delivered up nineteen horses—one had fallen down a precipice—their arms and ammunition, the baggage, one dog, and two tail-less Persian chickens is an epic. So much so that His Majesty the King-Emperor summoned Jemadar Tiwakli Khan from France, where he had re-joined the Regiment, and personally decorated him with the Order of British India.

As will have been observed, the 18th went to France, where they arrived in the December of 1914, having been preceded by the 19th a month earlier; the 18th were in the Meerut Brigade of the 2nd Indian Cavalry Division, the 19th in the Sialkot of the 1st. Both Corps were in all the actions of the Indian Cavalry Corps.

At Gauche Wood, near Epehy, during the German counter-battle of Cambrai, the 18th found themselves on December 1st,

1917, fighting dismounted tooth and nail mixed up with the 2nd Battalion of the Grenadier Guards, and thus for the first time in history were Indian Cavalry fighting as infantry with the finest infantry in the world. The fighting spirit and dash of all ranks of the 18th Lancers made the greatest impression on the Grenadiers—so records the History of the Grenadier Guards. In memory of Gauche Wood the Grenadiers later presented one of their Bugles to the 18th; in return the Indian regiment gave a silver statuette of a Bengal Lancer. The comradeship existing between the two Regiments is now cemented annually on Gauche Wood Day by the exchange of telegrams. The 18th also received a silver tray in the shape of a Maple Leaf—the emblem of Canada—from the Fort Garry Horse, a Canadian Regiment with whom they had fought side by side; while another silver salver was presented by the Royal Canadian Dragoons. Who after this dare say that among the armies of the British Commonwealth the Army of India is not an equal partner?

Not long after Cambrai, in which the 19th had also taken a part, the Regiments were on their way to Palestine, where the summer of 1918 was spent in the unhealthy Jordan Valley. The great cavalry chance of the war had come, and in the series of brilliant operations that followed in the autumn, the 18th and 19th played their rôle with élan. The 18th were with the Brigade selected to raid Nazareth—the enemy G.H.Q. from which the German-Turko Commander-in-Chief escaped in his pyjamas by a matter of two hundred yards: the 19th with the column through the Musmus Pass, making thereafter one of the outstanding night marches of the War. And so on to Damascus, where ended what was one of the greatest cavalry victories in history. 250 miles in just a week was a triumph of horsemanship; and the laurels went to "A" Squadron (Sikhs) of the 19th Lancers, who arrived at Damascus without losing a horse from sickness or causes other than battle.

The Regiments continued the pursuit towards Aleppo, but

the world-wide influenza epidemic of that time took its toll, and the 19th had to halt; the 18th being just able to continue reached that town on the day of the Turkish Armistice, October 31st, 1918.

The 18th returned to India, November, 1920, the 19th following early in 1921, to amalgamate at New Delhi on the 23rd August of the same year.

Red, with a blue pagri, was worn by the 18th from their birthday. The 19th on their foundation, were dressed in French grey with a scarlet pagri and cummarbund. French grey with red facings had been the uniform of the old Bengal regular cavalry regiments that had mutinied, and there being large stocks of the material available, the Regiment were fitted out with this; on their return from China, the kit was changed to dark blue with red facings and blue pagris. The facings of both Regiments subsequently underwent changes; the 18th from blue to white, and the 19th from red to light blue and later French grey. Upon the amalgamation the scarlet coat, white facings, and gold lace—the 19th had silver lace—of the senior regiment were adopted as the present uniform.

In 1936 Lord Wigram, Colonel of the Regiment, emerged from his retirement to be permanent Lord in Waiting to H.M. King George VI, and thus once more a King-Emperor found counsel and guidance in an officer of the Indian Army.* Lord Wigram is also Keeper of His Majesty's Archives, Deputy Constable and Lieut.-Governor of Windsor Castle.

* Sir Dighton Probyn, of Probyn's Horse, was Private Secretary to King Edward VII, both as King, and as Prince of Wales.

RISALDAR-MAJOR, MURRAY'S JAT LANCERS, 1910.

CHAPTER XXIV

20th Lancers

"Charasiah," "Kabul, 1879," "Afghanistan, 1878–80," "Neuve Chapelle," "France and Flanders, 1914–15," "Kut-al-Amara, 1917," "Sharqat," "Mesopotamia, 1916–18," "Persia, 1916–19," "North-west Frontier, India, 1915."

A REGIMENT formed in 1921 from the 14th Murray's Jat Lancers and the 15th Lancers (Cureton's Multanis), both regiments which prior to the Great War had been class regiments; that is to say, comprised of one class of men only and not of mixed classes, as was the case with the remainder of the Indian Cavalry, with the exception of Skinner's Horse. The 14th was comprised entirely of Hindu Jats, and the 15th of Multani Pathans and kindred tribes.

The history of Murray's Jat Lancers starts in the Aligarh District in August, 1857. The countryside around being in a very disturbed state, a small European force was sent out from Agra Fort on the 24th of that month to attack the rebels near Hathras. Thakoor Gobind Singh and other Jat landholders joined the column with 70 horsemen and 200 matchlockmen. Captain Murray of the disbanded Gwalior Contingent, and a volunteer, took command of the horsemen, and in the action

that followed, the Jats, particularly the mounted men, did such good and gallant service that subsequently Captain Murray was instructed to raise a levy 400 strong for local service in the district under the designation "Jat Horse Yeomanry". Murray was unable to make much headway in recruiting until after the fall of Delhi, and it was not until mid-October that active steps could be taken, Gobind Singh's men being used as a nucleus.

Very slow progress was made. Jats had not previously been recruited into the Bengal regiments, and while willing to help the Government to put down the mutiny as volunteers, were adverse to accepting permanent military service. The influential *zemindars*, however, subsequently enrolled themselves as officers, and bringing their relatives and tenants with them, the Corps was soon brought to strength; and discipline and drill became slowly the rule.

To the surprise of those who knew the Jats and the difficulties that were experienced in making them accept even local military service, the whole of the men, with the exception of one troop who wished to join the Police, petitioned to be made a General Service Regiment. By June, 1858, their offer was accepted, the numbers raised to 600, and the "Jat Horse Yeomanry Regiment" came into being.

Prior to their acceptance for general service the Corps had been employed patrolling the Ganges to prevent Rohilkhand rebels from crossing. In an action opposite Kutchla Ghat, Captain Murray with 200 men defeated a large body of rebel cavalry 600 strong, who had crossed especially to attack him.

By the end of September, 1858, the Corps had marched to Allahabad, joining the field force operating along the Nepal frontier. In the battle at Ruttonpore the Regiment behaved most gallantly, capturing several guns. They continued on field service until 1860 when law and order were finally restored.

For four years they remained at peace, being cantoned at Sultanpur, Oudh; and then, in 1864, the Bhutanese began to

give trouble. The Regiment joined the Bhutan Field Force and operated in the Dooars for almost two years. The Bhutanese rarely came down from the hills and there was not much that cavalry could do, but some dismounted parties of Jats gave very good accounts of themselves. The Regiment suffered severely from cholera, dysentery and malaria, and large numbers of men had to be permanently invalided; while the severe climatical dampness practically rendered the whole of the clothing and horse furniture useless, so that the whole unit had to be re-equipped at the conclusion of the campaign.

The Regiment were detailed for the Afghan War in 1878, but being employed in minor duties did not join the field force until January, 1879, remaining in the Kurram until events in Kabul necessitated prompt action. They marched—when owing to lack of transport other regiments were unable to do so—with each man leading his horse, which carried two extra sacks slung over the saddle, containing ten days' supplies and ammunition. One march was for 25 miles up the bed of a stream covered with boulders. They reached the column advancing on Kabul in time to fight the action at Charasiah. Many sorties from the Sherpore Cantonment at Kabul gave them much hard work, until they returned to India, 26th January, 1880. Field Marshal Lord Roberts in September, 1913, just a year before he died, wrote "I always remember the good work done by the Regiment (14th Jat Lancers) when we were hard pressed by the Afghans in the Chardih Valley on the 11th December, 1879. The retirement by squadrons was carried out as if on parade."

General Sir John Murray, K.C.B., the father of the first Jat regiment to be raised in the Indian Army, and which, under his command for twenty years had so brought out the soldierly qualities of the Jats that they were regularly henceforth enlisted, died 20th May, 1902. A memorial in the shape of a *Dharamsala* near the fort at Delhi was erected to his memory, largely from funds contributed by the Jats themselves.

In 1903 the title of the Regiment became Murray's Jat Lancers, the last change before the amalgamation.

The Regiment being on the North-west Frontier when the Great War commenced, saw little prospect of serving overseas. They, however, had their hands full with the Mohmand small war of 1915, the Regiment charging the enemy together with the British 21st Lancers on September 5th, 1915—a charge made through high crops, and one the subject of much discussion at the time.

Unexpected orders sent the 14th Jat Lancers to Mesopotamia in July, 1916, where as part of the 7th Cavalry Brigade, they took their share in the brilliant operations which commenced with the storming of Kut-Al-Amara and culminated in the capture of Baghdad. They were in support of Cassells' Brigade when he blocked the Turkish retreat. In the pursuit to Mosul the 14th Lancers caught the 14th Turkish Regiment nicely in the open, and charging home with the lance they put the whole of the enemy *hors de combat*. Heavy as the Turkish fire was, curiously, only one man was killed and five wounded, with fifteen horse casualties.

The 15th Lancers (Cureton Multanis) were formed as "The Multanee Regiment of Cavalry" from *risallahs* of horsemen of Pathan and Baluch tribesmen of the Derajat district, who had volunteered to serve in the campaigns against the mutineers of 1857.

After embodiment at Lahore, with Captain C. Cureton, 2nd Irregular Cavalry in command, they proceeded south and joined the forces in Rohilkhand and Oudh, taking part in many actions with distinction until 1859.

A wing was employed in the Mahsud blockade, 1860.

A year later the Corps became the 15th Regiment of Bengal Cavalry.

They earned honours for Afghanistan, 1878–80, but being employed on the Baluchistan frontier saw only hard work escorting supplies, without the glory of any great action.

On the 8th August, 1914, the 15th Lancers were mobilized for service, and became the Divisional Cavalry for the Lahore Division about to proceed to France. They were the first Indian Cavalry Regiment to land in France, but had little opportunity for mounted combatant work and were thus employed on foot, supporting the infantry in all the actions of the Lahore Division, including Neuve Chapelle. Leaving France with the Indian Infantry Divisions in December, 1915, the 15th proceeded to Mesopotamia in their company. On arrival at Basra they did not join these Divisions then on the Tigris' but proceeded to Persia, where they remained as part of the blockading cordon endeavouring to keep enemy propagandists and other mischief makers out of Afghanistan. In the small actions against the raiders and tribesmen the fighting was at times severe.

During the war, in 1917, after nearly sixty years as such, they ceased to be a class Mussalman regiment. Owing to the enormous numbers of recruits required to replace the war wastage, the man power of their existing recruiting grounds was insufficient to keep them up to strength, two squadrons were thus recruited from Jats.

The amalgamation of the 14th and 15th Lancers into the 20th Lancers took place at Sialkot, 21st September, 1920.

In 1937 the Regiment became the training regiment for the 3rd Indian Cavalry group with a permanent station at Lucknow.

CHAPTER XXV

The Central India Horse (21st King George V's Own Horse)

"Kandahar, 1880," "Afghanistan, 1879–80," "Punjab Frontier," "Somme, 1916," "Morval," "Cambrai, 1917," "France and Flanders, 1914–18," "Megiddo," "Sharon," "Damascus," "Palestine, 1918."

FOR upwards of forty years prior to 1857, the garrisons of the inaccessible regions of Central India were provided by contingents raised by the feudatory chiefs of the various states, but administered by the East India Company who provided the British officers. Originally little more than armed police, these troops gradually developed into trained forces due to their association with the regular corps, and through whom they became disaffected, despite the loyalty of their chiefs. After the outbreaks in the Bengal Army all, save a few staunch elements in the ranks, mutinied.

It was necessary to improvise units to put down the insurrection and so the faithful men of the 2nd Gwalior Cavalry, the Sikhs of the Bhopal Cavalry, and a detachment of Malwa Cavalry were formed into a regiment in 1858, under Captain

H. O. Mayne of the Hyderabad Contingent, for local service in Central India. This regiment took part in the final chase of Tantia Topi after the re-capture of Gwalior.

Meanwhile, in Hyderabad, early in 1858 Lieut.-Colonel W. F. Beatson, a veteran of thirty-five years' service, was ordered to raise two regiments of Irregular Cavalry. By the May, both regiments were ready for the service of hunting down the rebel bands then roaming Central India.

Eventually the insurrection in Central India having been brought to a close, a reorganization of the forces became paramount, and thus in 1860 Mayne's Horse and the two regiments of Beatson's Horse were amalgamated into one corps of three regiments, styled Mayne's Horse. To the Corps also were transferred another "mutiny unit" the Mainpuri Levy. The new Corps was to be primarily for local service in Central India, but the men were to be enlisted for general service anywhere in or out of India. The Corps was not, however, to be under the orders of the Commander-in-Chief, but of the civil Agent to the Governor-General at Indore.

There were many inefficients and Mayne soon weeded them out; as a result the Corps became much below strength, and thus Meade's Horse, a body raised by Captain R. J. Meade, the Brigade Major of the Gwalior Contingent, from loyal men of disaffected regiments at Agra in 1857, was transferred to Mayne's. This Corps had taken part in many actions in the pursuit of Tantia, and it was to Meade that the latter finally surrendered at Sipree.

Mayne, though a brilliant soldier, evidently did not get on with the Agent-General—Sir Richmond Shakespeare—who recommended that Mayne should be removed from his command. The Government of India concurred, Mayne was posted to the Madras Army, and as a final indignity his Regiment's title was changed and his name deleted therefrom. Mayne, broken in health and spirit, left his command only to die at Government House, Allahabad, at the age of forty-two: his

services in the Mutiny entitle him to no obscure place in the records of those times.

By this action, on the 8th September, 1860, were born at the almost inaccessible cantonment of Goonah, "The Central India Horse" of three regiments. It did not remain of three long, for in the following year (July, 1861) the Government decided to hold the jungles of Central India—the home of the dacoit and the tiger—with two regiments, and the 3rd C.I.H. was thus disbanded. The two regiments were quartered at Goonah and Agar, cantonments which they held as their own until the Corps were de-localized in 1921. Their principal duty was to preserve law and order in territories where those qualities were not held in any great respect, and to protect travellers from molestation passing up and down the Bombay-Agra Grand Trunk Road.

After Mayne's departure, the Corps was commanded by officers distinguished for their services in the Mutiny—Henry Daly of the Guides, Sam Browne, Probyn, and Watson—all of whom had raised the Regiments which still bear their names to-day; all, with the exception of Daly, were wearers of the Victoria Cross. Daly, however, received a Commandership of the Bath, an honour then much prized by officers of the Indian Services. Sons of all these distinguished men afterwards served in the Regiment, the command in France during the Great War being held by Colonel Browne, son of the famous Sam.

The withdrawal of the British Indian troops into the Sherpore Cantonment at Kabul in the winter of 1879–80 led to a composite regiment from the 1st and 2nd Central India Horse being despatched to join the relieving force. On the way, they had to cross the Kabul river, then in full flood and 400 yards wide; an achievement managed with the loss of only one horse; while some of the men crossed as often as ten times to swim horses over. Lord Roberts, as he afterwards became, was so greatly impressed by the Regiment that he ordered them up from

DUFFADAR
1ST CENTRAL INDIA HORSE, 1890

Jellalabad to Kabul to augment the cavalry selected for the famous march to Kandahar.

"Bobs Bahadur"—as Roberts was known to the Indian troops—took his force with the minimum of baggage some 300 miles in twenty-two days; they lived on the country and often had to cross long desert stretches where no water could be had: despite these hardships and the scorching sun, the troops arrived fresh and fit for action. The March has often been condemned by critics as unsound, since it was effected by a force without a base through a hostile country devoid of communications towards an objective probably in the hands of a recently successful enemy. But the March was justified by the results, and it was one of the greatest achievements of India's Army.

In 1897 a change was made in the organization of the Corps, the two Regiments were separated, and became distinct corps, having their own commanding officers; while almost all of the political duties that were performed by the old Commandant were taken over by civil officers. Previous to this the commanders of the individual regiments had been styled Second-in-Command.

The troubles on the Frontier in 1897 saw a composite regiment again formed which did sterling work in the Kurram Valley.

Both Regiments of the Central India Horse, which in the reorganization of 1903 had had the numbers 38 and 39 affixed, were honoured on the 1st January, 1906, by H.R.H. The Prince of Wales, afterwards King George V, becoming Colonel-in-Chief: upon his accession, the title of "Prince of Wales' Own" assumed previously, was changed to "King George's Own".

The increasing molestation of the King-Emperor's subjects —most of whom were Indians—in Eastern Persia, led the Indian Government to despatch the 39th Central India Horse to that country in October, 1911, for the purpose of protecting traders and others in Bushire, Shiraz, and Ispahan. A not very

agreeable task in a country where the inhabitants were not disposed to be friendly; and one officer and several men lost their lives in the many brushes with Kashguli tribesmen during the Regiment's nineteen months' stay.

It was but a short while to 1914, and in that October the 38th Central India Horse were mobilized as part of the Mhow Brigade of the 2nd Indian Cavalry Division to arrive in France in the December. They were decimated at Cambrai, but their bravery in extricating the 2nd Lancers from being surrounded was superb. On removal to Palestine they were in the van of the cavalry in Allenby's offensive. After minor affairs with the Syrian Arabs the 38th C.I.H. returned to India in February, 1921.

The 39th remained in Central India training officers, men, and horses, and sending them to serve the King-Emperor in four continents and against every kind of foe.

The old cantonments having been handed over to the Gwalior State Forces, the amalgamation into one regiment of regular cavalry took place at Quetta late in 1921.

The Corps was originally modelled upon that of the Guides, and thus like them have worn khaki or drab since their inception, the facings being maroon, a blue Ludhiana *pagri*, white breeches, scarlet *lungi*, and black puttees completing the kit.

Major-General H.H. The Maharajah of Rutlam, who fought with the Regiment in France, and H.H. The Nawab of Jaora have been Honorary Officers since 1908, a similar commission being given to H.H. The Nawab of Bahawalpur just after the amalgamation.

1939 saw the passing of the horse for the armoured fighting vehicle.

CHAPTER XXVI

The Indian Artillery

IN the early Armies of India the artillery was generally found from guns landed from the fleet manned by naval ratings detached for the purpose. As the fighting forces of the East India Company developed, properly organized artillery corps became essential; and by 1857 all the Presidency Armies had well organized batteries, horse and field, European and Indian.

The mutiny of all the Indian artillery of Bengal led to the decision to have no artillery in the Army in India manned by Indians, except two companies in Bombay, the Golandaz, originally used for the Aden and Jacobabad garrisons, and four batteries of the Punjab Frontier Force. These units eventually became the six senior batteries of the Indian Mountain Artillery, the first four retaining the additional title of "Frontier Force."

The 1st Royal (Kohat) Mountain Battery F.F. was raised at Bannu in 1851 from disbanded detachments of the horse artillery of Rajah Dulip Singh, the last of the Sikh rulers prior to the British annexation of the Punjab—all these detachments had seen service with the British at the siege of Multan. The Battery as such was employed in the frontier tribal operations of the Mutiny period, and many subsequent thereto; first major operations being in the Afghan War of 1878–80.

(*India Office photo*)

GUNNERS OF THE INDIAN ARTILLERY

The 2nd (Derajat) Mountain Battery F.F. was also formed from disbanded Sikh horse artillery in 1849 at Dera Ghazi Khan. A detachment saw service in Oudh and Bundelkand against the mutineers, and the whole Battery served with distinction in the Mahsud Waziri Expedition of 1860.

The 3rd (Peshawar) Mountain Battery F.F. was raised in the frontier Capital in 1853. All the frontier "shows" of the time demanded their services, and in 1871 they were brought right across to the other side of India to form the artillery for the Lushai small war. In this seven months' campaign in the hinterland of Chittagong the guns were carried on elephants.

The 4th (Hazara) Mountain Battery F.F. found nativity in 1851 at Haripur from Hazara gunners trained by Major Abbot for the defence of Hazara which District that officer was settling. Ambeyla Pass and the Black Mountain country gave them great opportunities before they were to be constantly engaged in the Afghan operations which began in 1878.

The 5th (Bombay) Mountain Battery is the oldest of all, being raised as the Bombay Foot Artillery in 1827, and as such fought with the Bombay columns attacking Multan in the 2nd Sikh War; and are the only battery to bear the honours "Mooltan", "Punjaub" and "Abyssinia" in which latter campaign (1868) they had been sent to support the Bombay Indian Brigade. They were thus the first of the present artillery of India to serve overseas, though one must not forget that the Viceroy's Body-Guard had horsed the artillery sent to Egypt in 1801.

The 6th (Jacob's) Mountain Battery is the second oldest, being also raised as Bombay Foot Artillery in 1843. They are the only Battery to bear a personal name, all the others having place names. The name arises from the Battery having taken over the guns of Jacobabad in 1875, which up till then had been manned by infantry men of Jacob's Rifles, now the 5th Battalion 10th Baluch Regiment.

Except the 5th, none of the Batteries bear honours prior

INDIAN FIELD ARTILLERY

"A Tank shot over open sights"

to the campaigns in Afghanistan, 1878–80, in which wars all of the other five Batteries then existing played prominent parts.

In their early years some of the Batteries had been organized as light or field artillery, but in the year 1876, all were put on a mountain basis, and thus ceased to draw their guns, which henceforth were carried, usually by parts on mule back, but sometimes by elephant or coolie.

The 1st took part in that very daring night advance on the Peiwar Kotal, the Battery Commander laying down his life; while the shelling by the Battery of the enemy's camp had such an effect as to be the decisive point in that victory. It proceeded with Sir Frederick, later Lord, Roberts to Kabul, where on arrival he personally offered his thanks. That great General held that his objective had only been reached through the aid of the Battery and the 3rd Sikhs, now the 3rd Royal Battalion Frontier Force Regiment. Addressing them under the walls of Kabul he said:

"So confident were the Ghilzais of their success that I hear they brought down their women to witness your discomfiture. You were able, however, to beat them off with heavy loss, notwithstanding their great numbers. You have set an example to the whole Force by showing what a few men can do when properly led."

After taking part in the heavy fighting around Kabul, they returned to India via the Khyber to be again recalled. They, with three troops of the Central India Horse, horses, mules, and men, swimming the Kabul River, 400 yards wide in full flood with the loss of but one driver and a cavalry horse—a feat never before recorded in our military history.

The 2nd formed part of Sir Louis Cavagnari's escort to Kabul; and in the following year fought at Charasiah, marching again to Kabul, fighting in most of the actions from the beleaguered Sherpore, and on the famous "walk" to Kandahar.

The 3rd operated through the Kojack Pass to Kandahar, covering the evacuation of that city in 1878.

The 4th were in the Khyber, fighting Ali Masjid with the force relieving Kabul, where they were left to form the garrison on Roberts' departure to the south; as a consequence many small affrays fell to their lot and they were always busy.

The 6th during the Afghan War were for a time a mixed British and Indian battery based upon Quetta; they assisted in the seizure of the Kojack Pass and marched to the relief of Kandahar.

The Burma War, 1885–87, requisitioned the 4th and the 5th. While, in 1886, two more mountain batteries were raised; the 7th (Bengal) at Rawalpindi, and the 8th (Lahore) at Mian Mir, the cantonment of the Punjab capital. These new Batteries also went to the Irrawaddy.

The 4th, after the advance to Mandalay, garrisoned Bhamo, and supplied the guns for the sorties therefrom against the Kachins and Yunanese on the Chinese frontier.

The 5th went to Suakin in 1896 to take part in the Egyptian Dongola Expedition.

By 1897 all Batteries were back on the North-west Frontier, where previously year in and year out they had been constantly employed against tribal turbulence. The mass outbreaks along the whole frontier brought "Punjab Frontier" to the 1st, 2nd, 5th and 8th; "Tirah" to the 1st, 2nd and 5th; "Chitral" to the 2nd and 4th; "Malakand" to the 8th.

The 5th played a prominent part in the capture of the Dargai heights, losing the Commandant, Captain de Butts, R.A.; the subaltern Lieut. Edlemann receiving the D.S.O.

To the 6th, however, fell the supreme glory of those campaigns. At Maizar, when escorting the Political officer, a section of the Battery, together with detachments of the 1st Sikhs, were treacherously attacked by Pathans pretending to be friendly, and with whom they had been fraternizing. The British officers were killed and the section surrounded by hordes of fanatical tribesmen. The guns were got into action and fought on by the Indian gunners under their Havildar-Major, and when their ammunition failed they effected an orderly retirement

to safety without the loss of a gun, although many of the gun mules had been shot down. No less than nine Indian Orders of Merit were earned by these indomitable men, and June 10th, Maizar Day, is to-day a day of celebration in Jacob's Battery.

Peace, more or less, reigned on the frontiers, and except for the expedition across the sea to Somaliland, 1901–04, in which the 8th took part, no actions for which honours were granted happened until 1914.

After the frontier wars of 1897, four more Batteries came into being.

The 9th (Murree) being raised not at the place, but at Abbottabad, in 1898–99, where its completion was delayed due to the requirements of animals and material for the South African War. Twelve months after completion a section was sent to East Africa for service in the Jubbaland Field Force operating on the Abyssinian border. A campaign of very different type and character to that of the Indian frontier; dense bush having to be penetrated through which progress was only possible in Indian file, while insect pests caused havoc among the animals—one large fly drawing blood at every bite and leaving a deep sore an inch in diameter. The Indian ranks willingly adapted themselves without demur to the difficult and trying circumstances, and each man received the Africa General Service Medal.

The 10th (Abbottabad) raised there in 1900–01, saw service in the Aden Hinterland in 1903 during the Arab dispute about boundaries; and again in 1904 in Tibet in the attack on Gyantse.

11th (Dehra Dun) and the 12th (Poonch) were babies born at Dehra Dun seven years old when somebody threw a bomb in Balkan town; a lot of people who should have known better got excited, and eventually before many weeks were out, India's Mountain Gunners were drawn into the quarrel.

The first to go were the 1st and the 6th, despatched to Egypt for the defence of the Suez Canal. They played a great part in

Mountain Artillery in Action on the North-west Frontier

repelling the Turk, the 6th capturing the two guns now in their possession.

Both took part in the original landing at Anzac Cove on April 25th 1915, with the Australian and New Zealand Corps, with whom they served and supported as all true artillerymen should. For eight months they were heavily engaged under continuous fire, day in night out; as a consequence, great friendship sprang up between the gunners of India and the men from 'down under'. Every year on Anzac Day—a great day in Australia—telegrams are exchanged between the Batteries and the Australian and New Zealand Regiments; the latter presented silver cups as a memento of the days of great gallantry when the 1st and 6th Batteries were the first artillery ashore; as well as the last guns to leave at the final evacuation of Anzac on the 20th December, 1915. The 6th fired 10,000 rounds from their ten-pounder guns, whose carriages had no arrangements to check recoil. These two units were the only Indian Artillery to serve in the European theatres, and both bear "Anzac", "Landing at Anzac", "Defence of Anzac", "Suvla", "Sari Bair", "Gallipoli, 1915", "Suez Canal", "Egypt, 1915–16".

The 3rd and the 10th went early to Mesopotamia fighting "Basra" and "Shaiba", remaining there principally on the "Tigris" until 1916, when they returned to earn "N.W. Frontier, India 1917", and the 3rd, "Baluchistan 1918". At Nasirizeh, the 10th Battery's escort was forced back by superior numbers and the enemy got in among the guns; they were promptly thrown out by the drivers and gunners, armed with swords and a few rifles. As a mark of distinction for the collective bravery shown, they have been allowed to wear bayonets on parade, a unique honour in the Artillery.

The 7th and the 8th landed in East Africa in 1914, both subsequently being present under Smuts at "Kilamanjaro", the 8th returning in 1917, the 7th to remain until the Armistice earning "Nyangoo" and "Nurungombe". The 2nd also arrived in this theatre in 1916 and took part in the latter action; they also remained to chase Von Lettow until the 14th November, 1918. Such were the communications and the countryside of the Dark Continent, that an action was fought in the East African theatre before Kasama the day following the Armistice. East Africa was a campaign in which wild carnivora rapidly made short shrift with the wounded. The 4th also joined this mixed force of Briton, Boer, Rhodesian and African for the last year of the operations.

The 1st and 6th after Gallipoli refitted in Egypt, and joined the forces of Maude to earn "Mesopotamia, 1916–18", as well as "Persia, 1918"; at which time a foretaste of the mechanization of pack artillery was gained, one section being transported in lorries as cavalry mobile artillery. The honour "Persia" is also shared by the 11th who had proceeded to the Euphrates in 1918, and by the 16th (Zhob) Battery raised outside of India at Bunder Abbas, South Persia, in 1918, the only artillery unit still remaining in the Army raised in a theatre of war.

The 14th (Rajputana) Battery was a *war-baby* lusty enough within a year of its birth to proceed to Mesopotamia, where, with the 5th who had been in the North-west Frontier frays of 1915, they proceeded in the last year of the war to take part in the victory at "Sharqat".

The 9th who bear the honour "North-west Frontier India, 1914, 15–17", got their oversea chance together with the 12th, proceeding to Palestine, 1918, taking part in the closing phases earning "Megiddo" and "Nablus". The 9th were for a time brigaded with the Ayrshire Territorial R.H.A. in support of the New Zealand Mounted Brigade, another example of Empire co-operation, which fired them to such an extent that although they had 50 per cent of their number down with sickness in the unhealthy Jordan Valley the Battery was never out of action. The Palestine honours are also enjoyed by the 19th (Maymyo) Battery, raised during the course of the war at Maymyo, Burma, May, 1918.

The 6th hold the record for any unit of the Imperial Armies, having been on active service overseas continuously for five and a half years.

The 7th and 8th, together with the 15th (Jhelum) Battery raised 1917, provided the 'screw gunners' in the 4th Afghan War, 1919.

The remaining mountain batteries were all raised during the Great War or subsequent thereto, and as yet have no emblazonments.

In recognition of the services of the Indian Mountain Artillery in the Great War, the title "Royal" was conferred on the 1st by the King-Emperor George V on the 31st January, 1922.

The personnel of the mountain batteries, both gunners and drivers, are Indians, and prior to 1927 the men were enlisted into the Indian Mountain Artillery and the batteries were units of the Indian Army—only their British officers belonging to the Royal Artillery. In that year, however, they were included for the first time in the Royal Artillery, and became, in effect, units of the British Army, the men being enlisted into the Royal Regiment. This course is a little difficult to understand, and it is only reasonable to suppose that with the Indianization of the artillery of India's Army, that these Batteries with their long and glorious records will grace the Indian Regiment of Artillery to which they rightly belong, and that that Regiment will bear the title "Royal" as do the artilleries of Canada, Australia, New Zealand, and Malta—South Africa has no separate artillery corps, her batteries being part of the South African Mounted Rifles.

We now come to the Indian Artillery formed on the 15th January, 1935, at Bangalore as a field brigade. An occasion marked by a telegram of good wishes from the Royal Artillery, who have presented a silver model of an 18-pounder gun to grace the Officers' Mess. Till that day no Indians had been enlisted as gunners in the horse, field, or heavy artillery; but

since the Great War—and in some cases even before it—they had been enlisted as drivers and farriers.

The field brigade consists of four batteries; the 1st of Madrassis, originally formed from the disbanded Madras Pioneers; the 2nd, Punjabi Mussalmans; the 3rd, Rajputana Rajputs; and the 4th, Ranghars.

It might be mentioned for those who think the pace of Indianization too slow, that it takes nine months to train an Indian Artillery recruit up to the standard reached by recruits at the Royal Artillery Depot in fourteen weeks. This is doubtless due to the lower educational attainments of the average Indian tyro, which a development of the country's educational system will in time no doubt rectify. Gunnery is highly specialized work, and guns badly manned are a two-edged weapon; a serious obstacle to victory, and destructive of essential morale in the forward combatant ranks.

From now on, officers posted to the Brigade will be Indians from the military academies of Woolwich and Dehra Dun; and in course of time all officers will be Indian. And why should they not be? The task of the British officers posted to the Brigade, and they have been carefully selected for their sympathy, is, as their Commanding Officer has said, to instil the right ideas into the young officer and mould them as the senior officers themselves were moulded in their youth; for it will be opportunity in the end that will make the man.

CHAPTER XXVII

The Corps of Sappers and Miners

THE inability of the infantryman to overcome an enemy entrenched behind apparent impassable obstacles or fortifications brought about the necessity of employing in an army skilled personnel able to use mechanical or technical means of assisting the foot soldier to victory. India, unlike most countries of the world, has not one corps of military engineers, but three, each preserving separate entities and bearing their own battle honours; unlike their sister corps the Royal Engineers whose honours are covered by their motto "Ubique". All three corps were formed originally from battalions of pioneers recruited for the old Presidency Armies of India and whose territorial designations are incorporated in their present titles.

The senior corps, now "The Queen Victoria's Own Sappers and Miners" were raised as pioneers in the late eighteenth century. Previous to that year bodies of volunteers from European and Indian infantry were formed for pioneer duties as the exigencies of a campaign demanded and were disbanded on its conclusion. After this the experiment was tried of enlisting voluntary labour on a temporary basis, but such voluntary labourers proved not only inefficient for military work, but were apt to throw down their working tools and bolt whenever the enemy appeared in sight. So, on the 30th September, 1780, were founded the Madras Pioneers, a corps to be officered entirely from the Infantry of the Line.

As military engineer science developed, this officering of what was rapidly developing into a technical corps by non-technical officers led to serious difficulties. There had already been established a service of Military Engineer Officers, but these were administered by the Presidency Chief Engineer, a civil appointment to the incumbent of which the Commander-in-Chief had no official access; while the Pioneers were under the Quartermaster-General: the result in war as the armies developed was not difficult to imagine. As a consequence the corps were re-organized and by a General Order of 1820 became a sapper and miner corps to be officered by engineer officers: owing, however, to a dearth of such specialists in the Madras Army the re-organization was not carried out until 1831.

On their becoming sappers they were allowed to retain the honours earned as pioneers "Carnatic" and "Sholinghur" their first, being for services under Eyre Coote against Hyder Ali (1781–82); "Mysore" (1791) and "Seringapatam" (1799) for the wars against Tippu Sultan. During this period the Corps had also been in minor actions against the Dutch in Ceylon and Malacca, and the French at Cuddalore.

The Madras Sappers are the only corps of the engineers to bear the special honour of "The Sphinx superscribed Egypt", an emblazonment earned in 1801–02 for services in the expedition under Baird sent from India to assist Abercomby to turn the French out of Egypt; the honour was, however, much belated, not being granted to the Corps until seventy years later.

When Wellesley—afterwards the "Iron Duke":—

'Against the myriads of Assaye,
Clashed with his fiery few; and won."

the Madras Pioneers were there in the brilliant attack by 4,500 disciplined troops that routed Scindia's army of 40,000 on the 23rd September, 1803.

Many small local campaigns employed the pioneers of the southern Presidency in the following years. In 1811 the Corps embarked for the conquest of the Dutch East Indies, thus adding "Java". Back in India again, 1817–19 saw the Corps engaged in the Second Mahratta War and the granting of the honours "Nagpore" and "Maheidpoor". For their services in the investment of the former fortress they received a special order of thanks from the Army Commander.

Now let us change from shot and shell, to cash and rations. Up to 1818 the men had been paid in pagodas, fanums, and cowrie shells, but this year saw the establishment of the more familiar rupees, annas, and pice. The Pioneer received then Rs. 7–13–4 per mensem, and a monthly pension if invalided of Rs. 3–14–8: the cash allowance for a Company's stationery and postage being Rs. 2–5–4 for the same period. The European Sergeant did himself well on Rs. 1–2–8 per diem, plus Rs. 14 per mensem if he were the Sergeant-Major.

Now back to active service. This time Burma, 1824–26, a campaign in which the success of the forces resulted largely from the services of the Madras Pioneers; the escalading of almost impregnable stockades appearing to have been their especial flair. "Ava", the name of the old Burmese capital, was thus added to their shield.

The first Burmese war was their last campaign as pioneers under infantry officers; the Directors of the East India Company having remarked that it was highly desirable that the Pioneers should be put under officers of engineers and regularly instructed in the art of sapping and mining, making pontoons and bridges, roads and surveys, fortifications and other buildings. So back

INDIAN OFFICERS OF THE QUEEN VICTORIA'S OWN MADRAS SAPPERS AND MINERS, WHO REPRESENTED THEIR CORPS AT THE CORONATION OF KING GEORGE VI.

to their regiments went the officers of the foot and gentlemen from Addiscombe, the Company's 'shop' took over, although infantry officers did not entirely cease to be posted to sapper corps until 1870, fifty years later.

The second corps now "King George V's Own Bengal Sappers and Miners" were formed at Cawnpore in 1803 as the Bengal Pioneers and were partly re-organized as sappers and miners in 1819, just prior to the issue of the General Order of the following year. As pioneers they took part in the second and third Mahratta Wars.

The junior corps started life actually before the two senior, having had a continuous existence as a company of Pioneer Lascars in the Bombay forces since 1777; officially, however, the date of formation of the Bombay Sappers and Miners is only recognized as 1820, the title "Royal" being bestowed a century later for their services in the Great War. As pioneers they fought in the Mahratta and Mysore wars, but unlike the Madras Corps bear no honours for those campaigns.

To write the history of the three Corps of Sappers and Miners from 1820 onwards is to write the history of the Indian Army, for, whenever, and wherever, Indian troops have taken part in a campaign of consequence, whether at home or in distant lands, a portion of one or another, sometimes of all three; have accompanied the troops in the field.

As sappers, Bombay—to describe the Corps by their Presidencies—were first to earn honours "Beni-Boo-Ali" (1821) for the campaign in Arabia; followed by Bengal three years later with "Bhurtpore".

Bhurtpore had been unsuccessfully attacked in 1805, a reverse almost entirely due to the failure to make full use of trained engineers. The fort was surrounded by a deep moat, and water from a jheel close by was used by the Jats to fill it. A cavalry patrol was sent to reconnoitre and measure the moat; they failed to do this properly, reporting its width in ambiguous terms as a result. The defenders subsequently flooded the moat, and the portable bridges with which it was intended to cross the gap in the final assault became too short, and the attack failed with heavy and unnecessary casualties. The lessons of two decades earlier had been learnt and in 1825 the capture of the fortress was largely due to the remarkable zeal and fine work performed by the Bengal Sappers and Miners.

In the years following Bhurtpore, the Bengal Corps were employed on development and construction work; much of the Grand Trunk Road below Delhi, many of the older hill roads, and what is now the Western Jumna Canal being the result of their handiwork.

The first Afghan War brought Bombay "Ghuznee, 1838", "Khelat", and "Afghanistan, 1839"; the blowing down of the Kabul gate of the first named fortress by the Bombay Sappers being not only a most gallant deed, but the decisive factor in the capture of the enemy stronghold. The Order of Merit, then the highest decoration obtainable by the Indian soldier, was awarded to six other ranks for the conspicuous bravery displayed.

In 1937, Shanghai, Hong Kong, the Woosung Forts, and Nankin, were very much in the public eye at the commencement of the Sino-Japanese "troubles"; but not quite a century previously, in 1842, the Sappers of Madras were capturing Shanghai, quartering Hong Kong, assisting the fleet to demolish the Woosung Forts, and marching to Nankin to conclude the first Chinese war which gave them the right to wear a Golden Dragon and "China" on their badges.

The second Afghan war (1839–42) saw Bengal advancing into Afghanistan from the then frontier town of Ferozepore. Crossing the Indus at Sukkur, where they constructed a bridge, from local material, capable of transporting the army across, and 1,000 yards long, within twenty-one days, they earned "Cabool, 1842" and took part in the gallant defence of Jellalabad.

Madras, instead of Bombay, captured the honours for the Scinde War of 1844, earning "Hyderabad" and "Meanee"

"MODERN TOOLS ON THE FRONTIER"
A MADRAS SAPPER BORING HOLES FOR ROCK BLASTING

'bagging' an enemy standard in the latter battle. While a year later Bengal was on the Sutlej at "Ferozeshah", "Sobraon", and three years thence with Bombay at "Mooltan" and "Goojerat" each gaining those honours as well as "Punjaub".

In another little trip to Burma in 1852, Madras, by storming the Shwe Dagon Pagoda and the Pegu stockades added "Pegu" to their roll. Four years later they found themselves with detachments from Bombay engaged against the Shah; both gaining "Persia" with the addition of "Reshire" and "Khooshab" for the junior.

The Great Mutiny, which can be divided into three phases:—

(i) The siege and capture of Delhi, 1857.
(ii) The relief of the Lucknow garrison, 1857, and the subsequent capture of Lucknow city, 1858.
(iii) The campaign to pacify the country, 1858.

found employment for all three Corps. Bengal was the only one to suffer from disaffection, a little less than two-thirds of the ranks deserting; the remainder, however, more than made up for this, in the gallantry displayed before Delhi. The desperate task of demolishing the Kashmir Gate in broad daylight thus ensuring the victory is an epic in the annals of war: three Victoria Crosses and fifteen Indian Orders of Merit were awarded to the gallant band thirty strong. Madras and Bengal supplied the engineers for the Lucknow relief force under Outram, the former being the first to plant the flag on La Martinere. Both Corps thus carry "Lucknow". Bombay were also not out of it, for immediately on their return from Persia they, with the co-campaigning detachments from Madras seeing service in Rajputana, both earning a Victoria Cross—Lieut. Prendergast of Madras and Lieut. Goodfellow of Bombay—and the honour "Central India".

After the Great Mutiny the three Corps all passed into the service of Queen Victoria and the technical officers into Her Majesty's Corps of Royal Engineers, from which Corps a certain

number of specialist warrant and non-commissioned officers are still drawn.

The troubles of 1857 were not soon over before Madras and their Dragon were once again in the Far East, scaling the "Taku Forts" and occupying "Pekin" where peace was signed in the autumn of 1860.

From China to Africa went the engineers of the south, being joined by their brothers from the west, to take part in the Abyssinian campaign of 1867 under the command of Sir Robert Napier, himself an old Royal Engineer. In this difficult country both Corps made roads, railways, and other works essential to the progress of an army. They were not behind in the fighting either, for a company of Madras aided by a rocket detachment of the Royal Navy prevented the turning of the British flank; while both Corps forced the entrance to the fortress in the assault on Magdala then considered almost inaccessible: The reward—"Abyssinia".

Both Madras and Bombay sent detachments to Malta and Cyprus in 1878.

And the next two years saw all Corps employed for some time in the Afghanistan campaigns made necessary by the treachery of Sher Ali, the Amir, and his son, Yakoob Khan, resulting in the murder of the envoy Sir Louis Cavagnari. The honours earned being "Afghanistan, 1878–80" by all three, with "Kandahar 1880" for Bombay, Bengal having four others "Ali Masjid", "Charasiah", "Kabul, 1879", and "Ahmad Khel". The result of these campaigns was a long peace with the border country not disturbed for forty years. In the serious reverse at Maiwand, a party of forty-four ranks of the Bombay Sappers covered the retirement of the Royal Horse Artillery, and afterwards, with some men of the 66th Regiment and the Bombay Grenadiers, carried out an orderly retreat until only eleven were left. The bearing of this little remaining handful of men was such that the enemy declined to go near them until they had all been shot down.

"Egypt, 1882", "Tel-el-Kebir", "Tofrek", "Suakin" borne by the senior Corps record their second and third visits to the land of the Sphinx in 1882 and 1885; in the latter year at Tofrek, or McNeil's Zareba as the action is sometimes known, they suffered very severe casualties.

1885 is also noteworthy, for in that year the rank of *Sepoy* was changed to Sapper, and submarine mining was undertaken at India's defended ports for the first time.

All three Corps fought in the actions on the road to Mandalay

Sappers and Miners Making an Irish Bridge

and the downfall of King Thebaw, which accounts for "Burma, 1885–87".

Early uniforms followed those of the rest of the army and were fashioned after the style of European troops, being a short and very tight dark green or blue coat ornamented with braid, with tight short white drawers. Red tunics with blue facings and blue trousers with a broad red engineer stripe were introduced upon the organization into sappers; thus copying the Royal Corps of Sappers and Miners, the then name of the Royal Engineers: the latter dress still remains the full-dress to-day. The subject is brought in here as khaki was issued to the Corps for the first time in Burma in 1886.

The next decade was one of great activity on the North-west Frontier and all Corps bear "Punjab Frontier" and "Tirah" for their services, Madras and Bengal having also "Chitral", and the former "Malakand". In the Tirah campaign Lieut.'s T. C. Watson and J. M. C. Colvin of Bengal were awarded the Victoria Cross for their conspicuous gallantry on the 16th September, 1897.

The close of the nineteenth century saw all the Indian Sappers and Miners with the allied forces in China, the last honour to be earned save one before the great conflict began in 1914: Bombay participated in the difficult expedition against the Mad Mullah which gained for them "Somaliland, 1901–4".

Bengal went with Younghusband on 'the roof of the world' to Lhassa in 1903; and with the Abhor expedition of 1911; both campaigns required heavy work in mountainous country before the army could proceed.

From 1914 to the end of the War, the three Corps were everywhere; and between them share almost all of the Great War honours granted for service in theatres outside Europe; while all three had representative units in France and Flanders with the Indian Corps from October, 1914 until their withdrawal. Their strength was increased to almost five times the normal, Bengal alone rising from 1,700 in 1914 to 10,000 on Armistice

Day, 1918. All participated in the Afghan war of 1919. During the world war their tasks were as varied as they were dangerous; there was nothing in fact which they were not called upon to do, and which they failed to do.

To-day all the Corps are organized into Army Troops, or Divisional Headquarters Companies, Madras alone having Field Troops, units organized for operations with cavalry; while Bengal and Madras have Printing Sections for undertaking the typographical work of the field army. Each Corps has a training battalion permanently stationed at the three headquarters, Bangalore, Roorkee, and Kirkee. The Survey of India, a department of the civil government, is also staffed to a large extent by officers and men of the Sappers; arduous and trying work in peace, but essentially necessary for defence in time of war.

All the martial races are to be found in their ranks, Madras also recruiting Tamils, Christians, and what are now known as the depressed classes of the South. The outcaste men who serve in Queen Victoria's Own on return to civil life form a new and exclusive caste known as the "Quinsap", confining their marriages to the daughters of those who have served in the Sapper ranks. They consider themselves above their own people, and thus does the oldest Corps play its part in the social uplift of the country.

H.H. The Maharajah of Mysore is Honorary Colonel of "Queen Victoria's Own", while H.H. of Jaipur is an officer of "King George V's Own", and H.H. of Malerkotla of the "Royal Bombay".

1933 began an epoch, for in that year Indian gentlemen were admitted as commissioned combatant officers for the first time; and thus has been opened what was for long a close preserve of the 'Sapper from Chatham' to the product of the Indian Military Academy at Dehra Dun. A wise and an inevitably just decision.

Success in war is dependent to-day almost entirely upon the

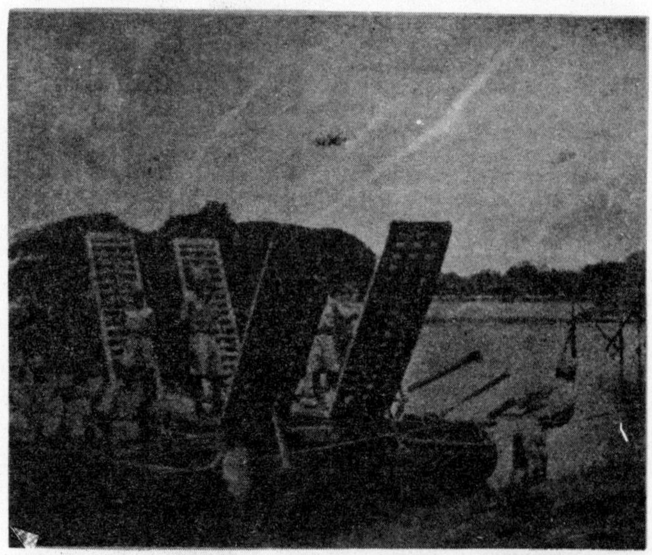

(India Office photo)

Sappers and Miners with Folding Boat Equipment

use of machinery, and the duties the Sappers will be called upon to do will not only be multifarious, but important. Where actual operations against an enemy are in progress, they will be required to overcome any obstruction to the progress of the army, by constructing bridges, railways, and roads. Or they will have to impede an advancing hostile force by demolishing all that will facilitate its progress, making obstacles to deny it passage, and constructing such defences as are requisite. They must provide accommodation for the troops in camp, including arrangements for water, sanitation, and essential services, from the supply of pumps to ice-machines. For the performance of these duties, not only must the military engineer be a trained and effective soldier, but the highest professional standard is necessary. With the records of bravery, loyalty, and devotion to duty, ever before them, one can hope that the traditions of a century and a half of some of the oldest corps in India's Army will be safe in the 'young entry's' keeping.

A Field Signal Office

CHAPTER XXVIII

The Indian Signal Corps

THE importance of the duties of the signal service of an army is paramount. Without speedy and reliable means of communication betwixt the military high command and the components of the field army, as well as between themselves colaterally, or with the co-operating services in the air and on the sea, success in war would be impossible.

Leaving aside the primitive, the first method of long-distance defence communication was by means of optical shutters displayed from specially constructed towers. A mode designed in 1767 and used in England when rapid communication between London and the naval ports became of prime importance during the wars with the French. A marvel of the age, for the daily time signal sent over the sixty miles from London to Portsmouth in 1816 was acknowledged in three-quarters of a minute. The system was introduced into India later, and some of the towers until a few years ago still stood along the trunk road between Calcutta and Barrackpore. From a military point of view it

lacked adaptability, but from it the French developed the semaphore method and its resultant code, now little used.

The telegraph became practical about 1835, and its first use by military in the field was during the war between Britain and France and Russia, when some twenty miles of land cable were laid in the Crimea. It was only partially successful as the field-mice developed a taste for the gutta-percha insulation and disconnection was frequent.

In the opening stages of the Great Mutiny the telegraph under civil control played a great part. But it was not until the Abyssinian campaign of 1867 that the telegraph was used by Indian forces in the field. Then, a party of the Bombay Sappers and Miners under a Royal Engineer subaltern laid a line supported on poles and kept Sir Robert Napier's communications intact during his advance to Magdala. This party too had their trials from the fauna, for vultures perching on the wire caused many breaks. It was during this campaign that the code invented by the American, Samuel Morse, was used for the first time.

The Germans developed the field-telegraph company as a tactical unit during the Franco-Prussian War of 1870–71, and about this time a similar unit in the British Army was introduced. To the first telegraph company formed was posted a Lieutenant Herbert Kitchener, R.E., who later became Commander-in-Chief of the Army of India—1902–09.

Like most scientific or technical branches of the military forces the signal corps evolved from the engineers—the Royal Air Force originally started as the Balloon Battalion Royal Engineers—and as far as India is concerned the present Indian Signal Corps developed from the three corps of Sappers and Miners. Experience had shown that if there was to be unification and standardization of methods the inter-communication requirements of the army could not be left to the fighting units. And so a special service called in its infancy the Indian Signal Service was started in 1911, the companies of which, although to all intents and purposes Sapper and Miner units, were independently administered, the personnel being found from those corps as well as from trained signalmen from the other combatant branches of the Army; the ranks then, as now, being British and Indian mixed. The organization was four divisional signal companies; and thus five divisions of the then field army had no organized signal unit, nor was there any depot for the training of recruits.

With the outbreak of war in 1914 expansion became vital, and had it not been for the potential reserve of telegraphists in the civil telegraph departments not only of India, but of Britain and the Dominions, the Indian Divisions in the field could not have functioned properly, nor would the great difficulties which arose through the constant adoption of new ideas issuing from the immediate experience of the War been surmounted as they were.

The first units of the new service were formed on February 1st, 1911, and the tirocinium of the Corps may be said to have been in the Abor Expedition on the North-east Frontier the same year.

During the Great War, Indian Signal units served in France, Belgium, Mesopotamia, Egypt, Palestine, Syria, Russia, Persia, the North-west Frontier; and in East Africa where animal life in the shape of the giraffe caused much wire breakage, due to the supporting air-line poles—the most economical method of field telegraphy—being too low to clear the neck of that unusual quadruped; while white ants put an effective end to field cables.

The knowledge gained in the war caused the creation in 1922 of a new and entirely independent corps under the style of the Indian Signal Corps. And all operations of India's Army since then have included in the force employed detachments of the new Corps.

The establishment in 1939 comprised 2,000 British and 3,500 Indian ranks, as opposed to a few British and some 350 Indian in 1914. Indian Commissioned Officers are now posted. The

Corps is organized into field-units for operating with the various types of formations; while a training centre for recruits is situated at Jubbalpore. There is an Army Signal School at Poona for wireless instruction, and where soldiers from Indian regiments are trained as instructors in signalling—internal communication within a regiment still being unit responsibility.

The present age is one in which the instruments and weapons of war and their methods are subject to much and rapid change. The methods of signalling used vary according to the type of, and proximity to, the enemy. They include carrier-pigeons, dogs, the telephone, radio, hand and automatic direct telegraph, flags, and the sun's rays reflected upon the distant station by a mirror; and last but not least, that most perfect machine of all, the human being, either as a runner on foot, or else borne by animal or mechanized means; and at times the most reliable of any.

H.R.H. The Princess Royal of Great Britain and sister of the King-Emperor is the Colonel-in-Chief of the Corps, and the Colonel-commandant, General Sir Robert Cassells.

The classes enlisted are Madrassis—who are slouch-hatted and comprise the whole of the Waziristan District Signals, thus proving that Madras men can serve on the Frontier—Punjabi Musalmans, Sikhs, and Dogras. Some of the units are all one class, in others they are composite; all have a certain proportion of British ranks.

Mercury, the Greek mythical messenger between the Deity and the Universe, is the talisman of the Signal Corps of the Armies of the British Commonwealth. He is depicted in the Indian Corps as standing on the globe bearing in his hands the Star of India. The uniform follows that of the Royal Corps of Signals, being similar to that of the Sappers and Miners, save that the red coats have black facings. The head-dress fringe of green, dark and light blue signifies the three physical divisions of the universe in which the Corps operates.

ASSAYE

CHAPTER XXIX

The 1st Punjab Regiment

" Sholinghur ", " Carnatic ", " Mysore ", " Seringapatam ", " Assaye ", " Leswarree ", " Nagpore ", " Ava ", " Bhurtpore ", " Burma, 1885–87 ", " Bourbon ", " China ", " Suez Canal ", " Egypt, 1915 ", " Aden ", " Shaiba ", " Kut-al-Amara, 1915, '17 ", " Ctesiphon ", " Defence of Kut-al-Amara ", " Tigris, 1916 ", " Baghdad ", " Mesopotamia, 1915–18 ", " North-west Frontier, India, 1915 ", " Afghanistan, 1919".

AS the homes of the majority of India's soldiers lie in the land of the five rivers, it is but fitting that the two senior corps of the Indian Infantry should bear that territorial name. But it must not be forgotten that the traditions of these regiments were founded, and the battle-honours of the eighteenth and nineteenth centuries won, by men of Madras.

The story of the modern infantry of India begins with the amalgamation into battalions of the independent companies of sepoys raised by Indian officers for the protection of the isolated trading posts of the East India Company in the Carnatic and on the Coromandel coast: The growth of their commercial ramifications, together with fears of an invasion of their acquired territories by Hyder Ali, Sultan of Mysore, and the prospect

REPRESENTATIVE DETACHMENT OF THE 1ST PUNJAB REGIMENT AT THE CORONATION OF KING GEORGE VI, 1937

of the challenge by the French for supremacy, led to the overhaul of the Company's military resources. So, by the amalgamation of existing companies, by drafting from established units, or by beat of drum were formed between the years 1759 and 1794, the ancestors of the corps of foot which after many changes of title became by 1901 the 2nd, 6th, 16th, 22nd, and 24th Madras Infantry.

The men of Madras for over a hundred years enjoyed an unblemished record as soldiers, not only at home, but in the early expeditions oversea. They stood well under arms, but unhappily the climatic conditions in which the campaigns of the closing years of the nineteenth century were fought, told on the stamina of the southern classes willing to accept service. The Madras regiments made every effort to tap new sources of recruitment, but suitable material equal to the virile peasantry of the north was not forthcoming, and by 1903 all the then existing Madras Infantry, with the exception of certain Carnatic battalions, were converted into "Punjabis" and sixty added to the existing regimental number.

The Carnatic battalions finally disappeared in the reorganization of 1929, and thus in India's Army now only the 1st, 2nd, and perhaps the 8th, Punjab Regiments remain as the descendants of the foot of the old Madras Army.

The senior battalion of the present 1st Regiment were just one year old when they assisted in the signal defeat of the French at Wandewash. From 1767 to 1769 they were fighting Hyder Ali, particularly distinguishing themselves at Trincomalee.

Both the 1st and 2nd Battalions were present at the reduction of Pondicherry, 1778. And when Hyder and the French again invaded the Carnatic in 1780 they, together with the 3rd Battalion, were in the final capture of the French fortress. All were in the many operations which ended in the enemy defeat at Sholinghur.

All the first four battalions were ardent fighters in the second

The Band and Bugles, 2nd Battalion

1st Punjab Regiment—1938

Rajput Indian Officer, 1st Punjab Regiment

Mysore War, 1790–92, a campaign almost abortive due to inept generalship.

Lord Cornwallis gave the Mysore ruler another chance which he failed to take and, as French intrigue made further war inevitable, Seringapatam was again laid siege to. In the approach march, the present 10th Battalion came into action for the first time at Mallavelly; the 2nd also joined the fight. All the battalions except the 5th took part in the severe and victorious assault on the 4th May, 1799, which finally ended Mysore hostility. The 2nd and 5th Battalions continued in the operations which extirpated the insurgents under the freebooter "King of the Two Worlds" Dhoondia Waugh, the 2nd being in at the death in Wellesley's final action at Conaghul.

Uniforms about this time were based upon the European model, an iron-framed cloth-covered imitation shako being the headdress; the coat was white, later red yellow-faced. Short drawers edged with blue were the first lower garments, but these appeared to offend the susceptibilities of those prudish times and white trousers became the order of the day. Not for over a century were shorts again to become fashionable.

In 1803 began the first of the great Mahratta campaigns which, besides bringing some measure of peace to the lower orders of Central India, also laid the foundations of the unification of the forces of the three Presidencies. Two armies were formed for the campaign, the northern under Lake, the southern under Wellesley. The latter opened operations by a march to save Poona, followed by the storming of the great fortress of Ahmednagar: with him were the 10th Battalion. The 1st later joined the column and these two battalions are responsible for the "Elephant" and "Assaye" now emblazoned on crests and colours. The 3rd and 5th Battalions were at the same time with Colonel Stephenson's supporting force. At Argaum the joint columns attacked the five-mile front of the Persian and Arabian mercenaries. Desperate fighting followed, the day

being saved by the 2nd Battalion repulsing an almost overwhelming charge of the Mahratta cavalry: the four battalions were also present at the storming of the impregnable Gawilgarh, the 5th especially distinguishing themselves.

The Regiment also bears the honour for Lake's victory at Leswarree; but none of the present battalions was there. The honour belonged to the old senior regiment of Bengal Infantry —the 1st Brahmins—who came into the Regiment as the 4th Battalion in 1921 to be disbanded in 1929: "Bhurtpore" was similarly their work.

During the French Wars between 1793 and 1908, shipping from Calcutta was the constant prey of enemy ships; and the Government of India were constrained to take action. An expedition was thus sent in 1810 to seize Mauritius and Bourbon, the French bases. The French defenders only had 2,000 men and some undisciplined slaves, while the Indian force numbered 11,000 bayonets; in consequence the islands were thus surrendered without any action of importance. Bourbon (Reunion) was subsequently returned to France.

The two senior, as well as the junior, battalions were all utilized in the final campaign to subdue the Pindaris and their supporting Mahratta princes. The 1st went to the rescue of the Nagpur Residency; the 2nd operated around Cuttack; while the 10th in the southern Mahratta territories found much work to do in reducing the numerous forts, large and small, of the various feudatories, including the taking of Sholapur.

The 3rd Battalion which had not been on major active service since 1799 went to Burma in 1824, for the first war made by the Honourable Company beyond the borders of India. The 5th and 10th Battalions also served in the campaign.

Events in Afghanistan overshadow practically all else in Indian military history in 1842. But despite the dark days owing to blunders in the north, an army from the south was sent to China. Madras men had no objections to serving overseas and so the 1st and 2nd Battalions sailed to invest Nankin.

For their services in this amphibious warfare the Regiment now bears the Golden Dragon.

The 3rd Battalion saw interesting operations between 1839 and 1844 at Kurnool, and against the Sawant Warree insurgents, Privates Cotapah and Veerapah receiving the newly constituted Order of Merit, as well as a gold medal from the Scottish City of Perth for protecting and saving the body of Lieut. Campbell of the Bombay European Light Infantry. The Victoria Cross had not then been instituted.

The Great Mutiny did not disaffect the Madras regiments, and the 3rd and 10th Battalions both provided a company for the Madras Rifles, a unit extemporised for service in Bengal during those troublous times.

The third Burmese War was the last noteworthy active service the Regiment was to see as Madrasis, the 2nd and 3rd Battalions earning the honours for the laborious and unhealthy subjection of widespread dacoity which followed the surrender of King Theebaw in 1885.

From 1903, when men of the Punjab took up the arms of the mustered-out southerners, none of the Regiment were required for any active service, the period until 1914 being one of preparation of new men under old colours, which produced a result in the following years the reverse of the example set by the proverb of vintage liquors.

The 1st and 3rd Battalions (62nd and 76th Punjabis) were soon on the banks of the international high-waterway of Egypt, a detachment of the 1st charging one of the few successful enemy parties to cross—the action being perhaps the last in which the sword was drawn and used with effect by infantry officers. Both later went to Mesopotamia, the senior to dwell awhile at Aden in mid-1915, the junior arriving in time for Shaiba.

The 2nd Battalion (66th Punjabis) went early to the Persian Gulf and advanced with Townshend to Ctesiphon where the 3rd Battalion also joined. Both made history in that battle:

Sikh Indian Officer, 1st Punjab Regiment

the 2nd, at one time unable to make headway and with every British officer a casualty, fought with *Dahs* brought from Burma: while the stupendous efforts of the 3rd Battalion were undeniably the cause of the enemy being unable to advance further on that fateful 22nd November, 1915. The retirement to Kut displayed their fine qualities; both suffered the hardships of the siege and the horrors of captivity—of the one hundred and twenty-one Rajputana Rajput survivors of the 2nd, only nineteen were ever seen again.

The 1st Battalion fought in the endeavour to relieve Kut, the 5th (82nd Punjabis) from the North-west Frontier joining the same brigade. Later, the 1st cordoned the Persian border.

The 10th Battalion (84th Punjabis) joined the Mesopotamia forces in April, 1917, migrating to Salonika and the Black Sea the following year.

When the Afghans attempted their invasion of India in 1919 companies of the 3rd Battalion were shut-up in Jandola. To their relief went the re-formed 2nd. The 5th also saw some service in these operations, as well as in the consequent chastisement of the Mahsud-Waziris, in which the combat was much more severe.

The Regiment has been represented by one battalion or another in most of the frontier operations since the Great War ended.

The full dress uniform is red with green facings. The 5th Battalion maintain one distinction in dress, having a dark blue fringe in the pallah: during the Great War this unit wore blue puttees, a fact which the official history records was the means of averting a serious reverse. In the confusion of an intricate trench battle in Mesopotamia things had definitely gone agley, and the units engaged became inextricably mixed. A brilliant brain-wave ordered the "*Kala Pattis*" to stand-fast. This order, rapidly taken up by the n.c.o.s, so cohesed the men that they held their ground and beat off the enemy counter-attack. The incident being considered of special merit, the distinction in

dress was returned to the Battalion in the form of the present ornamentation.

The 5th Battalion possess three superb figures of Buddha which are reputed to date from about the first century A.D., and are believed to be the best preserved statuettes of that era that exist. The figures are of great value and have now been scheduled as ancient works of Art and will thus remain as treasures of India for all time.

With the introduction of the Indian Territorial Force in 1921 an 11th Battalion was raised originally as the 1st (Territorial Battalion) 62nd Punjabis and has permanent headquarters at Jhelum, the home of the 10th Battalion.

All the Company Officers of the 2nd Battalion are now Indian as a result of the scheme of 1923.

On the day of his Coronation in 1937, His Majesty the King-Emperor honoured the senior regiment of his Indian Infantry by becoming their Colonel-in-Chief.

CHAPTER XXX

The 2nd Punjab Regiment

"Sholinghur", "Carnatic", "Mysore", "Maheidpoor", "Ava", "China", "Pegu", "Lucknow", "Burma, 1885–87", "Loos", "France and Flanders, 1915", "Helles", "Krithia", "Gallipoli, 1915", "Suez Canal", "Egypt, 1915", "Megiddo", "Sharon", "Nablus", "Palestine, 1918", "Aden", "Defence of Kut-al-Amara", "Kut-al-Amara, 1917", "Baghdad", "Mesopotamia, 1915–18", "North-west Frontier, India, 1915, 1916–17", "Afghanistan, 1919".

HALF-WAY through the eighteenth century the quasi-military forces of the East India Company in their southern territories were a little more than watchmen numbering not more than five or six thousand, with rather less than one man in three armed with a lethal weapon, the bulk relying upon the primitive spear, or more likely, a bow and arrow. The incessant intrigues and enmity between the Nizam, Mysore and Mahrattas, as well as the French, and their possible effect upon the Company's future developed the necessity for a more cohesed defence force, the early expansion of which caused the beginnings of the ancestral units of the 2nd Punjab Regiment.

The first four Battalions were raised as hostilities in the Carnatic increased between 1751 and 1776. All took part in

(*India Office photo*)

A Detachment of the 1st Battalion 2nd Punjab Regiment at Drill in Malaya

varying degrees in the diverse phases of those prolonged and almost continuous operations. The 2nd, 3rd and 4th fought under Eyre Coote at Sholinghur. This battle, although Hyder Ali's third defeat, did not end the strife and a year later both the 2nd and 3rd Battalions suffered capture and martial imprisonment for two years; the 2nd to Hyder, the 3rd to the French at the siege of Cuddalore. Both were re-formed by the time war broke out with France in 1793, and in the seizure of Pondicherry recaptured their lost colours.

The story of the 5th Battalion starts from 1798, two years after the 2nd had returned from fighting the Dutch in Ceylon and the spice island of Amboyna, and a year after the turban had been introduced into the Madras Army in place of the uncomfortable imitation European shako.

The dress of the southern sepoys at this time was a white jacket faced with blue, the turban and cumarband being of the same colour. Their gross monthly pay, from which there were many deductions, was the equivalent of five and a half rupees, acceptance of which entailed a penalty of death for subsequent desertion.

All the existing active units of the Regiment took some share in the first attempt of the East India Company's Government to settle affairs with the Mahratta federation: the 2nd Battalion are the only surviving unit of the force which took part in the successful enterprise against the Raja of Berar, better known as the Bhonsla.

In the second and final war, the 4th and 5th Battalions audaciously stormed the formidable batteries at Maheidpoor in December, 1817 and so put paid to the account with Holkar and the Indore Durbar.

It is rather amusing to read in the ancient digest that His Excellency the Commander-in-Chief took exception to the frequency of courts-martial in the 4th Battalion. He felt that such were unnecessary, or at least ought to be, in a corps of such character, respectability and gallantry in the field.

The 2nd and 3rd Battalions crossed to Burma in 1824, the former being rewarded by the Directors of the Honourable Company with the badge of a galley and the motto "*Khushi wa Tari*" on account of their willingness to embark on foreign service, which owing to caste restrictions Bengal regiments were unable to do.

Not to be outdone by the other two, the 4th Battalion volunteered to a man for service at the other end of Asia in the not very creditable war with China over the opium traffic. They fought well in the assault on the Ching-Kiang-Foo Forts, and on the conclusion of hostilities also received a motto, "*Tayar-o-Wafadar*" (Ready and True) in consideration of the marked forwardness that had been shown by the Corps to serve overseas. The Battalion also received the distinction of wearing the Dragon in company with the rest of the forces employed.

The 2nd Battalion again crossed to Burma in 1852 to lose over three hundred men in the victorious assault on Bassein.

During the Great Mutiny the 5th Battalion were brought up into Bengal. As the 27th Madras Native Infantry they played a loyal and distinguished part for two years. They were the only Madras Infantry in the Lucknow operations, being part of Outram's force at the Alambagh and in the capture of the Oudh capital. It was through them that all Battalions of the Regiment carry the much-prized honour on their colours of to-day. The 2nd Battalion played small parts in the suppression of Tantia Topee, while the 3rd and 4th released British troops by garrisoning Hong-Kong and Singapore.

King Theebaw's vicious attitude towards India which resulted in his downfall, and Upper Burma becoming a theatre of almost unparalleled dacoity, again called for the services of the 3rd and 4th Battalions. The latter saw little of the fighting, not arriving until late in the campaign, but the former were in the difficult assault on Minlah, and from 1885 were continuously on innumerable active operations for upwards of five years during and after the official termination of the war.

The state of Burma, and the North-eastern Frontier of India after the annexation, necessitated the establishment of a permanent local force, and to this the 3rd Battalion were allocated as the 2nd Regiment of Burma Infantry. Madrasis were then mustered out, Punjabis henceforth being enlisted for the ranks.

Internal trouble in the state of Manipur in 1890 led the paramount power to interfere. The small force, all Indian, sent to depose the ruler were strenuously resisted. There was much heavy street fighting, and in the end several British officers were treacherously murdered under pretence of parley. A small detachment of the 3rd Battalion with some Gurkhas under the command of Lieut. C. J. W. Grant, on learning of the disaster, gallantly attempted to relieve Manipur. More or less opposed all the way, Grant captured the fort at Thobal; and with a handful of young soldiers, the majority recruits of eight months' service, held out for thirteen days against all attacks of the greatly outnumbering Manipuri sepoys. An outstanding defence which earned the Victoria Cross for Grant and the Order of Merit for each of his eighty men.

After Manipur, the yellow-faced red uniform of the Madras line was changed to drab with white facings for the 3rd Battalion, a distinction abolished in 1922, when all battalions adopted red with green facings. Green facings had come into general use for the Madras regiments converted into Punjabis in 1903. The new regiments were, however, permitted to retain the colours and honours earned by their forebears.

As the 67th, 69th, 72nd, 74th, and 87th Punjabis, the five senior battalions fought the enemies of India and the British Empire in the world-spread campaigns of 1914–21.

The 2nd Battalion (69th Punjabis) the first to see action, reduced the Turkish defences in the Gulf of Aden, while on their way to guard the Suez Canal. After a spell in the Dardanelles, they went on to France, fighting with conspicuous distinction at Loos. Withdrawn to Egypt ostensibly for Mesopotamia, the Aden operations again detained them *en route* and they remained

The Band of the 3rd Battalion 2nd Punjab Regiment Marching through the Khyber Pass

in that trying, unhealthy, but not unimportant theatre until their return to India. No sooner landed, they were pushed up to join General Dyer's column for the relief of Thal, then besieged by the forces of a misguided Afghan King foolish enough to believe that India in the year 1919 was devoid of real soldiers.

The 1st Battalion (67th Punjabis) left Secunderabad post-haste in 1915 for the valley of the Tigris, in the hope that they would be able to assist General Townshend to consolidate what seemed to be the certain capture of Baghdad. It was ordained otherwise, and so the Battalion suffered the misery of capitulation to the enemy, after a spirited defence which will long illuminate the pages of the valiant deeds of the soldiers of India.

The 3rd Battalion (72nd Punjabis) were on the North-west Frontier and remained on that front, as important to India as any other, until the prospects of defeat in France resulted in Indian battalions being dispatched to Palestine in replacement of corps ordered to stem the contemplated break-through in the West. Thus came the honours for Allenby's victories over the Turks.

The 4th Battalion (74th Punjabis) in 1914 were in the far eastern garrison of Hong-Kong. They had the novel experience when ordered to support the force for the capture of the German Chinese treaty port of Tsingao to find themselves delayed through their transport striking an ice-field and becoming almost frozen in. The ship was in the end only freed by the men being brought up on to the stern decks, to jump in unison until the bow was clear. The 4th Battalion eventually came home to the Frontier in 1917, and shortly afterwards joined the 3rd in Palestine.

As the 3rd Battalion had watched the Mohmand line in the early days of the War, so the 5th kept an eye on the Mahsuds and other Waziri fraternity, until they too were required in 1917 for the larger arena of Mesopotamia. There, unluckily, they did not have much opportunity until the final operations of 1918. But in the Kurdish Revolt they fully made up for

lost time, doing great work around the Beziyan Pass. Marching 670 miles over mountain country in eleven days, they reached the Persian border in time to avert the contemplated Bolshevik Russian invasion. In the Arab Rebellion they also saw much fierce fighting in the operations around Hillah, 1920–21.

The 10th Battalion, the training battalion at Meerut, came into being as the 2nd/67th Punjabis at Loralai during the Great War as a frontier defence unit. The Battalion now no longer goes on service, but instructs the recruits for the other five.

Since the Great War, all the senior Battalions have actively shared in the frontier small wars—some operations of which far outweighed in magnitude any of the pre-war affairs. The 5th forms part of the new Indianised division.

The night defence of Kila Hari post in the Loe-Agra operations of the Nowshera Brigade in April, 1935, by a party of the 3rd Battalion is an epic. The small detachment attacked by an unexpectedly well-led tribal *lashkar* of Shamozais a thousand strong, were almost overwhelmed in the old-fashioned hand-to-hand fighting that followed. Revolvers, bayonets, stones, and knives all played their part in a fight of the utmost determination and severity kept up for over ten hours, and one almost unequalled on the Frontier, except perhaps for the desperate affair on the Craig Piquet in the Ambeyla operations of 1863. That such an assault should have been defeated with a shortage of ammunition, bombs, pyrotechnical lighting, and without effective protective wire speaks well for the leadership of the Indian officers and the discipline of the men. All ranks engaged proved themselves worthy of decoration; but this was not possible and so five Indian Orders of Merit and four Indian Distinguished Service Medals were conferred upon those who showed extra-special magnificent conduct in the beating off one of the most amazing attacks ever made by tribesmen on the North-west Frontier.

The 1st Battalion has recently garrisoned the new Empire defences in Malaya. The 5th will go down to history as the

first troops to reach Shanghai in 1927, when that then undefended great international trade centre was threatened with destruction by civil belligerent forces in South China.

The 2nd Punjab Regiment have served their King, their Country, and their Empire in more wider and varied fields than most of the other Indian Regiments.

NAIK—BOMBAY GRENADIER BATTALION—1801

CHAPTER XXXI

The 4th Bombay Grenadiers

" Mangalore ", " Mysore ", " Seringapatam ", " Egypt ", " Kirkee ", " Corygaum ", " Beni Boo Ali ", " Meanee ", " Hyderabad ", " Mooltan ", " Punjaub ", " Central India ", " Abyssinia ", " Kandahar, 1880 ", " Afghanistan, 1878–80 ", " Burma, 1885–87 ", " Somaliland, 1901–04 ", " Egypt, 1916–17 ", " Gaza ", " Megiddo ", " Nablus ", " Palestine, 1917–18 ", " Aden ", " Tigris, 1916 ", " Kut-al-Amara, 1917 ", " Baghdad ", " Sharqat ", " Mesopotamia, 1915–18 ", " East Africa, 1914–16 " " Afghanistan, 1919".

TO the uninitiated it may be as well to explain that a grenadier was originally a soldier whose particular duty it was to throw hand-grenades or, as we commonly say to-day, bombs. Later these bomb-throwers were organized into formations, and practically all typical infantry corps of the eighteenth century had their grenadier companies, composed of their most powerful men to whom were ascribed the post of honour on parade. As the grenade devolved into obsolescence, the grenadier sub-units became "crack companies" wearing a distinctive headdress or

badge and developing considerable *esprit-de-corps*. And later, whenever some special service was required, it became customary to form the grenadier companies into battalions for the purpose—Napoleon even went further and organized them into brigades.

Thus in 1778, when the Bombay Government desired their own nominee—Raghunath Rao—to be installed as Regent to the youthful Peshwa of Poona, the grenadier companies of the 2nd, 3rd, and 5th Battalions of Bombay Sepoys were temporarily amalgamated into a battalion to form part of the expedition to support him. The enterprise was a failure—the force was to all intents commanded by a committee—and the only bright spot in the ill-advised operations was the amazing gallantry of the grenadier regiment in the rear-guard actions from Talegaon when 2,600 Bombay troops were pursued, by not thrice, but thirty times that number of Mahratta soldiery. In the words of the enemy prince Scindia, the Grenadiers were "a red wall, no sooner beaten down than built up again".

In the ensuing year the remnants of the Talegaon companies were the basis of a battalion formed of one grenadier company from the existing six Sepoy Battalions and two grenadier companies of the newly-raised Marine Battalions. Whether this was intended to be a compliment for the bravery displayed at Talegaon no records exist; if it was, the usual lack of imagination was displayed, for the new unit became the 8th Bombay Native Infantry without the title of Grenadier. This deletion nearly caused mutiny; for a grenadier was something more than the common sepoy. The authorities, however, were adamant; and what is now the 1st Battalion of the Bombay Grenadiers remained for the time undistinguished in title from the rest of the line.

The Battalion were part of the garrison of Mangalore, then besieged in person by Tippu, Sultan of Mysore. From the 16th May, 1783, to the 23rd January, 1784, the small British Indian force of just under two thousand, resisted the attacks

of nearly seventy-five times that number of the Mysore Army, assisted by some French. Although men and ships were available, no attempts at relief were made, and it was not until men were dying in dozens or going blind through the salt ration failing, that the gallant commander capitulated with the honours of war. The defence of Mangalore has achieved no distinction in popular history, but as a military achievement it has been considered by some authorities to have been even greater than the defence of the rock of Gibraltar, when consideration is given to the means available for denying the enemy entrance. After Mangalore the restoration of the title of Grenadier could no longer be denied and it was restored. The White Horse of Hanover, then the mark of the grenadiers of the British Army, was adopted as the badge.

The title Grenadier was in those days only conferred upon a unit for prowess in battle: those to whom that word only conjures visions of London, the Guards, and the senior regiment of infantry of the armies of the British Empire, here may be told that an Indian regiment is the senior regiment of Grenadiers in the King-Emperor's service; the equivalent honour not being bestowed upon the First or Grenadier Regiment of Foot Guards until thirty years later, for the severe handling they gave the French Imperial Guard at Waterloo.

Mangalore was also unique in that it is one of the rare instances where an honour was given other than for a victory, and it is now only borne by the The Black Watch and India's Grenadiers.

The 2nd Battalion was born at Calicut in 1796 as the 13th Bombay Native Infantry. Upon the reorganization of the Bombay foot into two battalion regiments in 1798, they and the Grenadiers joined to become the 1st and 2nd Battalions of the 1st Bombay Infantry. On account of the 1st Battalion being Grenadiers the Regiment were given premier seniority in the new re-numbering.

With the dawn of the nineteenth century the French en-

(*India Office photo*)

AN OFFICER QUESTIONING RECRUITS WHO HAVE OFFERED THEMSELVES FOR ENLISTMENT

deavoured to secure a hold on Egypt. This was considered a grave menace to India and a British force under Sir Ralph Abercomby was despatched to assist the Turks to expel them. Lord Wellesley, then Governor-General, decided that India's safety demanded co-operation—in fact it was he who proposed the despatch of the expedition from England. Thus a force under General Baird, of which over one half were Indians, was sent to the Red Sea. After traversing many miles of arid and trackless desert it reached the Mediterranean in August, 1801, just too late to take part in the fighting. The French General upon hearing of the approach felt unable to withstand attacks from front and rear, and so capitulated.

The second-in-command of the force was Colonel Arthur Wellesley, but he was unable through illness to embark. Had he sailed, epic battles would not have been great victories of a Duke of Wellington, nor an "Iron Duke" have been numbered among Britain's Prime Ministers; for, the transport to which the great soldier was allotted, sailed to sea never to be heard of again.

All the Indian infantry for the first Egyptian Expeditionary Force, as we might describe it, were supplied by the present 2nd and 10th Battalions, the latter having been raised in 1800 as the 1st/7th Bombay Infantry. We are told by contemporary writers that the troops from England were much impressed by the smartness and bearing of the Western Indian soldier. Few incidents in the early history of India's Army can be as romantic as the appearance on the banks of the Nile of warriors from the lands of the Indus. Thus comes the special honour of "The Sphinx" to the 2nd and 10th Battalions, the Regiment bearing the honour of Egypt: though why on the introduction of the regimental system into the Indian Army, the former honour was not extended to all the Battalions as were all other honours gained by the other components, is one of the inexplicable mysteries which ever confound the student of Indian military lore and custom.

On their return from overseas the 2nd and 10th Battalions

saw much service in the general plan which aimed at keeping the Mahratta princes within the confines of their proper territories; while the 1st Battalion played a most honourable part in the unsuccessful first siege of Bhurtpore, when a Bombay column went to the aid of Lord Lake in 1804.

All Battalions were constantly engaged in local service during the ensuing years until 1817. Then the Peshwa—Bajee Rao—treacherously joining the hostile Mahratta combine against the Indian Government, contemplated destroying the latter's troops encamped at Kirkee. Bajee Rao's hordes fell upon two thousand Indian and eight hundred European troops. But never undaunted, the Bombay force advanced, ignored their envelopment by the Mahratta horse and, with the 2nd and 10th Battalions well to the fore, charged with such audacity that the enemy lost his nerve and bolted.

January 1st is always a great day for the forces in India, parades being held in anniversary honour of the Queen-Empress Victoria's assumption of the Imperial throne: to the 2nd Battalion of the Bombay Grenadiers the parade has a double significance, for on the same day in far back 1818 that battalion achieved entry into the army as Grenadiers.

In the closing days of the previous year Poona was threatened by the Peshwa's troops. The authorities there called to their support a detachment, then at Sirur, consisting of three hundred of the Poona Horse and the 2nd Battalion all under the command of Captain Staunton of the latter.

On the evening of the last day of the old year Staunton's force marched, and by the following morning reached the village of Corygaum, there to discover the whole of the Peshwa's army 25,000 strong. The Mahratta troops immediately sought action against this handful of soldiers, destitute of rations or water and exhausted by a trying night march, and there began the brilliant engagement, almost the first where the sepoys were unsupported by European troops, save for twenty-four Madras Artillerymen of whom twenty became casualties. Every inch

of the village was disputed with desperate valour. Despite their tremendous sufferings the men fought on against an enemy which gave no quarter—even the medical officers had to lead as combatants. By dusk both sides were exhausted, and so disheartened were the Peshwa's ranks at their failure to secure what seemed an obvious victory, that they slunk away and refused to face any more the indomitable men. In commemoration of the most momentous struggle ever to have been fought by Indian soldiers the sword and medals of Captain Staunton are annually trooped with the colours on Proclamation Day.

After Corygaum, both Battalions having achieved the same distinction, the Corps became the 1st Grenadier Regiment.

Pirates in the Persian Gulf, of which the leaders were the Beni Boo Ali tribe, sent the 10th Battalion to the coast of Arabia in 1821, where they stormed the tribal stronghold and destroyed the pirate fleet. Of the total 202 casualties in the force, 148 fell to the 10th Battalion.

The Regiment saw much minor, but sometimes desperate fighting during 1839, when communications with the army in Afghanistan were much harassed by the hostile Baluch tribesmen.

At Meanee, in the subsequent campaign which brought Sind within the Indian frontiers, the 1st Battalion more than held their own against tremendous odds; while at Hyderabad, so Sir Charles Napier records, the 1st Grenadiers advanced with the regularity of review right up to the enemy entrenchments. They did not, however, receive the Meanee-honour, owing to Sir Charles' anger with their Commandant for halting quite rightly tactically when Sir Charles thought otherwise. The honour for the battle thus comes from the 5th Battalion disbanded in 1930, who were originally the 12th Bombay Infantry.

The since disbanded 4th Battalion, then the 9th Bombay Infantry, brought the honours given to the columns which suppressed the rebellion in the Southern Punjab after the settlement of the 1st Sikh War.

1857 touched the Regiment lightly, the 10th getting most of the work, doing extremely well in the *mopping-up* operations in Central India during the following year.

The 2nd Battalion went to Abyssinia, 1868, but no fighting fell to their share, they being employed as pioneers constructing roads and railways to smooth the path of one of the best organized expeditions that ever left India's shores. The men suffered severely from the climate, and scurvy took a very large toll.

1880 found the 1st Battalion in Kandahar, and at the battle of Maiwand they suffered a most terrible ordeal. For four hours, when other regiments were falling back, they stood firm under the most appalling fire, and when forced to retire over ground devoid of any kind of cover not a man attempted to run, they being saved from complete destruction by a belated charge of a squadron of the mis-managed cavalry. After Maiwand the Bombay Army became much *bad-named*, but no slur could ever lie on the Grenadiers, nor should it lie on the other ranks of that Army, for the action was nothing but bad generalship, and but for the bravery of the troops might have been an utter disaster.

Beyond the guerilla Burmese War of 1885–87, where numerous detachments of the 1st Battalion were engaged with rebels for upwards of two years, no further serious operations came the way of the Regiment during the reign of Queen Victoria.

The 1st and 2nd Battalions were engaged in small wars in 1902; the 2nd around Aden; the 1st in Somaliland, a campaign in which two officers—Lieuts. Rolland and Carter—were awarded the Victoria Cross.

Africa has a strange calling for the Grenadiers, and it was to British East that the 1st Battalion proceeded in 1914. In the ill-conceived battle of Tanga—won and lost it is said in the Mombassa Club—"the sacrifice of the gallant Grenadiers was unavailing" so states the official record. Later, much further land service fell to the survivors in the defensive operations of the East African Protectorate, and at the defence of

101st Grenadiers
Naik—1914

the little fort of Jasin the musketry of the Grenadiers was so telling that the German besiegers thought the garrison to be South African Europeans. Before they left for Egypt and Palestine in mid- 1916 the Battalion had the gratification of participation in the final capture of their original objective.

The 2nd Battalion were introduced to the great conflict at Muscat in the Persian Gulf. They subsequently joined the force before Kut-al-Amara, where two companies in a most courageous attempt to relieve their beleaguered comrades lost almost all of their ranks killed or wounded, the survivors being countable on the fingers of the hand.

The 10th Battalion, by then the 113th Infantry, also went to Mesopotamia, fighting at Sharqat; thereafter to join in the Aden operations.

The duplicate unit of the 2nd Battalion, raised during the Great War, earned the honour for the 3rd Afghan War, 1919.

Half of the 1st were back again in Somaliland, 1920, the ranks receiving the first post-war issue of the Africa General Service Medal. The whole Battalion were in the Razmak 'war' during the following two years, afterwards going east to meet more 'trouble' in Hong-Kong.

During the reorganization of the infantry in 1929 it was very much feared that the ancient and honoured title of Grenadier would disappear from the Indian Army. It was known that disbandment of the Regiment was contemplated. But due, it is believed, to the personal intervention of the King-Emperor George V, whose father had been Colonel-in-Chief of the 2nd Battalion since 1876—hence their subsidiary title of King Edward VII's Own—the two original grenadier battalions were saved, while the 10th Battalion never 'grenadier' until brought into the Regiment in 1921, was linked with the 10th Battalion of the 9th Jat Regiment to make a composite battalion serving as a training unit for both corps. There is also a Territorial battalion, the 11th, originally raised in 1922 as the 1st (Territorial) Battalion Merwara Infantry, headquartered at Ajmer.

The red uniform of the Regiment is faced with white. Having regard to the fine unbroken existence of the Corps unequalled by any Indian infantry unit, one is tempted to suggest that the facings should be blue, the cuffs flapped in true grenadier fashion, and the Regiment put to the right of the line as the Indian Grenadier Guards and linked to the Household troops of His Majesty, as is the similar regiment of the Dominion of Canada.

The third permanent grenadier regiment of the King, the Canadian Grenadier Guards, was raised in 1807 with James McGill—founder of the famous university—as the Colonel, and first saw service in the war with the U.S.A., 1812-14; which began a long and brilliant record in all the Canadian small wars until, and including the Great War. The present title was bestowed, and the unit raised to the status of household troops by direct command of King George V in the year 1912. Why, therefore, should not a similar distinction be conferred upon India's Army, and why not upon the Bombay Grenadiers?

CHAPTER XXXII

The 5th Mahratta Light Infantry

"Mysore", "Seedaseer", "Seringapatam", "Beni Boo Ali", "Kahun", "Mooltan", "Goojerat", "Punjaub", "Central India", "China, 1860-62", "Abyssinia", "Afghanistan, 1879-80", "Burma, 1885-87", "British East Africa, 1901", "Megiddo", "Sharon", "Nablus", "Palestine, 1918", "Basra", "Shaiba", "Kut-al-Amara, 1915-17", "Ctesiphon", "Defence of Kut-al-Amara", "Baghdad", "Sharqat", "Mesopotamia, 1914-18", "Persia, 1918", "North-west Frontier, India, 1914-15, '17", "Afghanistan, 1919."

DURING the Great War, authoritative voice on the Eastern Front was heard to declare that one of the surprises of the campaign was the exemplary behaviour of the Mahratta as a fighting soldier. Though why this should have occasioned surprise is obscure. It is true that for many years prior to and after the formation of the Indian Army into one entity in 1903, none of the corps which comprised sepoys from the ancient land of Maharashtra had come into the limelight on the popular stage of the Indian frontier, the lack of engagement on which was entirely due to the geographical disposition of the units.

But, the story of how a community of village clerks and peasantry was transformed into a nation of formidable warriors, of Sivaji "The Mountain Rat", the attempts at a second empire, and the long series of wars with the British, are one of the romances of Indian history. Even the early rulers of Bengal obtained their soldiery from Maharastra which has always been the home of men cradled in stories of military prowess.

The modern Maharastra for the recruiting officer comprises the Concan and the Deccan, the former the Bombay coast strip, the latter the table-land before the Western Ghats, the soldiers from which, though in appearance not so prepossessing as the noble Rajput, are steady, laborious and persevering, and have long served with credit and distinction since the days of the East India Company.

The 5th Mahratta Light Infantry originates from 1768 when the native levies were first formed into battalions when the Company, then in Western India practically confined to the island of Bombay, were forced to organize protective defence against the possibility of French invasion from the sea, as well as attacks from the Mahratta chiefs or Mysore. So were born the 2nd Bombay Sepoys, later the 103rd Mahratta Light Infantry of 1903, and the present 1st Battalion.

The 105th Mahratta Light Infantry of the Great War were raised in 1788 as the 3rd Bombay Sepoys: they changed their number to 5 in 1824, and as the "*Kali Pȧnchwin Paltan*"—they once wore black feathers in their turbans—they remain as the 2nd Battalion.

From their foundation both Battalions were continuously employed in operations on the Bombay mainland, the Deccan, Gujerat and Mysore. The 2nd were with the force invading Ceylon and were present at the capture of Colombo, 1796.

The following year were born at Calicut the Travancore Regiment of Bombay Native Infantry, later the 10th, to become the 110th Mahrattas of 1903, and now the 3rd Battalion.

The two older Battalions played a most distinguished part

in the extinction of Tippu Sultan of Mysore, and the year 1799 is one held in great veneration by the 1st and 2nd. On the morning of March 6th a small force, of which these Battalions formed a part, was detached as a concentrated out-post on the hill of Seedaseer, some fifty miles from Seringapatam—a dangerous proceeding since Tippu specialized in overwhelming small forces and cutting up armies piecemeal. He tried his usual tactics and the force soon found itself confronted by overwhelming enemy numbers, who at once commenced a vigorous assault. For six hours the three units engaged—the third was the old Bombay Pioneers—held their ground against repeated and determined assaults, and when the ammunition was almost exhausted joined the counter-attack with such avidity that the enemy were scattered. The force was unsupported by European troops, and the action has always been regarded as establishing the renown of the Bombay Army. The anniversary has for some years been celebrated by a pageant depicting some part of the battle, and thus the Mahratta Light Infantry may perhaps later be established as the founders in India of the military spectacular tattoo. The two Battalions were also present at the fall of Seringapatam the same year.

A year later, in 1800, the present 4th, 5th, and 10th Battalions were raised; the 4th at Mangalore, the other two at Bombay. The present 5th Battalion were actually raised as the Bombay Fencible Regiment from local inhabitants, by public voluntary subscription of the Europeans in Bombay. Fencible was a popular term for the volunteer corps in England then being formed to meet the menace of a Napoleonic invasion, which in 1799 India also feared, owing to the establishment of French troops in Egypt. The corps was officered by leading officials and merchants, including the Governor; and thus became the forbears, not only of a battalion of the line into which the ranks were absorbed in 1803, but of the Bombay Battalion of the Auxiliary Force, India.

A Mahratta Sepoy—1773

For the next twenty years all the six Battalions were engaged in the innumerable expeditions against the Mahrattas, Pindaris, and the other turbulent folk in Central India. While in 1819 the 1st and 2nd Battalions were both engaged against Arabian pirates who offered stout resistance to the expedition to the Persian Gulf, the 1st Battalion suffering heavily when three companies were overwhelmed by the formidable Beni Boo Ali.

During the first expedition into Afghanistan the 2nd and 5th Battalions were deputed to protective duties in Baluchistan. During these subsidiary operations in which the political officers seemed to delight in ordering the troops here, there, and everywhere in the notorious terrible heat of the waterless Sind Desert, a detachment of the 2nd Battalion was sent to garrison the tiny fort at Kahun. From May until August no relief could be sent to the gallant defenders, then hemmed in by hostile Marris. The relieving column failed through treachery and thirst, and by September, after five months of siege, the fort was forced to capitulate, but with the honours of war. The tribesmen observed those honours. For the gallant conduct of the small detachment reduced through ambush at the beginning from 300 to 140 the honour "Kahun" was rightly awarded to the Battalion: while for the subsequent services of another detachment of the 2nd Battalion in the repulse of the three-day attack of the tribesmen on the town of Dadar, the 2nd Battalion were created Light Infantry.

A Bengal force, having been unable to subdue the fortress of Multan held by insurgent Sikhs in 1848, the 1st Battalion joined the Bombay Army column despatched by forced marches in support. The Battalion, after suffering heavily in the capture of the Multan Mandi, fought again at Gujerat, then went on to the frontier to be engaged for another year policing the Yusafai.

All Battalions were employed in the suppression of rebellion in Central India during 1857–58. Split into small detachments in that non-spectacular but remarkable achievement of the loyal troops employed, not the slightest sign of disaffection appeared anywhere amongst them and their steadfastness earned the highest commendation of the Government.

Bombay troops played little part in the post-mutiny expedition to China; nevertheless the two infantry battalions thereof selected to take part were the 1st and 2nd, although they were unfortunate enough to be left to garrison Canton. Rebels at Taiping, however, gave the 2nd Battalion their opportunity of adding "China" to the Regimental Colour.

The 1st and 3rd received the honour of becoming "Light Infantry" for their gallantry in Abyssinia, 1867–68. The honour of being Light Infantry, be it noted, arises in almost the same manner as Grenadiers, though being inferior thereto. In early days each battalion of typical infantry had a "Light" company composed of small picked wiry men, able to move swiftly to reinforce or surprise, as tactical opportunity offered; these too, became "crack companies", and later, the title Light Infantry came to be bestowed as a mark of honour. The 3rd Battalion particularly distinguished themselves in the storming of Magdala; while the 1st Battalion received as a memento from Sir Robert Napier an Abyssinian silver drum. Sir Robert in addressing the Battalions said:—

> "You traversed, often under a tropical sun, or amidst storms of rain and sleet, 400 miles of mountainous and rugged country. You have crossed ranges of mountains (many steep and precipitous) more than 10,000 feet in altitude, where your supplies could not keep pace with you. In four days you passed the formidable chasm of Bashilo; and when within reach of your enemy, though with scanty food, and some of you were for many hours without either food or water, you defeated the army of Theodore, which poured down upon you from its lofty fortress in full confidence of victory."

BANDSMAN
BOMBAY LIGHT INFANTRY—1845

The 2nd, 3rd, and 4th Battalions were in the second Afghan War, 1879–80; the senior in the relief of Kandahar, the latter two in the Bolan Pass. The 1st were also employed on the lines of communication.

The 2nd Battalion chased the elusive dacoit in Burma for two years, 1886–88, bringing honours for a most strenuous and discouraging campaign.

As the North-west Frontier trained the Punjab Regiments, so did Africa the Bombay. A landing party of the 5th Battalion made the Esa Somalis 'sit up' in 1891; while the 4th Battalion earned "British East Africa" for very trying services more than equal to the severest Frontier episodes for some months in 1901.

The great upheaval that began in 1914 not only resuscitated the reputation of the Mahratta as a fighter, but enhanced it. In that conflict no regiments were to suffer more the glories and trials of victory, or the agonies of honourable captivity, than the Mahratta Light Infantry.

The 1st, 3rd, and 5th Battalions, more familiar to Great War warriors as the 103rd, 110th and 117th Mahrattas, all formed part of the original expeditionary force to the Persian Gulf in 1914—the 5th Battalion had already been up there stopping gun-running in 1909–11. In the first battle for Kut they proved that the old Bombay regiments were worthy to stand shoulder to shoulder with the best troops of the Empire; confirmed it in the desperate battle of Ctesiphon when the brunt of the loss —half the fighting force—was borne by the infantry; and in the honourable retreat in the face of overwhelming Turkish forces to the glorious stand at Kut; a siege borne by all ranks with the same fortitude and courage which typified their behaviour in the victories before that catastrophe. The three Battalions were all captured at Kut; the campaign till then having already cost the 1st, 330 men, the 3rd, 1,170, the 5th, 410. For their conspicuously distinctive services from the original landing at Fao until the capitulation at Kut-al-Amara

the 5th Battalion, not then "Light Infantry" were in 1921 honoured by the title "Royal". This distinction, however, was not extended to the amalgamated Regiment in 1922, although the battle honours of all are carried by each, as is the title "Light". There are no Royal battalions in the other armies of the Commonwealth, only Royal regiments. The 5th, aptly, have since been selected for complete officering by Indian gentlemen, under whom the Battalion ably carries on the old traditions.

The 10th Battalion (114th Mahrattas) went to the unsuccessful relief of their three brothers shut up in Kut in December, 1915, but took no part in the brilliant battles that forced the Turk into retreat until the final *coup-de-grace* at Sharqat. They remained for the Arab Rebellion, playing the leading rôle in the action at Samawah, July 1st, 1920.

The 2nd and 4th Battalions (105th and 116th Mahrattas) remained behind as draft-finders; the former being on railway internal security duties, the latter on the Frontier. The 2nd proceeded to Mesopotamia in 1916, having their fire-baptism before Kut where they carried out a commended magnificent attack; Palestine called them in March, 1918, to earn all the infantry honours for the final decisive phase in that historical theatre.

The 4th also took the stage in Mesopotamia in April, 1917, but did not get much opportunity to prove their mettle until the severe actions of the Arab rebellion. At Hillah two companies were composited with two companies of Gurkhas, and the resultant battalion—officially styled "Gurattas"—carried out several successful fights.

The 5th Battalion re-formed from drafts in India during 1918, proceeded to the Persian operations in the September, remaining there until 1920.

The 1st and 3rd Battalions after the Turkish peace were sent into the force defending the North-west Frontier against misguided Afghan efforts in 1919; and in the Mahsud-Waziri

BUGLER
MAHRATTA LIGHT INFANTRY—1939

offshoot therefrom, 1921–24, the 1st fought an outstanding action at Palesina.

The 2nd Battalion endured great hardship, with that cheerfulness and fortitude which is characteristic of the Mahratta soldier, during the Burma rebellion, 1931.

There is a Territorial Battalion, the 11th, formed at Belgaum in August, 1921.

The red-coated black-cuffed Battalions of the Mahratta Light Infantry—the 5th Royal Battalion did not change to the customary blue, but wears blue fringes, lanyards, and hose tops as distinctive dress—since their birth in the days of the old "John Company" have served the Imperial Crown with steadfast loyalty and devotion. Gallant and dashing in battle, tenacious and unresisting when hard-pressed, no longer do the men of Maharastra stagnate in the backwaters of the western provinces; but since the Great War have given their full measure of service on India's land frontiers, keeping themselves ever ready to preserve the honour of their great Country and their Empire.

CHAPTER XXXIII

The 6th Rajputana Rifles

" Mysore ", " Seringapatam ", " Bourbon ", " Kirkee ", " Beni Boo Ali ", " Meanee ", " Hyderabad ", " Aliwal ", " Mooltan ", " Punjaub ", " Reshire ", " Bushire ", " Koosh-ab ", " Persia ", " Central India ", " Abyssinia ", " Kandahar, 1880 ", " Chitral ", " Afghanistan, 1879–80 ", " Burma, 1885–87 ", " British East Africa, 1898 ", " China, 1900 ", " Givenchy, 1914 ", " Neuve Chapelle ", " Aubers ", " Festubert, 1915 ", " France and Flanders, 1914–15 ", " Egypt, 1915 ", " Gaza ", " Nebi Samwil ", " Jerusalem ", " Tell Asur ", " Megiddo ", " Sharon ", " Palestine, 1917–18 ", " Basra ", " Shaiba ", " Kut-al-Amara, 1915–17 ", " Ctesiphon ", " Defence of Kut-al-Amara ", " Tigris, 1916 ", " Baghdad ", " Mesopotamia, 1914–18 ", " Persia, 1918 ", " East Africa, 1914 ", " Afghanistan, 1919."

THE Rajputana Rifles are the senior rifle regiment of India's Army and the 1st Battalion the oldest rifle unit.

One New Year's morning the writer was asked by a young R.A.F. officer, "What is a rifle regiment?" A bystander got in first, and with that terse conciseness which at once denotes the

RAJPUTANA JAT INDIAN OFFICER
6TH RAJPUTANA RIFLES—1937

'old soldier' of the line, replied: that they were a lot of uppish gentlemen to whom blacking appealed more than "brasso" and whose passion for doing everything differently was a complete anathema to the rest of the infantry, particularly when "showing off" before the aristocracy of governors, generals, and other exalted personages privileged, probably after a hectic "hogmany" to accept the salute of the marching-by troops, doubtless themselves suffering from the effects of over-celebration on a New Year's Eve: he did not exactly say all that, but as a court-martial witness avers "words to that effect".

It is true that rifle regiments all wear dark-green or black clothing, black accoutrements, and there is considerably less shine about them—not that they are any the less smart, far from it—than an ordinary infantry regiment. They too, drill somewhat differently, march in a different time, call their bayonets "swords", and in general have a whole heap of little differences which single them out from the rest of that great arm of the service which is the only one in the end that wins battles, despite what any new theorist may advance to the contrary.

How came all this about? Well, there are many authorities; but it is generally accepted that during the early wars in America the force suffered considerable annoyance from enemy sharpshooters who skilfully concealed themselves from view. To deal with this camouflaged sniping fraternity a select body of marksmen was formed, dressed in dark clothing to stalk and destroy them. The men were in effect hunters, and in the German army that name is still retained. As their value came to be realized and their uses developed, they were formed into bodies to move with the speed of light infantry but with greater concealment: hence, when later regimented, colours were not issued to them in order that their position should not be disclosed—colours being the rallying post were in those times always proudly displayed to the enemy—and so to-day rifle regiments do not carry these dual emblems of nationality and

brotherhood, but emblazon their honours of battle upon their badges. To-day the difference between the line and the rifles is one largely of tradition and ceremony, but in yester-years such were great realities. To be a rifle regiment was considered a noble privilege, and as the grenadiers took right of the line, so the rifles took the left; the flanks being considered the posts of honour.

The rifle regiment as a tactical unit found introduction into India about 1841, and the first unit to be converted was the 4th Bombay Infantry, now the 1st Battalion, 6th Rajputana Rifles.

The 1st Battalion's history of course goes back further, until 1775, when under James Stewart—an intrepid soldier concerning whom legends are still recalled by *pundits* of the Deccan—they were raised in consequence of the expansion of the East India Company's Bombay forces. All the Regimental honours for the campaigns in the Deccan and Carnatic, as well as the oversea expeditions to the French islands and to Arabia were the result of their efforts. They served in Wellesley's flying column in 1800 for the suppression of the adventurous Dhoondia; when by travelling light they learnt the first duties of riflemen. The afterwards great Duke commended not only their praiseworthy services, but the sensibility of their baggage train, then considered practically nothing, being twenty-four bullocks and fifty coolies; the officers having another fifteen of the former. It was on account of their good work at this time that one hundred years later they were given the title "Wellesley's Rifles".

The following years were busy ones for them until 1831, when they rested more or less for a decade until their conversion; thereafter to be brigaded for a long period with the British 60th Rifles; the senior British rifle regiment, born in America, and now the King's Royal Rifle Corps, whose red-faced green jackets were adopted by the 1st Battalion and later by all.

That the 1st Battalion inherited the first rifle tradition of

swiftness, is shown by their forced march to Multan in 1848—332 miles in twenty-five stages and only two days halt—and the second rifle characteristic of boldness, by their subsequent assault on the walled insurgent Sikh city.

They later served in the Persian War, 1857; hurrying back the same year to operate against the Bhils; and later in the Malwa field-force after Tantia Topi and other renegade leaders and soldiery.

The Battalion formed part of the besieged garrison of Kandahar relieved by Roberts' intrepid marchers, and under him fought in the enemy overthrow at the final battle for that city.

In 1898 a wing saw considerable service in Eastern Africa against the Somali tribes, then making much nuisance over their quaint ideas as to right and wrong. Other spells of active duty followed; some on the Frontier; another in Persian Baluchistan where the Ghilzais, angered by the effectiveness of the Royal Navy's interference with their intensely profitable arms-traffic, threatened destruction to the Indo-European Telegraph line. There for the moment we leave them.

The 2nd and 3rd Battalions of the Regiment were raised as line infantry in 1817 and 1818, to later become the 120th and 122nd Infantry of 1903.

The 2nd Battalion served in Sind and Afghanistan, 1839–41, taking part in Nott's victory at Kandahar. They joined the Persian Expeditionary Force of 1856, and in the storming of Reshire Captain J. A. Wood, seven times wounded, won the Victoria Cross; the first ever earned in an Indian unit. The standard captured in the battle, in which the Battalion played such a distinguished part, is still proudly preserved.

The 3rd Battalion earned their first honour in the eighty-fourth year of their age for services with the allied forces in China, 1900. Even then, to their dismay, they saw but little fighting.

The 4th and 5th Battalions were raised after the battle of Kirkee, from men belonging to the Poona Auxiliary Force who

had remained loyal. This force was established by treaty with the Peshwa and paid by him, but officered by the Company; the men were enlisted by the latter in their own territories, swearing to bear allegiance to the Peshwa so long as he remained loyal to the British. That worthy did not; and the battle of Kirkee, in which the 4th Battalion fought against the Peshwa's men, proved that the well-led disciplined Bombay sepoy was a match for the finest horsemen in India.

Just after the 4th Battalion became established a young man of seventeen was posted to them as a subaltern. His remarkable energy and ability quickly brought him to notice, and he became the Adjutant. Later he received the accolade and commanded the forces in the Persian War of 1856 in which the 4th saw their first major active service after Kirkee. Sir James Outram's services are commemorated in the title "Outram's Rifles" conferred in 1903.

Before, however, Outram's became a rifle corps, they were first honoured by being made Light Infantry, for their work in the terrible heat of the Sind Desert against the hostile Baluch clansmen during the first Afghan War.

The Battalion served with the reserve division in Sind during the second Afghan campaign, moving up to relieve Kandahar; but their force, delayed by many difficult obstacles, arrived to find they had been forestalled in their object by General Roberts.

For two exciting and arduous years, 1885–87, detachments of the 4th Battalion fought the brave and determined, but thoroughly undisciplined, fighters of the Burmese swamps and jungles: to leave with the epithet of the force commander, "Of all the good Bombay regiments I have had: the best."

During H.R.H. the Duke of Connaught's tenure in command of the Bombay Army he decided to form a rifle regiment as a *corps d'elite*, and to join the 1st Battalion, already rifles, he selected the 4th and 5th Battalions. So the red-coats were discarded for the rifleman's green jacket, the colours lodged, and the three battalions linked as the Rifle Regiment of Bombay Infantry.

The 5th Battalion first saw major action in Sind. In that campaign they did so well that Sir Charles Napier afterwards presented them with his portrait as a memento of their intrepidity at the battle of Meanee, doing so with the reflection that his name should be associated with them when those that survived the conquest of Sind shall be gathered to their comrades who fell in battle. And so to-day the 5th Battalion are "Napier's Rifles".

The 5th were on their way to Persia when news of peace sent them back to India, only to find that their comrades in Bengal had mutinied to an unexampled degree. The Battalion, whose honour was never sullied, joined in the Malwa and Central India columns, being present at the taking of both Jhansi and Gwalior, services which were so distinguished that they also joined the select company of created Light Infantry. One of the officers, Lieut. W. F. F. Waller, received the Victoria Cross for gallantry in the storming of the fortress capital of the Mahrajaha Scindia, himself loyal, when his "crack" Gwalior contingent was not.

The 5th served more as pioneers than light infantry on the lines of communication during Sir Robert Napier's hitchless campaign against the Lion King of Judah or more properly, perhaps, the Negus Theodore.

Like the 4th Battalion, the 5th also served in the Upper Burma Field Force 1886 to 1887; and as recorded, with the 4th became riflemen in the following year. They also joined the 1st Battalion in the Persian Gulf anti-gun-running expedition during the opening of the reign of King George V.

The 10th Battalion, unlike the five active battalions, were not a product of the old Bombay Army, but had their origin in the local contingent formed in 1853 to police the Shekawats, a predatory tribe inhabiting the country north-east of Jodhpur. The infantry of this civil force subsequently became the 13th

(India Office photo)

OFF-DUTY AT THE CINEMA—THE LADIES OF BOMBAY DISTRIBUTE CIGARETTES TO TROOPS BEFORE EMBARKATION

Bengal Infantry in 1851. Before then, however, they had combatted in the well-fought action at Aliwal. That the unit still remains is sufficient testimony to their services in the dark days which eclipsed the first army of Bengal. The Battalion were the first Indian regiment ashore at Tanga, German East Africa, 1914, where they suffered severely owing to the primary "bungling" which put the gallantry of the Battalion to no avail.

In 1914 the 1st Battalion were despatched to Bahrein island in the Persian Gulf to watch Turkish colonial activities. As was expected the Sublime Porte sided with the Central Powers and so that Battalion landed and fought the opening actions which sealed the fate of Basra. The 2nd soon joined them, and both were in the thick of the fighting of the "ever victorious army". In the first battle for Kut, Kipling's "Gunga Din" came to life in the person of Bhisti Mangla of the 2nd: going up and down the firing line, giving water to the men during the hottest part of the action, regardless of the danger; he well earned the mention which General Townshend made in his following despatch. Both garrisoned Kut; the 2nd Battalion holding the important Woolpress village outpost throughout the twenty weeks of siege. During their captivity the Sultan, as the Khalif of the Mussalmans, personally attempted to dissuade the Muslim officers from their allegiance to the King-Emperor; only to be affronted by the example set by the two senior Indian officers, both of whom belonged to the Regiment, who rejected the offered swords of honour with an emphasis that suffered them the further rigors of an Ottoman prison.

The 1st and 2nd Battalions were resuscitated from details and drafts in 1917, the former to go to Mesopotamia where they remained until 1919.

The new 2nd went to South Persia and Seistan where a young follower cook-boy won a Distinguished Service Medal for rescuing a wounded man in circumstances of great gallantry. Promoted to be a full-fledged sepoy he unhappily suffered death

a few weeks later when the whole of his company, officers and men, were entirely wiped out by the influenza scourge which ravished the world in the closing months of the War.

During the turn of our fortunes in Iraq, to use the Arab name, the 3rd Battalion were sent to reinforce Sir Stanley Maude. Again they were unfortunate in finding no great active employment. By 1920 they reached North Persia to hold the frontier in view of the possibilities apprehended from the newly formed Soviet Socialist Republics.

The 4th Battalion had to cool their heels in India until orders in 1916 despatched them to Allenby for the first phase in his advance which gave Jerusalem to the Allies. In the storming of the Prophet Samuel's tomb—Nebi Samwil, the key to the Holy City—and its subsequent defence, the work of the Battalion was outstanding.

The 5th Battalion, joyous at the news which suddenly put them into the Lahore Division then under sealed orders for France in 1914, fought in the indecisive struggle at Givenchy. The 5th remained on the Western Front, as the battle-honours show, until withdrawn to join the troops on the long road to Baghdad, and the after battles of Istabulat and Tekrit. After transfer to Palestine for the finale, they returned home, not to peace and quiet, but to disturbed Waziristan, where hostile tribesmen kept them on the go until 1924. The officer cadre of "Napier's" was among the first to be Indianized.

The 2nd Battalion received the title "Prince of Wales Own" for their war services, together with new colours duly inscribed in 1920. These were the last stand to be carried, for within a short two years the Battalion, together with the 3rd, laid-up their colours, cast off the red coat, and became part of the senior corps of Indian riflemen.

CHAPTER XXXIV

The 7th Rajput Regiment

" Delhi, 1803 ", " Leswarree ", " Deig ", " Bhurtpore ", " Afghanistan, 1839 ", " Khelat ", " Cabool, 1842 ", " Maharajpore ", " Moodkee ", " Ferozeshah ", " Aliwal ", " Sobraon ", " Chillianwallah ", " Goojerat ", " Punjaub ", " Lucknow ", " Central India ", " China, 1858–59 ", " Afghanistan, 1878–80 ", " Tel-el-Kebir ", " Egypt, 1882 ", " Burma, 1885–87 ", " Pekin, 1900 ", " China, 1900 ", " Macedonia, 1918 ", " Suez Canal ", " Egypt, 1915 ", " Aden ", " Basra ", " Kut-al-Amara, 1915 ", " Ctesiphon ", " Defence of Kut-al-Amara ", " Tigris, 1916 ", " Mesopotamia, 1914–18 ", " Persia, 1915–18 ", " North-west Frontier, India, 1915, '17 ", " Afghanistan 1919 ".

THE historian can find much for argument in the pedigree of the Rajput Regiment. The cold print of the Indian Army List gives the dates of the formation of the five active battalions as 1798 for the 1st and 2nd, 1804 and 1814, the 3rd and 4th, 1825, the 5th—we will leave the 10th for the moment. But the Regiment will have none of it, and claim to go back to 1778 to a common ancestor in the old 24th Bengal Infantry whose conspicuous services in the early Mysore War against Hyder

THE REGIMENTAL COLOURS OF THE SEVEN BATTALIONS OF THE 7TH RAJPUT REGIMENT BEFORE THE REGIMENTAL MEMORIAL TO THOSE WHO FELL DURING 1914–20, AT JHANSI

Ali, and later against the French, earned them the traditional distinction of capturing at Cuddalore that romantic figure of history, Bernadotte, the Corporal who became King of Sweden.

The old 24th, however, were disbanded in 1796 and the personnel transferred to the two battalions of the 7th Bengal Infantry which suffered an ignominious fate in 1857. The 24th were re-raised in 1803, and their second battalion was the foundation of the present 3rd Battalion, though this Battalion also claim descent from the 13th Bengal Infantry of 1778, which with the old 24th formed the two battalions of one regiment, the famous "*Kilpatrick ka Pultan*" of the Mysore and French wars in Southern India.

The 2nd Battalion of the new 24th Bengal Infantry, like the two battalions of the 7th, cut themselves from honourable military service in 1857, but the loyal remains—of which we shall hear more anon—were banded as "The Lucknow Regiment", and later became the 10th Battalion of the existing Rajput Regiment.

The history of the present 1st and 2nd Battalions actually begins in 1798. They are the only two units surviving to-day directly descended from the old Bengal units of the Grand Army of Lord Lake, which gave such sledge-hammer blows to the French trained mercenaries of the five Mahratta Princes, the 1st Battalion being the only unit now extant that took part in the memorable rout at Delhi in 1803 which restored the Moghul Emperor to his ancient throne. Both 1st and 2nd were at the Pass of Leswarree, which battle finished for ever the power of the French in Northern India; even if it did not finally settle affairs with the Holkar Prince of Indore. For their distinguished part in the campaign the 1st received an Honorary Standard inscribed "Lake and Victory". They are also the only survivors of that General's triumph at Deig.

The 3rd Battalion in 1828 became the 47th Volunteer Native Infantry, a title not indicating an embodiment of amateur soldiery, but to show that the Indian ranks had undertaken to

serve overseas whenever required. Many Bengal battalions on caste grounds declined this service, but some authorities hold that it was the complete loss at sea of two companies of the old 4th Bengal battalion whilst on return from Madras in 1770, that made a fatal impression on sepoy minds which afterwards became a very fruitful source of trouble.

Both 1st and 4th Battalions took minor parts in the war with Nepal, 1815: a campaign which in its early stages reflected no great credit on the leadership of the British Indian Armies until Sir David Ochterlony, an old subaltern of the old 24th, took the command. The war ended by bringing the valued friendship of Nepal and the first Gurkhas into the Indian Army and, the first lesson of caution in mountain warfare.

Despite the bravery of the troops, Lord Lake had been unable to subdue the Jat fortress of Bhurtpore in the first Mahratta War of 1803–4. In the many attacks the 1st Battalion behaved with conspicuous courage and suffered severe losses: but they managed to plant their shot-riddled Colours on the bastions. These Colours were afterwards condemned as unserviceable and ordered to be burnt; but before this could be done the sepoys got in first and tore them to pieces *pour souvenir*. Nothing was attached to this action at the time, but when Lord Combermere twenty-one years later was successful in reducing the fortress, the Battalion was again in the besieging force. Just before being ordered to make a difficult attack, the shreds of the old colours, which the men had preserved as sacred relics, were suddenly produced and tied round the new; and the sepoys solemnly averred to earn in the fight under the new colours as high a reputation as those who had fallen in defending the old. And right nobly did those gallant men make good their oath.

During the first Afghan War, 1839–42, the 1st Battalion were at the capture of Kalat, where they afterwards remained in garrison. The 2nd Battalion joined General Pollock's avenging advance through the Khyber.

Troubles incidental to the accession of a minor to the throne of Gwalior and the consequences of the army of that state attempting to establish a dictatorship, made paramount governmental intervention necessary. It was thought that the affair would be a picnic. But far from it; for the Gwalior troops opposed the forces under the Governor-General in person, and a fierce and bloody hand-to-hand encounter occurred at Maharajpore, at which the 1st Battalion and part of the 5th were participators.

Following the principle adopted by Clive that the sepoy line regiments should be assimilated in dress to that of the European regiments of the Company, all the Battalions from their inception wore red faced with yellow, and an imitation shako which later gave way to a cane-framed blue cloth-covered contraption called by courtesy a turban. In 1847 Kilmarnock caps were introduced, which became most popular amongst the soldiery, principally owing to their durability—some corps made them last for fourteen years. The Kilmarnock cap later evolved into the "pill-box" worn by the Gurka regiments to-day: in the earlier days, however, it was a larger article worn straight on the head.

In the Kilmarnock cap was fought the first Sikh War; and with the Army of the Sutlej sent to resist the Sikh invasion were the 2nd and 3rd Battalions. They did well at Mudki, one of the Indian officers of the 3rd capturing a Sikh cannon. Both fought heavily at Ferozeshah, as well as in the remaining battles of the war, which entailed the severest casualties yet experienced by the Army in India. The 4th, who had joined the forces just prior to the final victory at Sobraon, were specially praised by Sir Hugh Gough for their admirable services.

The 1st and 5th fought the actions of the second Sikh campaign now borne on the colours, the latter carrying the key position and capturing three standards at Gujrat, the battle which ended British strife with one of the bravest races of India.

Saugor, Hoshiarpur, Burma, Amritsar and Barnackpore, were

the stations of the existing active battalions, when the cataclysm of 1857 swept out of existence three-fourths of a regular army which could look back with pride upon a record of a century; the line of which to-day is represented only by ten battalions of which six form the Rajput Regiment.

The 1st Battalion, with every inducement to be false to their Colours in a district seething with revolt, deserted by all the other corps of the garrison, tempted by lavish rewards for the surrender of an overflowing treasury and an arsenal stocked with ammunition, stood firm and defended the important fort at Saugor from July, 1857, until the January of the following year. To quote the words of the Governor-General, "their fidelity was unexampled throughout the regular Native Army." As a special mark of approbation for their loyal and excellent services in circumstances of great difficulty and temptation, they were made a Light Infantry regiment.

The 2nd and 4th Battalions were disarmed as a precautionary measure which proved to be unnecessary, and after restoration served with much credit in the Punjab.

The 3rd were a sick and weary remnant *en route* from Rangoon to Allahabad, when events at Benares gave them much detached work. Despite their long spell in unhealthy Burma the men willingly volunteered to proceed overseas to China, where hostilities were threatening, and where they remained until 1860.

The 5th Battalion stood staunch at Barrackpore, immediately volunteering for service before Delhi. They too, however, were despatched to China to take part in the engagements in Canton during 1858–59.

Of the well-known and heroic defence of the Residency of Lucknow only a small share can be claimed by the ranks of the then Bengal Army. But to those few that can, to none is more honour due than to the loyal remnants of the 13th, 48th, and 71st Bengal Infantry, who vied with their British comrades in steadfastness and courage; who formed the nucleus of "The Lucknow Regiment" and brought the honour of a "Turretted

Gateway" as the now 10th Battalion's glorious share of the Regimental record. Indifferently fed, they were exposed to the severest artillery and musketry fire which daily reduced their numbers; and every man fully deserved the Order of Merit which he afterwards received, as did two of their officers—Colonels W. G. Cubitt and R. H. M. Aitken—the Victoria Cross.

The 1st Battalion achieved distinction in 1876 by becoming one of the senior royal battalions of the Indian Army; Her Majesty the Queen-Empress conferring the style of "The Queen's Own" and appointing H.R.H. the Prince of Wales, afterwards King Edward VII, Colonel-in-Chief; the yellow facings were then replaced by the customary royal blue.

In the Afghan and connected border campaign of 1878–80 all Battalions did good work; the 2nd with special distinction.

The third war on the Irrawaddy was distinguished by the 1st and 5th Battalion's gallantry at the storming of the Minla Redoubt, a particularly arduous task which ended the opposition of the Burmese Army. The 2nd and 10th Battalions also shared in the expedition.

Indian regiments being requisitioned for the outbreak in Egypt in 1882, the 3rd Battalion arrived on the Nile to take part in the victory at Tel-el-Kebir and the rapid pursuit to Cairo. The Battalion were afterwards honoured by becoming "The Duke of Connaught's Own", and H.R.H. is to-day the Colonel-in-Chief of the whole Regiment.

Although Rajputs of Hindustan had been enlisted into the ancestral Battalions right from their inception, Brahmins and Musalmans also stood in ranks; but between 1893 and 1897 the Battalions all became class units, thence recruiting only sons of the sun and the moon from the Gangetic plain. Rajputs are still enlisted, but to-day Punjabi, Musalmans and Pathans also go to make up the Regimental composition.

The allied columns in China for the relief of the Pekin Legations in 1901 included the 1st and 3rd Battalions; and the latter claim to be the first regiment to enter the celestial city.

OFFICERS OF THE 2ND BATTALION 7TH RAJPUT

REGIMENT ON ACTIVE SERVICE IN WAZIRISTAN, 1937

These two Battalions also did spells of active duty in the Persian Gulf, 1911–14, co-operating with the Royal Navy in suppressing the illicit arms traffic which was making things difficult on the North-west Frontier.

The 3rd Battalion returned to this hunting ground in 1914; the 1st, however, were sent farther afield to the Suez Canal, and did not return to the principal theatre for Indian troops until 1916, when they were almost annihilated at Dujailah: the Battalion later joined the Salonika and Black Sea armies until the end of the Great War.

The 2nd Battalion also went to Mesopotamia in 1915, remaining there until recall to the North-west Frontier in 1919.

The 3rd Battalion in Mesopotamia joined in all the battles of the first advance up to the by no means inglorious check at Ctesiphon, and the subsequent retreat to the final stand at Kut. Re-formed from drafts and by no means disheartened, the Battalion joined the Red Sea island garrisons and finally assisted in driving the Turks from the Arabian coast. The numerous awards for gallantry, both in the defence of Kut and the Turko-Arabian operations, manifestly indicate their magnificent spirit.

The 4th Battalion continually patrolled the North-west Frontier of India front of the Great War, until they too were ordered against the Turk and to the oil-fields of South Persia. But they found no great opportunity until the Kurds and Arabs became dissatisfied with the ideas of the peace-makers in far-off Versailles and rebellion mania prevailed over peace and order. The Battalion did especially well in the defence of Hillah.

The 5th as a unit were unlucky in the Great War; they early sent drafts to France, themselves going to Mesopotamia, but no chance came their way during their four months' stay in 1915, and later on column duty in Persia.

The 5th Battalion, however, brought the main share of the honours for the Afghan affair of 1919, when they fought from Chitral the advancing enemy columns on the Kunar River. In

these operations, the Chitral force—the mainstay of which was the Battalion—was completely isolated in the Himalayas and could expect no reinforcements in the event of a disaster.

The 10th Battalion were also retained in India until 1917 when they were sent into Persia, seeing wearing service without any compensatory glory both there, and afterwards in Baluchistan, 1919–20.

The Territorial 11th Battalion is stationed at Fyzabad, being raised there in 1921.

All the Battalions have seen service at one time or another in the trying operations, both frontier and internal, which have fallen the way of the Indian soldier since 1919. The 1st Battalion was an original selection for Indianization, and to use their own words "They wish for nothing better than to hand down unsullied the high traditions inherited from their forbears."

CHAPTER XXXV

The 8th Punjab Regiment

"Cochin", "Maheidpur", "Ava", "Afghanistan, 1878–80", "Burma, 1885–87", "China, 1900", "Loos", "France and Flanders, 1915", "Macedonia, 1918", "Helles", "Krithia", "Gallipoli, 1915", "Suez Canal", "Egypt, 1915", "Megiddo", "Sharon", "Palestine, 1918", "Tigris, 1916", "Kut-al-Amara, 1917", "Baghdad", "Khan Baghdadi", "Mesopotamia, 1915–18", "Afghanistan, 1919".

THIS Regiment is the third of those corps which by succeeding to a number, carries the honours earned by men unconnected with the present title. In this case the units which directly succeeded the old regiments of Madras actually went out of the regular line and became local units for police service in Burma; although by the time of that conversion their ranks comprised entirely men from the Punjab.

All of the early austral battalions formed towards the close of the seventeen hundreds from which descent is traced by the modern Regiment were actively employed in the numerous operations of the time in the Carnatic and the extreme south, but with the exception of the 2nd and 5th Battalions no honours were granted for these services.

The 5th Battalion earned, however, an unusual honour, granted in 1840, for service at Cochin in 1809. The Rajah of Travancore being considerably in arrear with his treaty payments to the East India Company, plotted to murder the Resident on account of the latter's suggestion to disband the expensive and somewhat unnecessary state forces as a means of finding the money. The conspiracy led to the reinforcement of the Resident's guard and so the present 5th Battalion, as the 17th Madras Native Infantry, and a small detachment of the 12th Foot, now the Suffolk Regiment, were sent to Cochin. The Travancore forces attacked them and the British commander recorded that "the small detachment of His Majesty's 12th Regiment behaved with great gallantry and showed a noble example to the 1st/17th Regiment who followed it entirely to my satisfaction." The Suffolk Regiment, however, did not receive any honour for their part in the affair; a point those who preach discrimination against India's troops might thus bear in mind.

The 3rd Battalion were styled "Light Infantry" in 1812 in honour of an achievement in a minor action in Mysore, when by a forced march of twenty-five miles they arrived to succour the retiring forces and turned the balance from defeat to victory. This Battalion also fought with some credit in the defeat of the Indore and allied forces at Maheidpoor by Sir T. Hislop in 1817 where 5,500 were victorious over seven times that number.

In 1824 Light Infantry Battalions as tactical units were introduced into the Madras Army, and the forbears of the 3rd Battalion became one of the four converted; the Directors of the East India Company instructing, "That, although the battalions of Light Infantry are liable, in common with other troops of the line, to be called upon to perform all duties to which Infantry are subject, yet, in order to maintain their dis-

A Madras Sepoy—1830

cipline more perfect in cantonment and to keep them in readiness for their more active service, they are not to be included in the rosters for fatigue and ordinary duties, unless in case of urgent necessity." The Board also recorded that the men were to be steady, well-limbed, active, of good caste and connection, and free from all natural or contracted complaints and above twenty years old. The battalions were also to be entitled to keep the right, and take precedence of all other then Indian Infantry. Based upon this ruling the present 3rd Battalion with its motto "Now or Never" should really be the 1st.

Madras Regiments at this time were apparently in the habit of misusing their colours; for in the year 1829 we find orders to the effect that such were not to be used for decorations at balls, and other festivals. In later years stories were current of orders being issued to the old loyal and gallant Madras Army that Commanding Officers were to cease drilling their battalions from their bungalow verandahs clad only in night attire. But whatever deterioration may have taken place after considerably more than a century of existence, we must not forget that it was the old Madras Army that principally ousted French power from India, fought victoriously all the early major campaigns and overseas expeditions that resulted in the unification of the territories from which began our present India, and the junior partner of the British Commonwealth—modern Burma. And it is with Burma that the 8th Punjab Regiment are most intimately connected.

For some thirty years prior to 1824 the eastern border of Bengal had been continually disturbed by invasions of Burmans. Much diplomatic representation had been made to the Burmese Court at Ava, but that arrogant Government viewed the failure of India to declare war as an indication of weakness. Aggressions continued to increase year by year, and when two Burmese columns invaded India with the object of seizing Chittagong a declaration of hostilities became the only remedy. The swamps and jungles of the Arracan littoral were an obstacle

too formidable for an advance by land, and so a sea expedition to disembark at Rangoon, followed by a speedy advance up the Irrawaddy on the capital became the plan. In this force were the to-day 2nd and 4th Battalions. The hopes of an early advance were soon destroyed and the force remained nine months at Rangoon. After many repulses, the war was finally ended by the destruction of the Burmese Army at Pagan; and the troops from Madras who formed the bulk of the force, returned to their homes.

The demands for troops to hold the communications with Kabul in the campaigns beyond the North-west Frontier from 1878 to 1880 brought Madras troops up to the Khyber. With them were the 2nd Battalion, who saw no fighting, but suffered so much from the cold climes of the north that the question began to be asked as to the suitability of Madrassis for service in what was then becoming the storm-centre of India: a question which was later finally answered by a total eclipse of southerners from the infantry corps.

The last major active service of any Battalion as Madras Infantry was in the Burmese War, 1885–87. Although King Theebaw had been deposed, and the official war ended, his disbanded self-indulgent soldiery, together with much undisciplined adventurous youth roamed the land. And, not only did the highways and byways become unsafe for the traveller, but no peaceful government could prevail. Pitched battles were impossible, and the fighting had to be carried out by small detachments constantly on the move; operations which greatly try the morale of troops when consideration is also given to the malarious and unhealthy state of Upper Burma, particularly in the rains. No less than 30,000 regular troops had to be drafted into the country before the lengthy guerilla warfare could be ended. These factors, together with the depredations of the aboriginal tribes of the Burmese frontier evolved the establishment of a special force to keep order on the borders and interior of Upper and Lower Burma.

Many Madras regiments in the new territory were now disbanded, including the ancestral battalions of the 8th Punjab Regiment, their places in the occupatory forces being taken by police battalions recruited from north of the Jumna. Between 1891 and 1893 these police units were brought back on to the regular military establishment of the Madras Army, receiving for a short while the old numbers of the disbanded line units; and those we are at present concerned with became by 1903 the 29th, 30th, 31st, 32nd, and 33rd Burma Infantry. The decision to abolish the Madras line in the united Indian Army of 1903 transformed these units into the 89th, 90th, 91st, and 92nd Punjabis; the 33rd, however, were not called upon to change their title, but only to add sixty to their number. It is through this connection that the present Battalions claim ancestry with the armies of the Carnatic.

As police units, and after their restoration to the Army proper, all saw considerable typical bush and hill warfare against the troublesome primitive tribes of the North-east Frontier.

The Burma battalions adopted the workmanlike drab of their colleagues on the opposite frontier, but with silver lace and buttons; with facings of blue, black, cherry, white, and yellow in order of battalion seniority. Later on, after the Great War and their amalgamation as a regiment of the Punjab, the blue facings of the 1st Battalion were adopted for all, as well as the badge of the mythical dragon of Burma, the Chinthe.

The 3rd Battalion went to China in 1900 narrowly escaping disaster on the way, a wing being ship-wrecked by the typhoon, which caused them to be too late for the early serious fighting of the allied expedition rescuing their ambassadorial establishments in the Imperial Manchu capital. The Battalion, however, found endless column duty in the severe cold of North China, and the wisdom of the change over of the ranks to Punjabis was aptly proved by the fact that no man left the ranks through sickness or fatigue.

The members of the 1st Battalion always prided themselves that they never did things by halves, and certainly during the great trial that began in 1914 they fully lived up to their ideal: for no other single battalion of the Armies of the British Empire served in so many theatres of the war as did the the old 89th Punjabis. Landing to destroy the Red Sea Turkish fort of Turba whilst in transport to Egypt—a minor fight in which three Indian Distinguished Service Medals were won—they proceeded to defend the Suez Canal—not letting the Canal defend them—and ended their first stage by arriving on the Gallipoli Peninsula where the first casualty occurred through an enemy shell scoring a bull by ejecting a sod of mother-earth right down the wide-open alimentary orifice of a wonder-struck sepoy. Within less than a month they were forwarded on to France into the Loos sector. No heavy fighting fell to their lot, but many casualties did on account of their constant work on patrol or in trenches.

Withdrawn with the Indian Corps to the sun and sand of the Tigris front, they did not mark-time as far as fighting was concerned. And in one of their many actions Lance-Naik Shamad Khan found himself on April 12th, 1917, in charge of a machine-gun section in a very exposed position covering a serious gap in the line. Undaunted, the Lance-Naik and his men beat off three counter-attacks, he continuing to work the gun single-handed after all his section, save two ammunition belt-fillers, were put out of action. For three hours the gap was held and, when the gun through enemy fire ceased to function, Shamad Khan and his two comrades held the post with their rifles until ordered to withdraw. The damaged gun, the other arms, equipment, and ammunition—save two shovels—were salvaged by this gallant n.c.o.: had it not been for his leadership and determination, to say nothing of his bravery, the action would have had a very different ending. Shamad Khan received the Victoria Cross, his two companions—such is the lot of the bravest soldier—nothing.

(*India Office photo*)

A Company of Punjab Infantry on the March in Northern India

The Battalion much depleted was sent back to the North-west Frontier and the Mohmand Blockade to refurbish and recuperate. They afterwards returned to Mesopotamia, proceeding in the last year of the War to Salonkia, where the great Armistice called a halt. We are told that the ranks received the news more stoically than their joy-making British comrades, the Subadar-Major inquiring if the men had permission to cheer. The Battalion's wanderings were, however, not yet over; and for the next year they marched and counter-marched all over Caucasia and through Georgia—though not the Georgia of the tune often played by the band of the Royal Marines which, upon occasion, headed their march. As far afield as Baku on the shores of the Caspian Sea, tramped the men until the shilly-shallying politicians finally decided their attitude towards Bolshevik Russia.

The 2nd Battalion were on the Euphrates from 1915 to 1918, fighting in the disheartening operations which failed to relieve Kut; and thereafter in the final victories. This Battalion also took part in the third Afghan War.

The 3rd Battalion also left the land of the Chinthe and the banks of the Irrawaddy for those of the Tigris in 1916 to take part in the re-capture of Kut-al-Amara, and the advance to the Calif's city. In the later battle of Ramadi they were much commended for their courage and endurance in the face of the enemy under extremely trying climatic conditions.

The 3rd Battalion, together with the 4th which had proceeded to Egypt at the end of 1914 to reach Mesopotamia in 1917, went over to Palestine to assist in the brilliant operations which caused Turkey to be the first of the enemy powers to seek an armistice with the Allies.

The Suez Canal Defences also received the 5th Battalion during the close of the first year of the War: they, however, went on to France, reaching the line just in time for Loos; to return again almost at once to Egypt in transit for Mesopotamia, where they spent almost two years before they, too, were sent to Asia Minor for service under Allenby. The 5th, which ha been selected to disprove the old fallacy that India is unabl to produce good officers, still bears the subsidiary title "Burma'

The 10th Battalion started life as a war-time formation—th 2nd Battalion of the 89th Punjabis now the 1st Battalioı After serving in posts on the Malakand during the aftermat of the War, the Battalion became the junior and training bat talion of the Regiment in 1912, and now turns Punjabi-Musa mans, Sikhs, and Gujars into efficient soldiers in the old canton ment of Lahore. The 8th Punjab Regiment are the only infantr to enlist the latter class who form the bulk of the graziers an shepherds of the Punjab and Kashmir.

SEPOY
BENGAL INFANTRY—1817

CHAPTER XXXVI

The 9th Jat Regiment

"Nagpur", "Afghanistan, 1839", "Ghuznee, 1839", "Candahar, 1842", "Ghuznee, 1842", "Cabool, 1842", "Maharajpore", "Sobraon", "Mooltan", "Goojerat", "Punjaub", "Ali Masjid", "China, 1858–59", "Kandahar, 1880", "Afghanistan, 1878–80", "Burma, 1885–87", "China, 1900", "La Bassee, 1914", "Festubert, 1914–15", "Neuve Chapelle", "France and Flanders, 1914–15", "Shaiba", "Kut-al-Amara, 1915", "Ctesiphon", "Defence of Kut-al-Amara", "Tigris, 1916", "Khan Baghdadi", "Mesopotamia, 1914–18", "North-west Frontier, India, 1914–15, '17", "Afghanistan, 1919",

ALTHOUGH bearing the name of that Indo-Sythic race which some ethnologists aver to be of gipsy origin, the ranks of the Jat Regiment to-day are not entirely composed of recruits from the industrious and dour yeoman peasantry of the South-east Punjab and contiguous territory, descendants of the "fire-born" and one of the thirty-six royal races of the Rajputs.

Prior to the Great Mutiny Jats were not generally enlisted into the Bengal Army, although their resolute and indefatigable qualities as soldiers were well-known—it was the Jats of

Bhurtpore (the modern Bharatpore) that successfully defied Lord Lake's many valiant attacks upon their capital in 1805—but during the later years of the nineteenth century their enlistment was carefully fostered into squadrons and companies of many existing corps; and in 1897 two regiments of Bengal foot, the 6th and 10th, were converted into Jat class regiments. They so remained, bearing the same numbers, until the fashions of the post-war period decreed in 1922 their junction with an old Bombay unit, the 119th Infantry, as the 1st, 3rd, and 2nd battalions of a mixed class regiment in which the Jat no longer predominated, but nevertheless to be styled the 9th Jat Regiment.

The senior battalion was raised in 1803 as the 1st/22nd Bengal Infantry from Brahmin and Rajput recruits, a little more than boys, at the time when the Mahratta forces under the Frenchman Perron seemed likely to make an intrusion into the East India Company's territory. The Battalion's tirocinium, however, was against non-co-operating Jats of the Hariana then recently made over to the Company as a result of the Mahratta War. Their chance came fourteen years later when the Bhonsla having besieged the Sitabaldi Residency, the Battalion marched sixty miles in thirty hours—no light feat in those days—to fight conspicuously in the resultant battle of Nagpur: and later in closing the bolt-holes in Malwa of those organized looters, the Pindaris.

The 2nd Battalion were one of the new units brought into being on the opening of the third Mahratta War. Established in 1817 as the 1st/10th Bombay Infantry, after only four months of training they were in serious action by the following February in the Konkan. The ranks we are told "judiciously included" many deserters who had been obliged to disappear from their old regiments due to ill-treatment of their families by the enemy Peshwa in whose territory they were living.

The 3rd Battalion came into the Bengal Infantry as the 1st/33rd in 1823 as a result of trouble with Burma over Arracan.

Only men willing to serve overseas were enlisted—a liability the average Bengal sepoy was unwilling to take. The new battalion, however, did not proceed to the land of the Pagoda, but to Penang, then recently exchanged with the Dutch for territory in Sumatra.

Both 1st and 2nd Battalions saw much service in those long and costly series of operations beyond India's frontiers which arose through fears of Russia's designs. Both joined the Army of the Indus in 1839 about to advance via the Bolan Pass to depose one Amir and instal another as part of India's foreign

(*India Office photo*)

"Cleaning Up in Camp"

policy. The 1st were left to garrison Quetta, the 2nd going on to Kabul, assisting later by covering fire the assault on Ghuzni by the British regiments of the column. The new Amir—Shah Sujah—having been duly installed. the 2nd returned home: meanwhile the 1st came up to Kandahar. As we all know, Shah Sujah was overthrown, and the British-Indian force at Kabul overcome with disaster, a Dr. Brydon being the only survivor to reach safety. Upon the news reaching Kandahar three companies of the 1st Battalion were sent to reinforce the small fort of Kelat-i-Ghilzai where, together with the now disbanded 2nd Bombay Pioneers, they defended their keep against Afghan hordes all through the severe winter of 1841–42. On the evacuation of Afghanistan the 1st returned to India via Kabul and the Khyber with the force objectively out to retrieve the previous disasters and secure the release of the unfortunate prisoners. A journey by no means without fighting from the second subjection of Ghuzni until the last skirmish at Ali Masjid. In recognition the Battalion were created Light Infantry, the men receiving a medal inscribed "Victoria Vindex".

The 1st Battalion crossed bayonets with the Gwalior troops at Maharajpore the year following their reception at Ferozepore by Lord Ellenborough—the Governor-General—who came up to the frontier to meet them. This reception was also meant to overawe the Sikhs then getting out of hand after Ranjit Singh's death. But the plan did not succeed, the Khalsa armies broke the old friendly ruler's treaty, and invaded British territory. As a result the 1st Battalion again became involved in heavy combat at the hard-fought victory at Sobraon in 1846.

Two years later occurred the murders which started the campaign that ended the Sikh dynastic rule at Lahore. A Bombay column from Sind went post-haste to assist in the subjection of insurgent Multan. With them were the 2nd Battalion whose deeds at the storm of the city are commemorated by their auxiliary title borne since 1903, "The Mooltan

Battalion" The 2nd also went on to reinforce General Gough, taking part in the *coup de grâce* at Gujrat.

When the long sown seeds of mutiny blossomed, with results that ended the career of the majority of the regiments that had consolidated modern India, the 1st Battalion were at Barrackpore in the first garrison to show insubordinate signs. Their action in the matter needs no explanation, they still stand on parade as proud as any soldiers of the King-Emperor. Public opinion in Calcutta, however, was so shocked at having close at hand a battalion that had not mutinied; that to appease the bankers, brokers, and merchants of the capital, the Battalion in company with the Governor-General's Body-guard were disarmed. They were, happily, not dishonoured, and the two corps together were the only Indian units to provide the future Viceroy's guards in the summer of 1857 and during the whole of 1858. Detachments of the 2nd Battalion, they being in the Bombay command, took active parts in the 'clean-up' in Central India in 1858. The 3rd Battalion being composed of men already considered to have broken their caste by service overseas—they had just returned from Burma—were not a very promising field for the seditionary fanatic, and the foolish actions of their brothers in Oudh caused them no second thought, as the former discovered in the subsequent operations.

The Mutiny ended; the war in China over the opium traffic, then in a state of suspension, was resumed in 1858; and so again overseas went the 3rd Battalion into the Far-Eastern fray to serve with honour at the capture of Shek Tsin.

It was twenty years before the two senior Battalions were required for any major service; and then Afghanistan. The 1st went through the Khyber with Sam Browne, took part in the storm of Ali Masjid, then an Afghan frontier fort, and the key to the Pass which the 1st were left to hold. The 2nd marched through the Bolan, and eventually found themselves shut-up in Kandahar. They did well in the sortie at Dej Kojah; and when the Kabul marchers gave them release fought with

Jat Indian Officer
9th Jat Regiment—1937

them the final action that ended the war. Major Waudby of the 2nd greatly distinguished himself in the campaign by his defence with two sepoys of the Battalion and three men of the Scinde Horse of the mud post of Dubrai for over three hours until the ammunition was exhausted and the little band overcome, Major Waudby fighting to the last with a hog-spear. For years afterwards the Afghan villagers tended his grave—a most unusual thing for them—averring that no braver man ever lived than he who lies buried there. Waudby Road in Bombay is their memorial, where a tablet remains erected, bearing the names of the gallant men.

The war of a fortnight followed by four years of guerilla warfare in the jungles and *jheels* of the land they were originally raised to fight in, required two years' hard service from the 3rd Battalion until 1890.

A year's service in China with the troops of the principal nations, then on duties protecting their nationals from the murderous preachings of the Boxer fanatics, came the way of the 1st Battalion in 1900–01.

The senior Battalion sailed to France and saw practically unceasing fighting in mud, water, snow and ice during the winter of 1914–15. They were recalled nearer home to rescue the 2nd who had gone with that force sent originally to protect the Persian oil pipe-lines, but whose subsequent victories caused discretion at G.H.Q. and Simla to go by the board, and ended in a sickening surrender in no way the fault of undaunted soldiers who had so nobly done their duty.

The 1st Battalion from France made spartan efforts to release their besieged brothers. Over the wet and muddy flats at Hannah they stormed the Turkish lines; but all to no end, since reinforcements there were none. And so out of the bloody struggle came unwounded one British officer, the Subadar-Major, a Havildar, and the Quartermaster; alone of the whole Battalion. These heroic attempts were again repeated at Sanaiyat, but fate denied the prize and Kut fell.

The 1st now had to 'lick their wounds', until reinforcements arrived and a new 2nd could be re-formed from drafts. Eventually, and still with stout hearts, both together fought for the final surrender of Baghdad.

The depleted units in Mesopotamia made heavy demands on the 3rd Battalion, who not only had to hold the Border against Mahsuds and other raiders, but to train hundreds of recruits and despatch them overseas. Eventually they completed the trio of the Regiment fighting the Turk in the Middle-East, departing after a year to confront him on the Black Sea where they took an active interest in the actions at Broken Bridge, Ismid, and Kandrah during 1919 and 1920.

After the War the 1st met the troubles of the Mohmand and the Khyber in 1919. Events which caused the temporary evacuation of Waziristan and commenced the long series of border wars in which all the active Battalions have played some part.

The Regimental training battalion is shared with the Bombay Grenadiers, and has the hybrid title of the 10th Battalion 4th/9th Regiments. The wing serving the Jat Regiment originated as the 2nd/6th Jat Light Infantry of the Arab Rebellion, but later were ordered to assume the identity of the disbanded 4th/9th Jat Regiment, the old 18th Musalman Rajput Infantry born as the Calcutta Militia in 1795.

The Territorial Battalion, the 11th, is quartered at Bareilly.

The 1st Battalion were raised to the honour of Royal in 1921. The 2/6th, a temporary unit with no war service or record, also assumed that honour by virtue of their connection with the 1st. Upon the introduction of the regimental plan into India in 1922 the blue facings of the Royal Battalion were adopted by all, the titles Royal and Light Infantry being, however, restricted exclusively to the 1st. A true regimental system would style all the Battalions by the more apt and euphonious title "The Royal Jat Light Infantry (The Mooltan Regiment)"

CHAPTER XXXVII

The 10th Baluch Regiment

"Aden", "Reshire", "Bushire", "Koosh-ab", "Persia", "Delhi, 1857", "Central India", "Abyssinia", "Kandahar, 1880", "Afghanistan, 1878–80", "Egypt, 1882", "Tel-el-Kebir", "Burmah, 1885–87", "British East Africa, 1896", "British East Africa, 1897–99", "China, 1900", "Messines, 1914", "Armentieres, 1914", "Ypres, 1914, '15", "Gheluvelt", "Festubert, 1914", "Givenchy, 1914", "Neuve Chapelle", "St. Julien", "France and Flanders, 1914–15", "Egypt, 1915", "Megiddo", "Sharon", "Palestine, 1918", "Aden", "Kut-al-Amara, 1917", "Baghdad", "Mesopotamia, 1916–18", "Persia, 1915–18", "North-west Frontier, India, 1917", "Kilimanjaro". "Behobeho", "East Africa, 1915–18", "Afghanistan, 1919.".

INDIA has no soldiers to-day specially enlisted as marines for service in her fighting ships, and whether future naval developments will require such a traditional force is doubtful, sailors nowadays being specially trained to amphibious warfare, as well as being much better behaved and more disciplined creatures than they were when marines were first introduced to fight and keep order in ships of war.

In past days several marine battalions existed in the East India Company's Army—one, the famous Marine Battalion of Bombay, actually saw action at sea against a warship of the U.S.A.; the only occasion upon which any portion of India's forces have fought against Americans. But the only survivor extant of these marine units, which at one time were engaged in the fights of the old Indian Navy from Macassar to Mauritius, is the 1st Battalion of the 10th Baluch Regiment raised in 1820 as the 2nd (Marine) Battalion of the 12th Regiment of Bombay Infantry: to serve immediately thereafter for three years in the Gulf of Persia.

The 2nd Battalion came into being five years after the 1st as the 2nd Extra Battalion of Bombay Native Infantry. A re-arrangement the next year (1826) made the two Battalions the 24th and 26th Regiments, numbers they were to retain with the addition of 100 in 1903 for almost a century.

Travellers by the P. & O. steamers are reminded by brochures and deck notices descriptive of the voyage that it was the Royal Dublin Fusiliers that stormed Aden in 1839 and brought that then piratical stronghold under the British flag. No one would wish to deny the successors of the old "Bombay toughs" their share of the honour and glory—historians tell us there was not very much—but nevertheless the steamers carrying His Majesty's Mails should also give credit that together with the Dublins—a name they did not receive until after the transfer of the Company's forces to the Crown—were the present 1st Baluchis.

The first units ever to bear the present regimental name were brought into being in 1844 and 1846 by Sir Charles Napier for local service in Sind as the 1st and 2nd Belooch Regiments. Both were raised in Karachi, which the 2nd Battalion had occupied while the 1st were taking Aden. Napier's two battalions are now the present 3rd and 4th. The present 2nd Battalion were also brought up from the Presidency a little later for service in Sind; so that right from the earliest days that Province has been connected with the Baluch Regiment.

The 2nd and 4th Battalions fought together at Khushab in the Persian War, 1856–57. A short campaign executed with boldness and brilliancy, but somewhat overshadowed by the later events of the Indian Mutiny, in which the 3rd Battalion was destined to be the only unit in the Bombay Army to join the field force before Delhi.

Their home-town had the 3rd Battalion in garrison when the news of the Mutiny reached the Sind Capital. Bartle Frere, the Commissioner, being a man of action awaited no orders but despatched them forthwith. Marching across the Sind Desert in May—to-day troops must travel in carriages fanned and iced—they dealt with the doubtful Bengal Cavalry at Jacobabad, marched on to Multan and Ferozepore where they picked up the long-awaited much-needed siege artillery and escorted it to Delhi. They arrived fit and well to join the final assaulting columns, and afterwards shared in the numerous operations for the re-conquest of Oude, being continuously on service in Hindustan until 1859. No wonder their reputation transformed them from a local corps into the regular line.

Sir Hugh Rose employed the 1st Battalion in many of his brilliant actions in Central India, in particular at the siege of Jhansi when the sun played as much havoc with the participators as did the lethal weapons of the time.

The 2nd Battalion were employed against the aboriginal Bhils, and the 4th on the North-west Frontier; the latter were also brought on to the regular establishment for their stout work.

Towards the end of the Mutiny period the turbulent Baluch became more effervescent and the Roland of Sind—John Jacob—raised a battalion of riflemen to 'swizzle-stick' the tribesmen. They were raised on the *silladar* principle, by which, for a higher wage they found themselves in kit and equipment: tradition has it that they even made their own rifles—a double barrelled muzzle-loading contraption of Jacob's own invention sighted up to 800 yards and complete with a sword bayonet.

THE PIPERS AND DRUMMERS, 10TH BATTALION,

10TH BALUCH REGIMENT—1938

Headquartered at Jacobabad, in more ways than one the hottest place in India, their foes received no peace—John Jacob saw to that—and they quickly displayed fighting qualities more than equal to the regular regiments so that they too were established as Jacob's Rifles and eventually became the Baluchi 5th Battalion.

Indian regiments have done much for greater Empire defence and all take pride over the varied fields in which they have served; but none but the 4th Baluchis can say, we have been on duty to protect the fellow creatures of our Empire in the islands of Japan—a unique distinction earned by a detachment sent from Shanghai in 1864, when the Nipponese threatened destruction to Queen Victoria's subjects, British, Chinese, and Indian.

The 3rd Battalion achieved prominence in Abyssinia during 1867. With the advancing column they took part in the only serious conflict of the campaign—the first in which quick-firing breech-loading rifles were in part used by the troops—and were one of the three Indian infantry units to take part in the storm of the capital perched high upon the almost inaccessible rock of Magdala: services which created them Light Infantry.

Empire politics gave the 2nd a sea-voyage and a month's holiday in Malta when Russia and Turkey made war in 1878.

Quetta having been ceded to India by treaty in 1878, the 4th Battalion were there in garrison when the Amir Sher Ali declined to receive an envoy from Calcutta and gave preference to one from Moscow. War being an instrument of policy in those days, hostilities naturally followed. And so the 4th Battalion advanced with Biddulph's force to Kandahar, going on to show the flag as far as the Helmund River. The 3rd and 5th Battalions were also actively employed against tribesmen endeavouring to sever communications with the field-force.

The treachery that followed the Treaty of Gandamak brought the 5th Battalion up to Kandahar, and they had the extreme misfortune to suffer heavily in the disaster at Maiwand where hopeless mismanagement lost us the day and the Bombay Army undeservedly its reputation for many a month—even the great Lord Roberts for once passed mis-judgement.

The 1st and 3rd Battalions hurried up to retrieve the fortunes; but kismet forestalled them. Roberts had marched from Kabul, collected the 4th then besieged in Kelat-i-Gilzai, and with the relieved Kandahar garrison fought the grand finale of the war. After Kandahar all four battalions were together for the first time.

Professional active operations of the Battalions after their return to India were greatly varied: the 4th joined the Egyptian Expeditionary Force of Wolseley in 1882; the 3rd in the subjection of the Karens in Upper Burma 1886-89—the medical officer, Surgeon J. Crimmin, I.M.S., winning the Victoria Cross; the 1st and 3rd, East Africa, 1896-99; the 2nd and 5th with the International relief expedition to China, 1900; the 3rd again in Africa—Somaliland and the Mad Mullah—1909.

Up till 1891 the first two Battalions had been regiments of Bombay Infantry recruited in the Presidency proper; but from that year onward only frontiersmen were enlisted, and the style altered to Baluchistan Infantry. The fusion that resulted in one army for India numbered them 124, 126, 127, 129, and 130 until they were finally regimented as one in 1922.

In the stupendous competition set by Lord Kitchener in 1906 to test the efficiency of the units of the Indian Army, the 5th Battalion won.

The Western Front honours for 1914 were all earned by the 4th Battalion; although in fairness to the 1st it must be stated that they, detained in India, sent three complete double-companies of their four to France, two of which joined the 4th Battalion, which by September, 1915, had sustained over 1,500 casualties.

The 4th Battalion were the first Indian regiment to attack the Germans, the first Indian Army officer to fall belonged to

them, and they are the only Indian corps to bear the honour "Ypres, 1914".

The night of the 30th/31st October, 1914, was perhaps for the British Empire the most critical day of the War. The whole weight of the German Divisions—including the Prussian Guard—was hurled against a thin line of hurriedly scraped earthworks in a desperate attempt to win the Channel ports of Northern France. With the weak and exhausted dismounted cavalry force holding Hollebeke were the 129th Baluchis. No ground they gave in the desperate fighting, and their machine-gun detachments were annihilated save for one lone soul—Sepoy Khudadad Khan—who fought his gun on until he fell severely wounded; happily surviving to be the first Indian soldier to earn the King-Emperor's most coveted decoration, the Victoria Cross.

The 4th Battalion having well-established their reputation as a fighting corps were later withdrawn from the European theatre and sent to East Africa, where they further enhanced their fame by earning the praise of the German commander, General von Lettow Vorbeck.

Prior to the war some of the Baluch battalions commenced to enlist Mahsuds, those virile undisciplined tribesmen that have cost the Indian treasury since the Great War ended many lacs, if not crores, of rupees. The Mahsud sepoys gave no trouble in France or Africa, in fact we are told they produced more corpses per shots fired than any other class. They did great work as scouts and snipers; one gentleman returning from a lone exploit to report that he had marked down a German outpost, shot the officer, the corporal, the sentry, and the bugler—the rest did not wait—"and, Sahib, here is the officer's pistol, the corporal's stripes, the sentry's rifle and the bugler's bugle."

The 1st Battalion expanded into three during the war, the regular battalion serving in Persia and Kurdistan. The 2nd/124th, raised overseas at Bushire in August, 1916, were

the first of the 'new army' Indian units to see active service: and very heavy active service it was, particularly before Kut and at Tekrit, until their departure to Palestine in 1918. The Battalion, one of the few 'war-babies' to reach manhood remains as the 10th and Training Battalion of the Regiment with a permanent home at Karachi.

The regular 2nd Battalion joined the Aden defences 1915-16 and thereafter served in Mesopotamia until 1917.

The 3rd Battalion replaced war-worn units on the East African front from September, 1917, until February, 1918; while a 2nd line formation fought in the finale of the Palestine theatre.

To East Africa also went the 4th Battalion, February, 1915, to September, 1917. At Salaita Hill they were instrumental in saving certain South African units from disaster and earned much praise from the men of that, in the past, not very sympathetically Indian-inclined Dominion. The 4th, too, joined the Palestinian campaigners.

In the third Afghan War the 1st and 3rd/124th (1st Battalion), the 1st and 2nd/129th (4th Battalion) all participated. From that time until the present day the North-west Frontier has never completely settled down, and all five active battalions have been continuously represented in the operations. While the 3rd joined in suppressing the Burma Rebellion of 1931.

The Baluch Regiment no longer recruit Baluchis; these for a variety of reasons having ceased to form part of the classes now enlisted. But the traditions of the only red-trousered infantry of the Empire—the coat is rifle green—are ably carried on by Pathans, Dogra Brahmins and the sturdy Punjab-Musalman; led in the 5th by Indian commissioned officers.

All the Battalions, except the 2nd, bear subsidiary titles intimately connected with the Royal House of Windsor. The 1st are "The Duchess of Connaught's Own", the 3rd "Queen Mary's Own", the 5th "King George V's Own"; a truly Royal regiment, though they are not classed as such, their facings not being royal blue or red, but cherry.

THE DRUM-MAJOR
3RD BATTALION 11TH SIKH REGIMENT (RATTRAY'S SIKHS)

CHAPTER XXXVIII

The 11th Sikh Regiment

" Defence of Arrah ", " Lucknow ", " Behar ", " China, 1860–62 ", " Ali Masjid ", " Ahmed Khel ", " Kandahar, 1880 ", " Afghanistan, 1878–80 ", " Tofrek ", " Suakin, 1885 ", " Defence of Chitral ", " Chitral ", " Malakand ", " Punjab Frontier ", " Samana ", " Tirah ", " China, 1900 ", " La Bassee, 1914 ", " Armentieres, 1914 ", " Givenchy, 1914 ", " Neuve Chapelle ", " Ypres, 1915 ", " St. Julien ", " Aubers ", " Festubert, 1915 ", " France and Flanders, 1914–15 ", " Helles ", " Krithia ", " Suvla ", " Sar-i-Bair ", " Gallipoli, 1915 ", " Suez Canal ", " Egypt, 1915–16 ", " Megiddo ", " Sharon ", " Palestine, 1918 ", " Tigris, 1916 ", " Kut-al-Amara, 1917 ", " Baghdad ", " Sharqat ", " Mesopotamia, 1916–18 ", " Persia, 1918 ", " North-west Frontier, India, 1914, '15, '16–'17 ", " Tsingtao ", " Afghanistan, 1919."

THE Regiment of Ferozepore, the Regiment of Ludhiana; these were the dignified old-world styles of the original two senior battalions of the nowadays eleventh regiment of India's Infantry, formed in 1846, it having been decided that two

'Seik' regiments should be raised on the left bank of the Sutlej to garrison the newly acquired Jullundur Doab and the extended spheres of influence resultant upon the settlement of hostilities with the Lahore Durbar. So were enlisted for the first time in India's service the bearded warriors of the disbanded Khalsa soldiery.

The Regiments were not, however, all Sikhs, a leavening of Musalmans being admitted, and the framework provided by transfers of Oudh Rajputs from the regular Bengal regiments. And, except for some periods, none of the present Battalions of the Sikh Regiment were composed exclusively of followers of the ten *Gurus*: while to-day only the senior three comprise entirely Sikh ranks, the two junior recruiting as well the now almost regimentally ubiquitous Punjabi Musalman. The greatest care has always been taken by the Regiments to see that none but those, who have accepted the *pahul* are taken for the Sikh vacancies, and it has thus been sometimes quite truthfully said that had it not been for the Sikh regiments, the martial Sikhism of Guru Gobind Singh would have long since lapsed into its former religious fervour rather than martial zeal.

The new corps although to be 'local', its enlistment was for general service, and the men were uniformed and accoutred as the regular foot. The idea of localization, however, seems to have been abandoned quite early, and by May, 1857, they were to be found cantoned down-country at Mirzapore and Benares.

In the great Hindu city were the men of Ludhiana. The authorities in haste resolved to disarm the disaffected 37th Bengal Native Infantry. By mistake the Regiment of Ludhiana were cannonaded, and in the confusion it was thought they had mutinied. The detachment at Jaunpur, hearing of the shooting-down of their comrades, got out of hand and some lost-tempered individual shot the Adjutant. The Regiment were not in the least tainted and an impartial court of inquiry proved the egregious blunder which nearly turned friends into foes, and almost led to the extinction of one of the most loyal

fighting corps from the Indian Army. All the guards at Benares were later found by the Regiment of Ludhiana, and due to their steadfastness the rebels obtained no real hold. The Sergeant-Major—Peter Gill—won the Victoria Cross for facing a party of desperadoes when armed with only his sergeant's short sword. And Sepoy Chur Singh received a Jemadar's commission for saving his commanding officer's life.

The Regiment of Ferozepore on the outbreak forced-marched to Allahabad, arriving to save the fort, then without a proper garrison. They joined Havelock's movable column, and were the only Indian troops with him when began the advance to Cawnpore, which gave the first check to the insurgent soldiery. The Regiment fought in all the actions to relieve Lucknow, and led in person by Havelock and Outram, with the Seaforth Highlanders, were the first troops to run the enemy gauntlet and reach the beleagured Residency. As a reward for their superb gallantry every Indian received a step in rank, and the Subadars the 1st Class of the Order of Merit.

It was found impracticable to leave the Residency, and so relievers and relieved joined for a further two months of defence until Sir Colin Campbell's rapid move enabled withdrawal to be effected. The Corps were left with Outram's force at the Alambagh, to hold that out-post until the rebels in Lucknow and Oudh generally could be effectively dealt with. The colour-staff proudly borne and broken by shot at Lucknow, and which in the following capture was planted upon the highest dome of the Kaiserbagh as a signal of victory, still does duty, although the attached colour has long since been replaced.

During the advance to Cawnpore the old unsuitable semi-full-dress fighting kit was discarded and permission accorded to fight in the loose shirt and turban; while at the close of the campaign the right to wear the red *pagri*—then a special mark of esteem—was bestowed upon the Ferozepore corps.

So much for the Great Mutiny services of "The 14th King George's Own Ferozepore Sikhs" and "The 15th Ludhiana

Sikhs", the 1922 pre-amalgamation titles of the 1st and 2nd Battalions 11th Sikh Regiment.

Eighteen months before the Mutiny, Captain Thomas Rattray had recruited at Lahore the Bengal Military Police Battalion of Sikhs, Muslims and Dogras for service in the aboriginal tract of the Santal Parganahs of Bihar. More popularly known as "Rattray's Sikhs", when the Dinapore Brigade mutinied they were the only troops, save the Lincolnshire Regiment, between Calcutta and Benares. Every man volunteered his services, which were availed of to the full, and no corps did more—few as much—as did "Rattray's Sikhs" to restore the authority of the established government in Bihar. Two officers —Lieuts. Daunt and Baker—received the Victoria Cross. The outstanding epic was the defence of the house at Arrah from July 29th to August 2nd, 1857, by eleven civilians and fifty Sikh soldiers of "Rattray's" against two thousand armed mutineers and a vast rabble of insurgent peasantry.

With the passing of the Company's armies to the Crown "Rattray's" became a regular dismounted corps, the 45th of the Bengal foot and of the later Indian Infantry, eventually to become the 3rd Battalion of the Sikh Regiment in 1922.

The 2nd Battalion were early volunteers for service in the second Chinese War and in 1860–61 were to be found in Shanghai defending that city in open skirmish with the Taiping rebels.

All three senior Battalions served in the second Afghan War, 1878–80.

The 1st were detailed for the Khyber; and in the fight for Ali Masjid the Battalion suffered the bulk of the casualties. Their career in the field-force was most unfortunately cut short at Dakka by an epidemic of "typho-malaria" which caused much distress and their withdrawal to more salubrious surroundings.

The 2nd Battalion formed part of the South Afghanistan Field Force destined to make an unprecedented march of 400 miles through most difficult country to occupy Kandahar for many months.

The 3rd Battalion also took part in the fight for Ali Masjid, and afterwards in the Bazaar Valley.

The Treaty of Gandamak and the establishment of our embassy at Kabul—the refusal of which was the ostensible cause of the war—brought about a six months' peace, rudely shattered by the murder of India's envoy and his escorting Guides.

On the breaking of faith in Kabul, Roberts—who the year before had operated, coincidentally with the Khyber and

THE OLD COLOURS OF THE 1ST BATTALION 11TH SIKH REGIMENT

Kandahar columns, from the Kurram to the Shatargardan Pass—at once marched on Kabul. The 3rd Battalion joined him, fighting in the disputory struggle at Charasiah which gave us occupation of the Afghan capital.

The 2nd and 3rd Battalions met for the first time in Kabul; to part when the relief of Kandahar became imperative and "Bobs Bahadur" took the 2nd with him south-westwards to the finale of Afghan operations for forty years. The 3rd formed the rearguard of the troops sent back to India.

The 2nd Battalion joined the Suakin Expedition in 1885. Part of the plan for the subjection of the Sudan Dervishes was the constructions of defended stockades or *zarebas*. At Tofrek, where the first of these *zarebas* was to be situated, the force sent to make the establishment was heavily attacked. A fight followed in which the 2nd Battalion bore the brunt of the impetuous assault, their gallantry and good discipline being the factor which saved the column from complete destruction.

The 10th and 4th Battalions came into being as a result of the Russian 'scare' in 1887, the numbers of two old Bengal units disbanded a few years previously, the 35th and 36th—the old loyal Mainpuri and Bareilly Levies of the Mutiny—being resuscitated as Sikh units.

Upon the death of the Mehta of Chitral in 1895 an uncle usurped the throne. His levies laid siege to the fort at Chitral. Here eighty-eight men of the 1st Battalion and 400 Kashmir State troops set up a defence of forty-five days which, through its sublime heroism, merited special recognition accorded by the battle-honour subsequently bestowed upon the two regiments. The 2nd Battalion joined the relief force, the troops of which were heavily engaged in the forcing of the Malakand Pass.

The Regiment bear a further honour peculiar to them conferred in recognition of the gallant defence, by two companies of the 4th Battalion, of Fort Gulistan on the Samana range in 1897. Hard by also was the detached post of Saragarhi, a little

mud-brick blockhouse set on the knife-edge of the Samana for the purpose of visual communication between Forts Lockhart and Gulistan. Here nineteen men and two cook-boys of the Battalion defended the post for over six and a half hours against some seven thousand Orakzais. No help could reach them in time, and they perished to a man. An episode which excited the admiration of the world, while the Government of India erected in the Sikh cathedral city of Amritsar a special memorial to those twenty-one valiant Jat Sikhs of the old 36th who made such a terrible sacrifice in the defence of their frontier before a cruel, fanatical, and unrelenting foe.

The attack on Gulistan was one of the incidents which led to the Tirah campaign. It is impossible to record the many brilliant episodes of the Regiment during that Afridi campaign. All the Battalions were heavily engaged, except the 1st: the latter were officially part of the avenging force, but had the misfortune to be sent to an area where, while there was much outpost and convoy, no actual fighting occurred.

1901 found the 1st Battalion with the International force in China. The same year were raised as the 47th Sikh Regiment of Bengal Infantry the 5th Battalion, whose first overseas service was on identical duty from 1905 to 1908.

The two senior battalions as well as the 5th all again left India's shores for foreign fields in 1914.

The 1st to the active defence of the Suez Canal—it was their patrol which discovered the attempt to mine the fairway—and thence to six months' continuous fighting on Gallipoli from the second battle of Krithia until the evacuation.

The 2nd and 5th brigaded together pioneered the Indian troops in France, the former being the first Indian troops to land on the mainland of Europe for soldier's work. The battle-honours show their record, but of course, do not mention that the 5th were singled out for special praise in the House of Commons for their gallantry at Neuve Chapelle: or that Lieut. J. G. Smyth of the 2nd received the Victoria Cross, hard-earned

at Festubert on the 18th May, 1915, for leading a bombing-party over exceptionally dangerous ground—to effect his purpose he had to swim a stream whilst under fire from all arms the whole time.

After withdrawal from France the 2nd were sent against the Sennusis endeavouring to invade settled Egypt from the west during the winter of 1915–16. A backwater in which heavy fighting fell to their lot; after which they returned home to the Mohmand blockade for the rest of the War.

The Mesopotamia theatre concentrated the 1st Battalion from the Dardanelles, the 3rd from the Derajat, the 4th from China—where in 1914 half of the Battalion with the 2nd South Wales Borderers allied with Japanese troops had successfully laid siege to the German treaty port of Tsingtao, and thus earned an honour exclusive to these regiments—and the 5th from France. The 3rd and 4th arrived in time to make one last effort to relieve Kut. These two, the following year, set an example which other battalions would find it difficult to live up to. They suffered over two thousand casualties between them, and had to be withdrawn to the Lines of Communication. In the words of Sir Stanley Maude their magnificent fighting qualities paved the way for and contributed greatly to the success of the 2nd Battle for Kut. After recuperation the 3rd took part in the final operations, the 4th proceeding to the North Persian Force.

The 5th Battalion later served in Palestine, and for their war services became "The Duke of Connaught's Own" on the occasion of H.R.H.'s visit to inaugurate the "Montfort" reforms of 1921.

The 10th Battalion did not serve overseas during the Great War, but sent representatives to fill the depleted ranks of many battalions.

The 2nd Battalion, together with the 10th, were very active during the third Afghan War, forming part of the 'flying brigade' despatched through the Khyber in 1919. They fought in the battle at Bagh Springs, and Sikh Hill near Dakka is named after them.

It is impossible here to detail all the operations of the several Battalions on the frontiers or overseas since the Great War ended. But mention must be made of the dogged determination and valour shown by the 3rd at the successful relief of Rumaitha during the Iraq Arab Rebellion.

Two companies of the 1st, together with their equipment, were flown from the rail-head to Kirkuk during the Kurdistan revolt of 1923. A journey which took less than one hour whereas the march would have taken a week. This was the first time in history of the transportation of a large body of troops by aircraft to meet the urgent needs of a serious military situation.

King George V became Colonel-in-Chief of the 1st Battalion when he was Prince of Wales. The 2nd received the distinction of "Royal" on the occasion of the Silver Jubilee, 1935. The 5th is happily proceeding towards complete Indianization. The 10th, numerically the junior of all, takes precedence on parade after the 3rd by reason of the seniority of their old number, a curious anomaly the result of the regimental system devised in 1922. Another Regimental unusuality is that while the red coat is common to all the Battalions, the facings are not, those of the 2nd being emerald green, the 3rd white, the others the yellow of the Khalsa and the old Bengal Infantry.

THE SUBADAR MAJOR
THE GUIDES INFANTRY, F.F.—1938

CHAPTER XXXIX

The 12th Frontier Force Regiment

"Pegu", "Mooltan", "Goojerat", "Punjaub", "Delhi, 1857", "Ali Masjid", "Kabul, 1879", "Ahmed Khel", "Kandahar, 1880", "Afghanistan, 1878–80", "Chitral", "Malakand", "Punjab Frontier", "Tirah", "Pekin, 1900", "Somaliland, 1901–04", "Suez Canal", "Egypt, 1915", "Megiddo", "Sharon", "Nablus", "Palestine, 1918", "Aden", "Tigris, 1916", "Kut-al-Amara, 1917", "Baghdad", "Sharqat", "Mesopotamia, 1915–18", "North-west Frontier, India, 1914, '15, '16–'17", "Afghanistan, 1919".

AS the name implies the Frontier Force Regiment has its ancestry in the units of that famous body originally raised for police and other duties in the extended spheres of Indian governmental influence which followed the defeat of the communalistic Khalsa Army.

The orders of the Governor-General in 1846 raised a "frontier brigade" for service in the Trans and Cis-Sutlej States, as well as a "Corps of Guides" for general service. The four regiments of the 'brigade' were consecutively numbered and styled "Sikh Infantry"; they have now descended as the first four battalions of the 12th Frontier Force Regiment, the 5th and 10th being

the Infantry of the Corps of Guides whose story up to the Great War has been told co-jointly with that of the Guides Cavalry. Actually, the two Battalions from the Guides are the senior, and have precedence over the others on parade; a curious arrangement which arose through the Guides generously allowing, when the present Regiment was formed in 1922, the Sikh Battalions to preserve their old numbers—tradition dies not less hard with Indian troops than with any others.

The Sikh Battalions were in their early days a little more than armed police, although trained as far as possible on the regular model. The men were recruited largely from the disbanded contingents raised in the Phulkian States and the Malwa which had come under British protection after the Sikh War. At the time many soldiers of fortune were at large and the Governor-General conceived the 'brigade' as a means of giving them employment. The composition was not all Sikhs, half being Musalmans and other classes—a composition still kept up.

The 2nd Battalion were raised in the Kangra Valley and comprised mostly Dogras, who in those days wore beards, and thus mistakenly received the nomenclature "Hill Sikhs"; Gurkhas as well as some Pathans were also in the ranks, and the unit later became known as the "Hill Corps", a title which stuck until 1903.

The insurrection which broke out in Kangra co-incident with the Multan rebellion and other defections now styled the 2nd Sikh War gave the 2nd Battalion their baptism of fire, and the first campaign reward to the Regiment.

After this all the Battalions had much 'frontier' work to do, in which the standard began to attract notice; they could not, however, owing to the terms of their service be employed on the Bengal side of the Jumna. But the restriction all the same did not curb the desire of the men to see action anywhere, and the 4th Battalion volunteered for the Burma War of 1852. Acceptance of their offer made them the first Punjab corps to ever cross the sea, and the colours they carried then are now one of their proudest possessions.

The 4th spent two arduous years in Burma where their behaviour was closely, as well as jealously, watched by the other units in the field. At the relief of Pegu the 4th Sikhs were ordered to lead the assault, Lord Dalhousie then Governor-General, remarking: "They are still untried, let them prove that they can fight as well for us as they did against us." The Sikhs went into the fray, but the Burmans would not wait, the bearded warriors and their blue steel were too much for them.

The Sikh Infantry were transferred to the Punjab Irregular Force in 1851, and until the smouldering fires of mutiny broke out six years later were constantly engaged somewhere against marauding tribesmen of the north.

None of the Sikh Battalions showed any renegade spirit in 1857, and all took some part in the suppression of the insurgent Poorbeah sepoys. The 1st Battalion were in Rohilkhand, where on the 31st March, 1859, at Jerwah they held their own for some hours against a force of more than ten thousand trained rebels. The 2nd Battalion in the Murree Hills chased the Mardan mutineers—"The Lost Legion" of Kipling. The 3rd served in Oudh and the Nepal Terai. A detachment of the 4th made a most valiant stand near Ludhiana aided by a cannon worked by the Deputy Commissioner against four mutinous regiments including cavalry, only withdrawing inevitably when ammunition became exhausted; the whole Battalion served in the assault and capture of Delhi.

When Quetta was being developed as a cantonment in 1877, the 4th Battalion were in garrison. There, on the 26th July, was won the only Victoria Cross ever earned on the parade ground; when Captain Andrew Scott went to the rescue of a Royal Engineer officer, the survivor of two, who, when inspecting barrack building operations, had been treacherously attacked by a gang of armed Pathans. Scott fought the band with a bayonet until they could be overpowered. Sepoy Ruchpal

Singh, who in the first instance armed only with a small axe rushed to the rescue to be instantly cut-down, received a posthumous Order of Merit.

The first three Battalions all crossed the frontier in the second campaign against the Kabul government. The 1st from Ali Masjid to Jellalabad. The 2nd in the Kandahar force, which when *en route* for home were held up by insurgent Afghans at Ahmad Khel: here the Battalion holding the key position were forced to form company squares to ward off the desperate charges of the Ghilzai tribesmen, who never before had displayed such bravery. The 3rd Battalion were in the defence of the Shatatgardan, and later the Sherpur Kabul cantonment. The 2nd and 3rd Battalions were both selections for the march to relieve Kandahar, the former afterwards punishing the Marris for their treachery in attacking the army's communications.

The 4th Battalion became involved in the Chitral relief in 1895, being present at the storm of the Malakand.

Two years later the 1st Battalion, then in the Tochi Valley, sent a detachment with the small mixed force engaged upon a political reconnaisance to Maizar, where tribal treachery afforded a prominent instance of proof of the discipline and training of the Frontier Force. The steady and orderly withdrawal under the Indian officers after all their British superiors had been killed or wounded was a military feat well deserving of the applause it received from the whole Empire.

In the Pekin allied relief columns the 1st Battalion fought side by side with the 14th American Infantry at Peytsang-Yangstan, their charge being the winning factor in which the brunt of the fighting fell on the Sikhs. The white ensign of the Royal Navy hoisted at the British Legation on the arrival of the relievers was afterwards presented to the Battalion by Lieutenant, now Admiral of the Fleet, Sir Roger Keyes, R.N., whose father had commanded the Punjab Frontier Force, 1870–78, the gallant hero of Zeebruggo himself having been born at Kohat.

(*India Office photo*)

AFRIDIS FROM THE KHYBER WAITING TO ENLIST AT A TOWN ON THE NORTH-WEST FRONTIER—FIGHTING TO THESE MEN IS SECOND NATURE

Somaliland is a long way from the North-west Frontier, and it was in this benighted territory that the 2nd Battalion were operating in 1903—the year of the dissolution of the "Piffers". A religious agitator of the type not unknown beyond India's administrative border was inciting murderous raids upon the peaceful tribes. One of the Mullah's tactics against the troops was the poisoning of wells with every conceivable kind of putrid fetid matter from corpses downwards. The cleaning of these was a constant trial of the troops, and many men were overcome by the fumes. Four members of the Battalion received Royal Humane Society's Medals for their daring in rescuing men incapacitated whilst on this most unpleasant duty.

The Somaliland campaign was one of great privation for the men—the water was so hard the food would not cook properly—but it was relieved by another of those small glories of the soldier of Ind. On April 17th, 1903, Captain H. C. Vesey and forty-nine men of the 2nd Battalion were by request included in a small column sent to cover the retreat of the King's African Rifles. What followed no one will ever really know, for none of this detachment survived. It seems that the column, forced to form a square against an unexpected outnumbering savage attack, were overwhelmed. The bodies of the Sikhs were afterwards found in an unbroken line, the only face of the square that had not been broken by the lances and broad spears of the Somali horse and foot. Such valour was the talk of the day, and is worthy of record with the other famous isolated defences for which "Jack Sepoy" is unexcelled.

The Battalion fought in the pitched battle that broke the Mullah's power, and after an absence of eighteen months returned to the Frontier with the new title of the 52nd Sikhs F.F. in June, 1904. But although the men received medals in recognition of their work, not for thirty years was the unit allowed to embroider the campaign as an honour upon the colours.

During 1911 the Crown Prince of Germany visited the Khyber areas, where three years before the 3rd and 4th Battalions had fought in the small war against the Zakha Khel and Mohmands. At Landi Kotal, the Kaiser's heir expressed some doubts as to the speed with which the Khyber Movable Column, then based upon Peshawar, could reach and man the Pass. To settle the point the call was sent to Peshawar; and His Imperial Highness was amazed to meet the 2nd Battalion two hours later marching towards Jamrud, the Pass entrance, in full fighting kit. The whole affair was entirely unrehearsed and in no way prearranged, a matter which the Crown Prince flatly declined to believe. What his *rapport* to the "Wilhelmstrasse" was, we do not know, but we do know that the Great German General Staff entirely under estimated the Indian Army.

At last, the German "Der Tag" dawned. And to rudely shock the Central Powers went forthwith the 1st and 3rd Battalions in the Frontier Brigade to serve on the Suez Canal, at Aden, on the Tigris, and in Palestine.

Before the 2nd and 4th were to fight overseas, the former at Sharqat and in the Arab-Kurd Rebellion; they had seen action on the India front. The 2nd in the Tochi Valley against the Khostwals, 1914–15; operations in which an officer—Captain Eustace Jotham—sacrificed his life in an attempt to save a trooper who had become unhorsed; an endeavour which earned a posthumous Victoria Cross. The 4th in 1917 operated against the Mahsuds in the Shahur and Khaisora Valleys, where Subadar Hukam Dad with thirty Punjabi Musalmans held an unconstructed piquet at Barwand all night against continuous attacks of greatly outnumbering tribesmen.

Although overshadowed by the more popular 'fronts' the operations on the north-west borders of India were of vital importance, and were as much a part of the Great War theatres as France or Mesopotamia—the equivalent stars and medals of the world campaign being earned by the Indian units that took part. And thus although disappointing to them, it was but

meet that the premier unit of the Frontier Force should be left to hold the border until they, too, could be spared for overseas. So, between 1914 and 1917, the Infantry of the Guides lived up to their past in the affairs, not only of outposts, but of pitched battle, against the Bunerwals, the Hindustani fanatics in Sitana, and the Mohmands. Eventually, in response to the final call on India's manhood, the 1st Battalion Guides Infantry were despatched to Palestine to take a part in the two principal battles before the Turkish Armistice.

A second battalion of the Guides Infantry was raised in January, 1917, by an officer of the 51st Sikhs by means of drafts from the 52nd and 54th; this battalion, which also served at Megiddo and Sharon, is now the 10th Battalion. Having regard to the geneity, no better selection could have been made for the training unit of the Frontier Force Regiment.

The active Battalions have all distinguished themselves in the war areas of the Frontiers of the last two decades. Space does not allow much detail.

The 5th Battalion head the post-Great War roll of the fighting services as far as decorations earned by the valour of individuals go; though it would be unfair not to say when making comparisons of this kind, that opportunity plays a great part. It would be equally unfair not to mention the Loe-Agra operations in which on the 29th September, 1935, occurred one of the stiffest encounters seen on the Indian Frontier for many a long month. There the 5th Battalion fully upheld the past glories of the Guides and created new records for the Frontier Force Regiment. All the British Officers, and many of the Indian, were either killed or wounded during a fight of the greatest severity against myriad Mohmands in which Captain Geoffrey Meynell earned the Victoria Cross. The fog of war had descended, and Meynell, who was the Adjutant, had gone forward to discover that while the men were maintaining a disciplined accurate fire they were being slowly overpowered by the weight of numbers of the tribesmen. He at once took command, and

Pathan

Sikh

Punjabi Musalman

Dogra Rajput

Types of Men Enlisted into the Frontier Force Regiments—1938

there ensued a fierce hand-to-hand struggle. Cut off from communication to his rear, he encouraged the men to fight on with him to the last. A decision which caused the infliction of heavy casualties on the enemy and resulted in overtures for a cessation of the fighting. Captain Meynell, it may be noted, had also won a Military Cross in a Frontier affair of 1932.

The evolution of the uniform of the Regiment is interesting. When the Sikh Battalions were formed the Governor-General insisted that no Sikh should be enlisted who would not wear a *topee* or rather the regulation headdress of 1847, then the large Kilmarnock cap; as a consequence the turban did not come into general use until 1861. Similarly interesting, the Guides were the only infantry other than Gurkha regiments to enlist Nepalese subjects, who though normally detesting the turban—Nepali folk-lore despises the *pagri* as an idler's badge—showed no objection to wearing it whilst serving in the senior Frontier corps. Gurkhas were mustered out in 1922, but only eleven accepted transfers into Gurkha corps: Gurkha Guides were undoubtedly fond of their Corps, and Gurkhas are also undoubtedly fond of military service, so one wonders whether the *pagri* and all its imputation in Nepal may not have been the cause of the refusal to accept transfer to the pill-box and slouch-hatted regiments of their own kin.

The early dress was also the idea of the Governor-General, being of similar fashion to the regular Bengal Infantry, namely, red coatees yellow-faced, with black trousers—in the 2nd, or Hill Corps, the coat was green. The Guides always wore khaki or drab, an invention of Lumsden their first Commandant. Khaki was adopted by all Battalions about 1861, the differential facings being yellow, scarlet, black, and emerald green in numerical order of the Battalions, with red for the Guides, a colour later adopted for all in 1922.

The 1st Battalion are "The Prince of Wales' Own", a title conferred as an honour in 1921. The 3rd Battalion similarly receiving "Royal" in 1935. The 4th is earmarked for the

Indianized Division. The 5th and 10th retain their old title as a subsidiary, as well as the Cypher of Queen Victoria bestowed when the Guides became one of the first Royal regiments of India. The 11th Battalion headquartered at Nowshera came into being in March, 1922, as the 1st Territorial Battalion, 51st (The Prince of Wales' Own) Sikhs. All the Battalions are regarded as Light Infantry, a distinction granted in the early days to all 'Piffer' units not rewarded by the style of 'Rifle'

SEPOY
FRONTIER BRIGADE INFANTRY, 1861

CHAPTER XL

The 13th Frontier Force Rifles

"Delhi, 1857", "Lucknow", "Peiwar Kotal", "Charasiah", "Kabul, 1879", "Afghanistan, 1878–80", "Tirah", "Punjab Frontier", "China, 1900", "La Bassee, 1914", "Messines, 1914", "Armentieres, 1914", "Festubert, 1914, '15", "Givenchy, 1914", "Neuve Chapelle", "Ypres, 1915", "St. Julien", "Aubers", "Loos", "France and Flanders, 1914–15", "Suez Canal", "Egypt, 1915–17", "Gaza", "El Mughar", "Nebi Samwil", "Jerusalem", "Megiddo", "Sharon", "Palestine, 1917–18", "Aden", "Tigris, 1917", "Kut-al-Amara, 1917", "Baghdad", "Mesopotamia, 1916–18", "Persia, 1918–19", "North-west Frontier, India, 1917", "Baluchistan, 1918", "East Africa, 1916–18", "Afghanistan, 1919".

LIKE those of the preceding regiment, the first four battlions of the Frontier Force Rifles were in their origin Punjab Infantry units of the Trans-Frontier Brigade created by Lord Dalhousie in 1849. No 3rd battalion exists, none having been created when the Regiment was formed in 1922: the old 3rd Punjab Infantry was disbanded in 1882.

The 6th Battalion was raised six years earlier than the others

by orders of Sir Charles Napier as the Scinde Camel Corps. Designed to be of value in war, the Corps proved to be an expensive item in peace, and so after ten years of existence, reorganization into a light infantry battalion armed with rifles was effected, and the administration transferred from the Bombay to the Punjab Government, the unit becoming the 6th Regiment of the Irregular Force—the 1851 development of the Frontier Brigade—with the style of "The Scinde Rifle Corps".

The "Irregular Force" became "Frontier Force" in 1865, administration by the civil government being terminated in 1886. The Force, as such, was abolished in 1903 when the titles of the Punjab Infantry battalions became "Rifles"—a compliment for their services—and the names of early commandants joined therewith; which made the 1922 family of the present Frontier Force Rifles, the 55th Coke's Rifles (1st Battalion), the 56th Punjabi Rifles (2nd Battalion), the 57th Wilde's Rifles (4th Battalion), the 58th Vaughan's Rifles (5th Battalion), and the 59th Royal Scinde Rifles (6th Battalion); the 10th Battalion were the old 2nd/56th Punjabi Rifles, a war formation of 1917.

The first recruits were largely found from 'demobilized' soldiers of the Sikh Durbar. The Force was permanently stationed in the Frontier Districts, the regular regiments not in those days, as they do now, sharing the arduous and wearing duties on the barren rocky hills and sandy stretches devoid of greenery, the home of the most turbulent and bravest tribes in the world; who entirely without mercy, they were then, as today, ever ready to harass and destroy the frontier outposts; the epic defences of which by isolated detachments of sepoys under Indian officers are now history seldom equalled in the stories of other nations.

Up till 1857 one or another of these "Wardens of the Marches" had been engaged in upwards of fifteen expeditions into tribal territory. And when the call came to subdue the disloyal men of Hindustan, the "Irregular Force" were more than true to their salt. On the thousand mile march from the Frontier to Delhi, in the heat of the Punjab summer, went the present 1st, 2nd, and 4th Battalions. Under John Nicholson the first two, with the Guides Cavalry, dispersed the rebels at Najafgarh and so made possible the safe arrival of the heavy artillery needed for the assault on the capital. After Delhi fell, the 2nd and 4th went on to join Sir Colin Campbell. And at the capture of Lucknow the latter vied with the now Argyll and Sutherland Highlanders in the storm of the Sikandarabagh: a brilliant feat of arms noted for the conspicuous gallantry of Subadar Gokal Singh. In the same assault Sepoy Mukarrab Khan of the 4th prevented the heavy gate-way from being shut by pushing his left arm, on which he carried a shield, between the closing doors. His hand was savagely hacked at, but he withdrew it and inserted the other, which was almost severed from his wrist. But he gained his object, the doors could not be closed by the rebels, and in swarmed the 4th. It was the Queen's Colour of the 2nd which the afterwards great Lord Roberts borrowed to place on the roof of the Mess House, thrice to be shot down, thrice to be replaced.

All Battalions were employed in the Mutiny subjection campaigns in Oudh, Rohilkhand, and in the Nepal terai, the 6th Battalion remaining on the Frontier where anti-tribal military operations were necessary all through that time.

The next important border campaign was in 1863 against the Hindustani fanatics and the tribes harbouring them. Known as the Ambeyla campaign, the operations gave the troops the fiercest fighting seen on the Frontier until recent years. All Battalions except the 2nd took part, the 6th making a conspicuously gallant charge during the attack on the "Eagle's Nest Piquet". But it was around the "Crag Piquet"—three times lost and won—that the most gallant deeds of the campaign will be associated. Here, with the 1st Battalion, Lieutenant H. W. Pitcher earned the Victoria Cross for gallantry and leadership which was the

admiration of all spectators; while Lieutenant A. P. Davidson, rather than retire from his post, died with his men, fighting to the last with a heroism that drew the admiration of the enemy chief.

As would be expected, the various columns and phases of the 2nd Afghan War absorbed all Battalions except the 6th, again unfortunate to be left keeping watch and ward; they, however, joined the others in replacement of the 5th in the following punishment of the Mahsuds who had been guilty of raiding and other outrages on the army's communications. The 1st Battalion were in the Kandahar column, the 2nd and 5th via the Kurram to Kabul, the 4th on the Khyber line.

The Frontier began to quieten down after the Amir Abdur Rahman assumed the Kabul throne, and no major operations were necessary until the general outbreak of 1897. The Orakzai had, however, given much trouble in 1891, and all the five active battalions joined the two expeditions in the Miranzai valley during which much fighting, some of it severe, took place before the clan could be brought to complete submission.

An unprovoked attack on the boundary commission endeavouring to fulfil treaty obligations with Afghanistan led to the despatch of a large punitive force into Waziristan. The 2nd, 4th, and 6th Battalions took part, and the medal-clasp earned, "Waziristan 1894–95", was the last to be issued with the India General Service Medal instituted in 1854, which had from then been granted for every Frontier campaign, and for which no less than twenty-one clasps had been struck, some wearers in the Frontier Force possessing as many as nine or ten.

Participation in the general rising on the border in 1897 as far as the Frontier Force Rifles were concerned began at Maizar in the Tochi Valley, where a small mixed escort to a political officer were suddenly treacherously attacked. The heroic devotion and sacrifice of Subadar Sundar Singh of the 1st Battalion and a dozen or so of his men, in holding a garden wall in order to cover the orderly withdrawal of men and guns under the

SUBADAR-MAJOR
"COKE'S RIFLES, P.F.F."—1870

supervision of the Subadars—all the British officers having become casualties—was one of the feats in an incident of bravery, training and discipline, which no military corps in the world could surpass. The 1st, 2nd, and 5th—the other two were holding the Mahsuds—all shared in what became the greatest campaign of the North-west Frontier until after the Great War.

Before that cataclysm began, the 4th were to see service in China at Fu-sing during the Boxer trouble of 1900–01.

The two Indian Divisions ordered to France in 1914 took with them the 4th, 5th, and 6th Battalions—the 2nd Battalion also left, but were detained on the Suez Canal. As we all know, the fighting was more severe than they had ever experienced and for which they were not equipped. The reserve system, too, broke down, and could not keep the units at fighting strength after the severe casualties incurred in holding the more or less impractical line of trenches until what time the Armies of the British Empire could be mustered to relieve them. As a result the 1st Battalion, detained on the Frontier, sent nearly half their men to France as reinforcements, and with them was Jemadar Mir Dost who received the Victoria Cross for not only conspicuous gallantry, but great ability when leading his platoon on April 26th, 1915, during the second Battle of Ypres; he afterwards collected various parties of the men who were without leaders and kept them under his command until ordered to retire: later on the same day he again displayed remarkable courage in helping to carry eight British and Indian officers into safety whilst exposed to very heavy fire.

After France the 4th Battalion garrisoned Egypt; then went to East Africa where they did very well in General Smuts' offensive and Van Deventer's pursuit of Von Lettow; they returned to India to occupy the Peiwar Kotal during the Afghan War of 1919.

The 5th also sojourned in Egypt, and thereafter served in Palestine from the third Battle of Gaza until the final victory of Allenby.

The 6th left the Western Front, and the 2nd the Suez Canal, for Mesopotamia and the attempts to relieve Kut. They both remained for the capture of Baghdad. After Tekrit, they too, joined the Palestine force at the Battle of Sharon and the subsequent advance to Damascus.

The 1st, after comprising part of General Benyon's expedition against the Mahsuds in 1917, went to East Africa, returning in 1918 to serve in the Baluchistan force against the Marris and Khetranis. A 2nd Battalion of "Coke's", formed in April, 1918, joined the Persian Field Force in the attack on the Kamarij Pass and the advance to Kazarun.

The 6th Battalion, in 1919, were able to place a Victoria Cross to their credit, awarded that year to Lieutenant W. A. McC. Bruce, but earned by him as far back as December 19th, 1914, at Givenchy. He, with a small party of surviving Sikhs and Punjabi Musalmans during a hotly disputed dawn attack captured an enemy trench, and with these few men, himself badly wounded in the neck, staved off repeated German counter-attacks, until in the end Lieutenant Bruce was killed and his six survivors overwhelmed and captured.

The collective services of the 6th Battalion in the Great War were considered to be so exceptional—at Neuve Chapelle the Battalion had been fought on in splendid fashion by the less than thirty-year-old Subadar-Major—that they were created "Royal" by His Majesty in 1921. H.H. The Mehta of Chitral is an Honorary Officer.

The 1st Territorial Battalion of "Coke's Rifles", raised by Raja Sher Mohammad Khan in August, 1922, is now the 11th (Territorial) Battalion with headquarters at Campbellpore.

"Coke's Rifles", the now 1st Battalion, were the only unit of the Frontier Force whose original uniform was not khaki. They adopted an indigo-dyed drill coat piped with red, long trousers of similar material with a red stripe, the pagri also being of the same colour. The whole outfit was more nearly black, and very conspicuous they must have been against the

dun-coloured hills of the border—no wonder the gallant Pitcher had sixty killed in his attack on the "Crag Piquet". And it was not until 1891 that khaki was adopted for field-service. The facings to the drab uniforms of the 2nd, 4th, 5th and 6th prior to the amalgamation were black, blue, emerald green, and scarlet. The present regimental full-dress is now rifle-green with scarlet facings.

In the matter of armament anything seems to have been good enough for the "Frontier Brigade". The 1st Battalion, and probably the others also, were first issued with worn-out and condemned Baker flint-lock muskets in which the great difficulty was to make the bullet fit tightly into the muzzle-loading smooth-bore barrel. The Scinde Rifles had Brunswick models: these had a two groove rifling, and fired a spherical projectile having a narrow projecting ring of softer metal to take the grooving; they fouled very easily and no dependence could be placed upon their accuracy at any range. Percussion arms were issued generally about 1852. The Enfield Rifle—the introduction of which was said to be one of the many causes of the Great Mutiny—was not re-issued until 1861–71; this had three grooves and was of shorter pattern than the Brunswick. The Snider, the first breech-loading rifle, a combination of the old Enfield with Snider's breech mechanism, came in 1874–75. The Martini-Henry in 1892–93. Magazine rifles—Lee-Metford and Lee-Enfield—from 1901, and the short Lee-Enfield in 1905. The last was the weapon of the Great War generally for the Indian Army, and an improved pattern is still the Infantry small-arm of the Armies of the British Empire. The old policy of keeping the Indian rifleman one pace behind the others rightly ceased a quarter of a century ago.

The Frontier Force Rifles have a Colonel-Commandant in Field-Marshal the Lord Birdwood, while the Colonel-in-Chief is His Majesty the King-Emperor George VI who so honoured the Regiment on his Coronation Day, 1937.

CHAPTER XLI

The 14th Punjab Regiment

"Taku Forts", "China, 1860–62", "Pekin, 1860", "Abyssinia", "Ali Masjid", "Ahmed Khel", "Kandahar, 1880", "Afghanistan, 1878–80", "Tel-el-Kebir", "Egypt, 1882", "Punjab Frontier", "Malakand", "Pekin, 1900", "China, 1900", "Ypres, 1915", "St. Julien", "Aubers", "France and Flanders, 1915", "Macedonia, 1918", "Suez Canal", "Egypt, 1915", "Megiddo", "Sharon", "Nablus", "Palestine, 1918", "Basra", "Shaiba", "Kut-al-Amara, 1915, '17", "Ctesiphon", "Defence of Kut-al-Amara", "Baghdad", "Khan Baghdadi", "Mesopotamia, 1914–18", "Merv", "Persia, 1915–19", "North-west Frontier, India, 1915, '17", "Narungombe", "East Africa, 1916–18", "Afghanistan, 1919".

ALTHOUGH the honours carried by the 14th Punjab Regiment commence with the Chinese War of 1860, the Regiment is the senior infantry descendant of the emergency corps raised in India's principal recruiting ground during the anxious days of the Great Mutiny. All, except the 4th and 5th,

Drum-Major and Pipers, 10th Battalion, 14th Punjab Regiment

Battalions were created by *en bloc* transfers of companies from units of Punjab or Sikh Infantry raised in the previous decade, or from the old-time famous Punjab Police Battalions. The 1st Battalion raised at Phillaur became the 7th Punjab Infantry, the 2nd Battalion formed at Nowshera the 8th, the 3rd Battalion at Multan the 11th, the 4th at Peshawar the 16th, the 10th Battalion at Mian Mir the 17th, all infantry products in 1857 of the energetic Punjab Administration headed by John Lawrence. The 5th Battalion, later to be famous as the only Pathan regiment, were a product of the adjoining province as the Shahjehanpur Levy of 1858.

All first saw service either on the North-west Frontier or in Mutiny operations, but none then to achieve any great historical fame save the 10th Battalion who, with marked gallantry and with considerable loss, stormed the fortress of Rampur Kassia with a trained insurgent garrison of 4,000 or more, on the 3rd November, 1858.

The Battalions, however, had proved their worth, and on the assumption of control of the armed forces of India by the Crown were brought into the re-organized Bengal regular army in 1861. And after a re-shuffling of numbers the Battalions became the 19th, 20th, 22nd, 24th, 40th and 21st Regiments of Bengal Infantry, numbers they were known by, although the titles became in 1903 Punjabis—the 40th, Pathans—until the infantry organization of 1922 constituted them one family.

The early kit was khaki drill, but later, when a form of coloured full-dress was adopted, scarlet coats and blue breeches were worn by the 1st, 3rd, and 4th, drab by the 2nd and 10th; with facings of blue for the 1st and 3rd, emerald green for the 2nd and 5th, while the 4th adopted white, the 10th scarlet: scarlet green-faced is now the present uniform.

The 2nd Battalion, which now bears the name "Brownlow's" after Field-Marshal Sir Charles Brownlow who raised them when a subaltern, volunteered *en masse* in 1860 for the resumption of hostilities with China suspended during the Mutiny period.

"Attestation Parade"

In the Indian Army recruits are not attested until after they the Colours, prior to joining their active service unit. The two

10th Battalion, 14th Punjab Regiment

have completed their recruit training, they are then sworn-in on igures in white are the Regimental Priests.

They took part in the storm of the Taku Forts unexpectedly defended by the Chinese with great determination, and in the advance to Pekin, where they were the first corps, British or French, to plant their colours on the city walls—the latter country had allied in the expedition whose object was to enforce ignored contractual obligations. The 3rd Battalion also volunteered for service, and being sent to garrison Shanghai had constant brushes with insurgent Chinese headed by Taiping until May, 1862.

The 10th Battalion were brought across India to the north-east for the operations in Assam in 1862–63; as were the 1st Battalion two years later to provide reinforcements for the trying and unhealthy war against the Bhutanese. The 3rd also operated on the eastern borders of Bengal against the predatory Lushai tribes who had been committing murderous assaults on the local inhabitants in 1871–72.

On the other frontier, the 2nd Battalion had joined in the Ambeyla Pass fighting of 1863, being noted for their remarkable steadiness in the defence of the "Eagle's Nest" piquet, and later in the recapture of the "Crag" piquet, great military events of the time which earned the Battalion a mention in despatches.

The 10th Battalion were one of the two Bengal battalions sent to Abyssinia in 1867–68; unlike the other—the now disbanded Sikh Pioneers—they were detained at the base, and on road protection duties; dull unpleasant hard work with no compensatory battle.

There were many 'frontier' affairs and similar work until the long and expensive series of operations known as the 2nd Afghan War, 1878–80. In these, the 2nd Battalion were present at the opening affray at Ali Masjid; the 1st at Ahmad Khel; the 10th in the Shutargardan Pass; the 4th in the Bazar Valley, the Kabul garrison, and the famous march culminating in the battle at Kandahar. The 3rd remained in the Khyber guarding the road, on which they witnessed the spectacle of a battalion returning after the operations with all the officers except the Colonel and the Adjutant under arrest. Discipline among officers in those days was very different from now, when it is exceptional to find an officer in arrest. But tempers then appear to have been more easily lost, and there is a story of the period which tells of a fire-eating colonel who drew his sword and chased his major off parade, while the ranks stood easy and betted on the result.

The selections for the Indian contingent sent to Egypt in 1882 included the 2nd Battalion, who fought in the hot dusty thirteen hour desert defeat of Arabi Pasha at Tel-el-Kebir: an action in which the attack was made at dawn after a daring night advance across a desert devoid of cover or landmarks, the columns being guided by a naval officer who steered by the stars. On their return the Battalion became "The Duke of Cambridge's Own". We are told that the Subadar-Major after the somewhat easy defeat of the Egyptians had no great opinion of the Fellaheen as soldiers, nor could he understand why after the campaign he should receive the very rare honour for a then Indian officer of a Companionship of the Eminent Order of the Indian Empire. This old officer had given long and meritorious service to the State. He was already a Sardar Bahadur, and a recipient of the Indian Order of Merit, both of which brought him in a monthly cash allowance. But he could never understand why the higher honour of the C.I.E., which as generally referred to sounded to him much like the Urdu for ink, brought him no cash return.

The "Mad Fakir" of Upper Swat during the general upheaval on the Frontier in 1897 expressed his intention of driving the garrisons of Chakdara, Malakand, and other posts back into India. Reported to possess supernatural powers so that his enemies' bullets would not harm him, and accredited with an army of all arms concealed in his homeland hills, he made things most unpleasant for the troops. He very nearly succeeded in a night attack on the Malakand Camp, and would

have done so had it not been for the stubborn defence of the garrison. Here Lieutenant W. G. Costello of the 3rd Battalion, then attached to the 4th, won the Victoria Cross; and two sepoys the Order of Merit, for rescuing from an area overrun with tribesmen a dangerously wounded havildar. As the disaffection spread the 2nd Battalion became engaged at Shabkadr; the 3rd in the Malakand relief, the Mohmand country, and the Tirah; the 10th against the Utman Khel, in Buner, and the action at the Tanga Pass.

The scene of the early overseas Regimental exploits was visited by the 4th Battalion in 1900, when they took part in the allied relief of the almost despairing citizens of the great nations besieged in Pekin by the Boxers. The 2nd Battalion—the Regimental pioneers in China—were also despatched to the same scene, but arrived too late for the serious fighting except for the capture of the Peitang Forts. The 2nd, however, took part in the 'mopping up', and on one occasion prepared to attack a defended town from which the enemy seemed to be defiant enough to join battle; but upon the accompanying field-battery firing a single shell, thought better of it, and came out to greet the advancing troops with offerings of chicken and hard-boiled eggs.

Up to now we have said nothing of the exploits of the 5th Battalion, for the very simple reason that up till 1888 the unit saw no service at all until the Black Mountain Hazara punitive operations of that year. In 1890, however, original organization of the Battalion was terminated and a decision made to fill the ranks with Baluchis. This proved more difficult than was supposed, the Baluch Sirdars would only assent to their men being enlisted if they themselves commanded the companies, together with other stipulations impossible to fulfil unless efficiency and discipline were to be largely set on one side. The fiat thus went forth—a battalion of trans-frontiermen. So at Peshin in Quetta, Major "Ali Baba" Graves of the 42nd Gurkhas raised the famous "Forty Thieves" and the only all Pathan regiment in

the Indian Army. The soubriquet was most unfairly bestowed; nor was it quite 'cricket' as they moved from station to station for the civil police to lay at their door every conceivable undetected crime committed for miles around. Though it may be true that, in the more peaceful garrisons, the thump of the *dhol* and the wail of the *srinai* of the *Pukhtan* musicians playing the Regimental March of "Zakhmi Dil" with its definitely Rabelaisian verses, did cause a certain amount of uneasiness amongst the 'locals'. The linked-battalion system of 1901 destroyed the all-Pathan character, as from then Punjabi Mussalmans and Dogras were to be admitted in order to facilitate changes in the event of major war.

The first service of the 40th Pathans, the now 5th Battalion, was on the "Roof of the World" in the expedition to the forbidden city of Lhassa in 1903–04. Operations undergone in the severest climatic conditions of cold, which earned what is now becoming a very rare war medal.

Four battalions, the 1st, 3rd, 5th, and 10th—then coincidentally mobilized for service together—all shared in the Frontier war of 1908, when the Zakha Khel incited the Mohmand tribes to disturb the peace which had reigned on the border for a decade. The first campaign of the India General Service Medal 1908, commenced by King Edward VII which was to last for frontier campaigns for just thirty years.

All the Battalions fought outside India during 1914–20.

The 1st Battalion after doing duty in the Persian-Afghan cordon in Seistan moved on to support the moderate Menshevik army in Russia against the communalist Bolsheviks, to experience burning heat, perishing cold, snow-covered sands, treachery, the defeat of allies, armoured-train fighting, an enemy led by Austrian prisoners, an action which brought one of the rarest of honours for real battle—"Merv"—still more fighting; all in unique circumstances, but completely eclipsed by the events enjoying the more popular publicity of the time.

The 2nd and 3rd Battalions were early arrivals in Mesopotamia, where they were later joined by the 4th who had been an original selection for France but had been detached for the defence of the Suez Canal. The 2nd Battalion took part in the battle of Kut-al-Amara, the fighting at Sannaiyat, and later the advance to Baghdad, Istabulat and Tekrit; whence they accompanied the 7th Lahore Division on its transfer to Palestine for the great advance through the Vale of Sharon in September, 1918.

The 3rd and 4th Battalions suffered heavily in distinguishing themselves at Ctesiphon; and during the retreat to Kut, where they both played a gallant part in that memorable siege, paying a heavy toll from the gross ill-treatment of the prisoners by the Turks who had hitherto been regarded as a chivalrous enemy.

A re-formed 3rd raised at Basra occupied Bushire, and afterwards stormed Spin Baldak Fort in the Afghan operations of 1919.

The 4th Battalion re-formed in India returned to the Euphrates to hit hard at Khan Baghdadi, where the entire enemy force was surrounded and captured, a return compliment for Kut. The Battalion subsequently went farther afield to Salonika and South Russia, eventually halting near Ismid where the 10th Battalion were forming part of the garrison. A detached company occupied a post some 3,000 yards north of the defences which were held, as a politically passive defence against Kemal Pasha's Nationalist forces. Owing to the supporting measures lacking co-ordination when this company was withdrawn, it walked straight into an ambuscade which it could not avoid and suffered severely. How any individual escaped alive was beyond the comprehension of onlookers, and if there had been seventy Mahsuds instead of 700 Turks not a soul would have got away. This little fight caused over 25 per cent casualties, earned a Military Cross, two Orders of Merit, and a Distinguished Service Medal. The Battalion subsequently

THE DETACHMENT OF THE 5TH BATTALION 14TH PUNJAB REGIMENT AT ADDIS ABABA, ABYSSINIA—1935

carried out other small operations before leaving for India after an absence of six years.

The 5th Battalion journeyed from far Hong Kong to France early in 1915. What the French thought of their National Anthem played on the shrill reed-pipe of the Pathan hills accompanied by the thump of the tribal tom-tom, history does not record; but to march through Marseilles headed by the *Pukhtan* band so performing was certainly a stout effort. With turbans replaced by Balaclava caps, their first casualties were suffered marching through Ypres to support a line stricken by the first taste of chemical warfare, and into an action which cost them fifty per cent of their strength. After Loos the "Pathans" were sent to East Africa, where they continued to give a very good account of themselves. The Ordnance Service seems to have neglected them, and patrols had to be selected from those men which had boots; others were clothed in captured German uniforms, one poor individual had to parade in a ration sack, while none had any socks or underclothing.

The 10th Battalion spent the first four years of the Great War in the various minor operations on the disturbed North-west Frontier. In 1918 they received welcome orders to proceed to Egypt and Palestine, and advanced with Allenby on the opposite flank of the force to the 2nd Battalion.

The "Forty Thieves", who were also in the Afghan War of 1919, have what is probably one of the most unique of war memorials. It is a monolith of stone depicting a rifle cartridge, overlooking the Indus gorge at Attock. A typical natural symbol of the thoughts of the trans-frontiersman in his rugged homeland, it leaves upon the traveller by road to the Frontier an impression not easily forgotten. The war memorials of the other Battalions are situated at Ferozepore.

An 11th Territorial Battalion came into existence in 1922 largely through the efforts of Mr. F. L. Brayne, M.C., I.C.S., an intrepid Punjab social rural reformer and originator of the

Gurgaon 'experiment'. Headquartered at Delhi the unit bears the sub-title "Gurgaon".

Since the Great War ended the 1st and 5th Battalions have done further oversea tours in Palestine and Iraq, while the 3rd formed part of the Shanghai Defence Force, 1927. But the greatest recent oversea service rendered by India's Army, and an international service at that, was when the 5th Battalion were ordered in August, 1935, to despatch a company as legation guard to Addis Ababa, the Abyssinian capital, during the Italo-Ethiopian war. Only Sikhs were sent, as the advisers on the spot considered them most likely to create the best impression, as the British Minister's escort was a Sikh detachment of the 8th Light Cavalry. After the Emperor's departure the town fell into the hands of the mob. And it was almost entirely due to the detachment of the 14th Punjab Regiment that many of the legation personnel and refugees of other nations were able to reach safety. Some sharp fighting was seen during these rescue efforts, but no casualties occurred to the 14th, and it was recorded that a little aiming drill would do the Abyssinian soldier no great harm. After eulogies in the House of Commons, and the assumption of control by the Italians, the Detachment departed in the December, being inspected on their leaving by the newly-created Viceroy of the House of Savoy. Their smartness evoked much comment, particularly from the invaders, and India can be proud of the service rendered to the great nations of the world by this small body of her forces virtually isolated for over fourteen months in circumstances for which there was no precedent.

HAVILDAR MAJOR
25TH PUNJABIS—1904

CHAPTER XLII

The 15th Punjab Regiment

" China, 1860–62 ", " Ali Masjid ", " Peiwar Kotal ", " Charasiah ", " Ahmed Khel ", " Kabul, 1879 ", " Kandahar, 1880 ", " Afghanistan, 1878–80 ", " Burma, 1885–87 ", " Chitral ", " Somaliland, 1901–04 ", " Loos ", " France and Flanders, 1915 ", " Suez Canal ", " Egypt, 1915 ", " Megiddo ", " Sharon ", " Palestine, 1918 ", " Tigris, 1916 ", " Kut-al-Amara, 1917 ", " Baghdad ", " Mesopotamia, 1915–18 ", " Persia, 1918 ", " North-west Frontier, India, 1917 ", " Kilimanjaro ", " East Africa, 1914–17 ".

THE 15th are truly a Punjab regiment; for all the Battalions were founded in the Province during 1857, when the fastest means of communication between the Governor-General in Calcutta and the Administration in Lahore was by steamer to Madras, telegraph to Bombay, and thence by any means possible to the north. As we know, Sir John Lawrence, although a civilian, had military genius which manifested itself so ably when by a stern and resolute spirit he overawed the Bengal regiments in his Province, and so won over the populace, that the peasantry and tribesmen flocked to the standard of order,

and enabled formation of the irregular corps which developed into some of the famous regiments of to-day.

Originally the Lahore, Peshawar, Rawalpindi, Ferozepore and Jullundur Punjab Battalions; then the 17th to the 21st Punjab Infantry of 1861, they now constitute the 1st, 2nd, 3rd, 4th and 10th Battalions of the 15th Punjab Regiment created in 1922.

The Battalions in their infancy must have been somewhat like the new armies of Britain of the era of 1914. Mostly recruited from police, tribal levies, and other hardy material; equipped with muskets from the disbanded mutinous regiments, their original training was nil. Uniform, although officially khaki, was just as the men fancied, and we read that for the first two years owing to constant service little or no opportunity presented itself of doing any drill.

Save the 10th, who stayed to hold the border, all the Battalions as soon as completed went south to teach the Poorbeah sepoys the sanctity of oath and contract; not, however, before the 2nd had a primary successful attempt with the 51st Native Infantry. This renegade Bengal regiment had been disarmed, but finding opportunity of seizing the piled muskets of a newly raised unit commenced to give fight, an action which caused several casualties to the Battalion before the mutineers were completely wiped out in a pursuing fight to the Khyber, through which they attempted to escape. A terrible lesson, but necessary at the time.

The 1st Battalion joined Coke's Brigade in Rohilkhand, and were at Sirpura. The 3rd were in Oudh. The 4th in the Jagdespur jungles of Bihar. The 2nd, after a bout with the Yusafai, came down to the campaign in the Nepal Terai.

Enlistment of the men was not yet for general service, and on volunteers being required for oversea war service in China, the 3rd Battalion willingly undertook to go. They spent a year there on detached duty in Chusan in 1860. The campaign medal, curiously, was not issued to the men, but in 1882 the

Battalion were allowed the inscription on the colours—a reversal of the normal usage.

In the summer of 1861 the 3rd Battalion were at peace in Lahore Cantonment when cholera broke out among the British troops with excessive mortality. Great difficulty was experienced in getting persons to attend upon the unfortunate patients in hospital, owing to the panic of the local inhabitants. The men of the Battalion on hearing of this, freely offered their services for this not very pleasant humane work, for which they were afterwards thanked by the British Government in a special despatch.

The 10th Battalion saw action at the capture of Dewangiri in the Bhutanese War, 1865. They also took part in the Jowaki Expedition of 1878.

For the Second Afghan War the 1st and 2nd Battalions were in the corps which occupied Kandahar, the 2nd returning as soon as the garrison was established to take no further part. The 1st marched with Stewart to Kabul, fighting against the impetuous attack at Ahmad Khel on the way; and then marched back again under Roberts to the conclusion with Ayab Khan outside the walls of Kandahar. The 3rd Battalion in the Khyber lost their Commanding Officer and Adjutant at Ali Masjid, after which they did some of the cleaning up in the Bazar Valley, thence to Kabul; to return after Abdur Rahman had assumed the Afghan throne. The 4th and 10th Battalions joined the Kurram force, the latter taking part in the cleverly designed attack of Roberts on the Peiwar Kotal; the 4th went on to Kabul, receiving honours for the defeat of the Afghan stand at Charasiah where they were in reserve: they later saw much sortie and outpost work, particularly at Lataband, an important connecting post with the Khyber.

After Upper Burma had been annexed in 1885, the country, contrary to expectation, did not settle down. The attempts of pretenders to gain the deposed Theebaw's crown, together with increasing outrage and disorder, which grew bolder as the weak

strength of the troops occupying the country became apparent, necessitated more reinforcements from India. Among the first to be sent were the 2nd and 3rd Battalions who arrived there in the spring of 1886. There were no pitched battles, all the operations being effected by detachments. And while casualties from fighting were very small, those from sickness were high, the Dogras and Sikhs losing almost thirty per cent of their strength—the Pathans, interestingly, only seven. After a year's very strenuous work the Battalions were relieved and returned across the Bay to recuperate.

The Orakzai clans began to disturb the North-west Frontier in 1891, and two expeditions were necessary into the Miranzai Valley. The 10th Battalion with the first expedition met no opposition, but had a hard time during the severe winter owing to the constant marching and road-making. It was the treacherous attack on their working parties on the Samana ridge which showed that no lasting result had been achieved, and thus caused the second expedition in which the 3rd Battalion also joined: on this occasion much fighting, some of it severe, occurred before the clans made their submission.

The 1st and 10th Battalions enjoyed the honour for the Chitral relief, the former taking part in the storm of the Malakand Pass, while the latter was in the reserve brigade.

The Dongola Expedition on the Nile employed an Indian Brigade at Suakin on the Red Sea whose duty was to draw off the rebel Dervish chief, Osman Dinga, whilst were started the first of the campaigns which eventually led to the recovery of the Sudan. The 2nd Battalion had six months' severe service in those sandy wastes in which the water was invariably the colour of red clay, the rations almost without vitamin, the barrack accommodation non-existent; while to add to the discomfiture, Osman failed to give battle and so relieve the monotony of service, or the minds of the Muslim troops who were much exercised by the fact that Mecca lay to the east and not to the west.

A TYPICAL INDIAN OFFICER—PUNJABI MUSALMAN—OF A PUNJAB REGIMENT

Africa also employed the 3rd Battalion who went over for the Somaliland campaign in 1903–04.

The 1st, 4th and 10th Battalions were all mobilized for the "Nowshera Manœuvres", the public nomenclature for the pending operations against the Zakha Khel and Mohmands in 1908. One of the prime movers in the raid which led to the punitive expedition was ex-Sepoy Multani of the 2nd Battalion. This daring adventurer in the raid which 'lifted' a lakh of rupees from Peshawar City, was accompanied by a bugler who on the falling in of the City Emergency Column sounded the "No Parade", an action which just delayed the troops sufficiently for the gang to make a perfect get-away. It was from this that the practice started of never dismissing troops on the Frontier otherwise than by whistle or verbal order.

The honour of being the first unit to leave India for, and the first to see action in, the Great War belongs to the 10th Battalion. Sailing from Karachi on the 17th August, by the 6th September they were engaged near the one-time man-eating lion infested town of Tsavo, repelling an attempted German invasion of what is now Kenya. The Battalion split up into detachments from then, and for two years saw many bush fights and contests on the border and in the subsequent invasion of German East Africa. They had a year's respite in India, and then again sailed for service in Palestine, and in the post-Armistice internal security operations in Upper Egypt.

In the immediate pre-Great War period, internal disturbances in China had caused a large increase in the garrison of Hong Kong; and here the 1st and 2nd Battalions were serving in August, 1914. Both were brought home in 1915 to the Tochi; the 2nd to be withdrawn for duty in Mesopotamia consequent upon the investment of Kut-al-Amara. The Battalion formed the advance guard in the long and difficult night march to Dujailah on March 8th, 1916, and fought in the following all-day battle designed to turn the Turk's right flank and relieve the starving defenders: a battle which very nearly

succeeded. But, as it did not, Kut fell, and except for a short lull thereafter, the Battalion was in almost continuous fighting until the fall of Baghdad.

About this time a 'hush-hush' enterprise under General Dunsterville started for Asia Minor with the object of providing a nucleus for organizing the Armenians into defending their own country. It never reached its destination and the problem of preserving communication with it into North Persia was a great tax on the Mesopotamia command. On this duty the 2nd Battalion remained until the end of the war.

The 1st left the Tochi Valley—a war front—in 1917, also for the Euphrates and Tigris. They just missed the victory at Sharqat, but in high hopes departed for Salonika, to be again disappointed by the Armistice. They later served on the Black Sea and in Anatolia against the Turkish Nationalists in 1920.

The 3rd Battalion from the North-west Frontier gave battle on the Suez Canal in February, 1915. By September they were fighting in the Loos offensive in France, from which they were withdrawn to join the relief of Kut. They did fine work at Beit Aiessa. After Baghdad the Battalion went to Allenby's command for the decisive Battle of Sharon, at the end of which they found themselves at Semakh on the Sea of Galilee, where four years before the Turk had rehearsed his plan for crossing the Suez Canal, in the repulse of which the Battalion had played a prominent part.

In 1914 the 4th Battalion were in Ceylon where serious rioting all over that island detained them from a war theatre until November, 1915, when they too joined in the Kut relief force. After being held up for almost a year opposite Sunniyat the Battalion shared in Maude's victories at Baghdad. Their war-services ended in Palestine, where, after having taken under heavy barrage of fire eight lines of trenches in sixty-two minutes in the attack on the 19th September, 1918, north of Jaffa, the Battalion saw no more fighting as their enemy was put to full retreat; which gave them a long march in his wake up the

Levantine coast.

The 4th Battalion left Egypt in 1920 for Waziristan. In one of the Frontier conflicts that followed, Sepoy Ishar Singh, when the convoy protection troops were attacked near Haidari Kach, won the Victoria Cross. He was No. 1 of a Lewis Gun section, and early in the action fell severely wounded in his chest. In the hand-to-hand fight that followed the officers and all the havildars of his company were killed or wounded, the Lewis Gun being seized by the enemy. Calling up two other men, Ishar Singh got up, charged the tribesmen, recovered the gun, and although bleeding profusely he again got into action. A reinforcing officer ordered him back to the ambulance. Instead of doing this he greatly helped the medical officer, pointing out where the wounded were and carrying water to them, making innumerable journeys to the river and back for the purpose. On one occasion when the enemy fire was very heavy he took the rifle of a wounded man and joined in the fight. He also shielded the medical officer with his body while a casualty was being dressed. It was not for three hours before he finally submitted to be evacuated, then being too weak from loss of blood to object. Gallantry and devotion to duty which inspired all who saw him and which is beyond all praise.

The 2nd Battalion was in Waziristan, 1921–23; in fact, all the active Battalions have done their 'whack' in the frontier strife of recent years. The 2nd were also in the Burma Rebellion, operating 1930–32, in the same area they had fought over forty-five years before, while a son commanded the same company as his father had in an action on almost identical ground. All ranks earned much praise and many well-merited decorations for their work in Burma; an episode which was to be repeated in the Loe-Agra-Mohmand operations of North-west Frontier from 1935 onwards.

Ambala is the headquarters of the 11th Territorial Battalion raised in 1922.

In 1914 the 2nd and 4th Battalions were dressed in drab

with scarlet facings, the other three in red, with white facings for the 1st, emerald green the 4th, and blue—originally light—for the 10th. The dress is now red, buff facings being adopted for all: the Regiment might thus be described as "India's Buffs" since they are the only Indian Army regiment to wear the colour for ever associated with one of the oldest and foremost fighting regiments of the British contemporary. The custom of wearing regimental pattern mufti clothing when off duty originated in the 2nd Battalion in 1910. The smartness of the men when off-parade, as well as their obvious comfort, led Simla to adopt the idea for the whole of the Indian Army.

The 3rd Battalion enjoy the honour of wearing on their appointments the Royal and Imperial Cypher within the Garter. The services for which this was bestowed are now unrecorded, but it is generally accepted that Queen Victoria did so for the humanity displayed during the Lahore cholera outbreaks of 1861. Thus does peace have its rewards for the soldier no less worthy than those of the battle-field.

No story of India's Army, however brief, would be complete without some mention of those faithful humans upon whom the soldier relies so much for his domestic services in peace and war, and who, while not sharing with him the glories and excitement of action, bear all the vicissitudes and hardships of the campaign. Without the *bhistee*, the *langri*, the *syce*, the *dhobi* and the *mehta*, as well as the all-important *bearer*, no regiment would be happy, few would succeed. The old grey-bearded mess-bearer of the 2nd Battalion—Allah Jowaya—is a case in point. Of great age, he could still keep up the regimental spirit, muster the Followers into his "Platoon", and at their head call upon them to endure the long hot and trying marches in Waziristan. Proudly wearing his medals for a long ago Egyptian campaign, in spite of his years he never spared himself, and at the end of an exasperating and heavy day was always ready to minister to those whose comforts were his especial care. The hewers of wood and drawers of water that

trail in the wake of the Indian fighting man may not be gifted with the warrior's physique, but they have stout hearts, and fully deserve the campaign medals they receive to mark their services. Medals were first issued to Followers after the close of the Burmese War in 1889, and were then and afterwards struck specially in bronze; King Edward VII stopped this differentiation, and with the issue to the troops of the India General Service Medal 1908, identical silver medals were distributed.

CHAPTER XLIII

The 16th Punjab Regiment

"Afghanistan, 1878–80", "Burma, 1885–87", "Chitral", "Tirah", "Punjab Frontier", "Malakand", "La Bassee, 1914", "Messines, 1914", "Armentieres, 1914", "Festubert, 1914", "Givenchy, 1914", "Ypres, 1915", "St. Julien", "Aubers", "Loos", "France and Flanders, 1914–15", "Macedonia, 1918", "Suez Canal", "Egypt, 1915–16", "Megiddo", "Nablus", "Palestine, 1918", "Aden", "Tigris, 1916", "Kut-al-Amara, 1917", "Baghdad", "Mesopotamia, 1915–18", "North-west Frontier, India, 1915, '16–'17", "Behobeho", "Narungombe", "Nyangao", "East Africa, 1917–18", "Afghanistan, 1919".

YES, but who were they? That is the question which is still asked when referring to the present infantry regiments of India, getting on for two decades after the old well-known numbers had been melted in the remoulding of 1922. Here is the answer anent the 16th Punjab Regiment.

The 1st Battalion began their existence at Ludhiana in 1857 as the 22nd Punjab Infantry. Under the Crown, from 1861,

they became the 30th Bengal Infantry, not losing that number until the merger sixty-one years afterwards.

The 2nd Battalion, raised under the orders of General Van Courtland, the Deputy Commissioner of Ferozepore, were also a Great Mutiny product. They became in 1861 the 31st of the Bengal foot.

The 3rd Battalion were originally the Allahabad Levy formed from lowly men who desired to prove their staunchness to an established order, when many of their social superiors set a bad example. They became the 33rd Allahabad Regiment when the East India Company was liquidated. Few records exist as to the soldierly qualities of the more humble strata of Hindu society, but the Regimental composition was gradually changed, and by 1883 the enlistment of men from the menial classes ceased. For a time, 1891–99, the Battalion was entirely composed of Punjabi Mahomedans, their thence title until 1903. They then became simply Punjabis—as did the senior Battalions.

The 4th Battalion originated in the welding of the loyal remnants of the disaffected Bhopal, Gwalior, and Malwa contingents into a battalion for local service in Central India. They were brought into the Line in 1903 as the 9th Bhopal Infantry.

The 10th Battalion were the old 46th Punjabis raised as Bengal Infantry in the last year of the nineteenth century.

The need for troops to restore order during the Great Mutiny was so pressing that although the recruits of "Van Courtland's Levy" were unused to fire-arms, and there was a dearth of instructors, they were considered ready for service by June, 1857. And so off went "Bloomfield's Sikhs"—a title they had received from the name of their Commandant, Captain C. G. Bloomfield—with the Hariana Field Force commanded by the General-cum-Deputy Commissioner, to assist in the successful squashing of the rebellion around Hissar, and that of the insurgent Jodhpur Contingent at Karnal. The 1st Battalion joined Sir Colin Campbell before Bareilly in 1858, remaining with the Oudh columns until 1861. The 3rd Battalion sallied forth as required from Allahabad and dealt with rebels and rioters still making trouble after the main operations had closed down.

In the year 1864 the Bhutanese made a series of raids into Indian territory which forced an unwanted war. With the invading column from Jalpaiguri were the 1st Battalion. After storming the fort at Dhalimkot the expedition met with little opposition save for the stand of the Bhutia warriors at Chamurchi. Bhutan was annexed. The punitive force proceeded to disperse. The Bhutanese, however, disliked the idea of being part of India, and so they attempted to efface the newly established garrisoned posts. War-like operations again became necessary and fresh troops were required. And so the 2nd Battalion were sent to join in the war which did not finally end until 1866.

The 4th Battalion became after the Great Mutiny a local corps for service in Bhopal under civil control. Until 1878 they were stationed by detachments in the jungles of Central India and the army saw little of them. They were mobilized and brought out of their seclusion to man the North-west Frontier during the first phase of the 2nd Afghan War, but except for this they remained localized until their conversion into a Line battalion in 1903.

The other Battalions were not required for the initial Afghan operations which had brought out the 4th, but when the massacre at Kabul again set India's forces in motion the 1st and 2nd were promptly despatched to hold the Khyber—a duty which did not pass without much detached fighting.

The campaign in Burma gave the 3rd Battalion onerous work for three years from 1887. As with all the regiments employed, sickness caused great depletion of strength.

All the first three Battalions saw much Frontier service in the "nineties". The 1st in the Isazai Expedition of 1892—

Colour Party
2nd Battalion, 16th Punjab Regiment—1938

coercive measures necessitated through a breach of engagements by the Black Mountain tribes. The 3rd against the Mahsuds for their unprovoked attack on the Indo-Afghan Boundary Commission in 1894. The 2nd—who had in 1895 mobilized for the Chitral relief—in the defence of the Malakand posts two years later: the incident where a Jemadar and twenty-five men held out for six and a half hours against a determined enemy who had set fire to the *serai* they were defending being considered worthy of special mention in despatches. The 1st were in the Tirah Expedition the same year. The 3rd in the Tochi Valley in 1899.

1902 saw the 1st Battalion in China, and 1904 the 3rd in Somaliland. Both were good advertisements of the efficiency of the Indian Army. The 1st in China attracted world-wide publicity for the excellence of their shooting in the International Rifle Meeting organized by the German units of the garrison. The Battalion won four out of five of the first prizes; and out of the twenty-four prizes allocated to the soldiers, twelve went to the men of the then 30th Punjab Infantry. The foreign press made much of the incident, pointing out that those who had sought to discourage the Indian troops in China would now have to admit that the Sepoy was indeed a very capable marksman. The International force in the Far East at the time comprised representatives of Britain, Germany, Italy, Japan, Russia, and the U.S.A., each of whom sent competitors.

The call for trained musketeers for a more serious "rifle-meeting" was soon to come. To it, in the autumn of 1914, were sent the 3rd and 4th Battalions; the former to halt in Egypt, the more fortunate latter going on to France to arrive in time to earn the "1914 Star". Neither were the first unit of the present Regiment to pass through the Suez Canal—that had been done by the 2nd Battalion when on their way to Malta and Cyprus in 1878, they being the first Punjab corps ever to do so.

The Indian Army was not prepared for a war in Europe. It

is said that some plan had been prepared only to be filed. But whatever the truth of that story, the 9th Bhopal Infantry had no place in it since they had no linked battalions on which they could draw for men and material to complete their war establishment and maintain them in the field. Still they had been selected, and despite much petty trouble, they reached Karachi ready to embark on the day that the British Expeditionary Force in France first went into action.

By October 27th, 1914, the Battalion reached the scene of fighting, severe and critical, in Flanders. The Germans were making a stupendous effort to drive the remains of the British Regular Army into the English Channel. Near Neuve Chapelle the line had nearly gone, and to the aid of the remnants of an unconquered British battalion were sent the 9th Bhopal Infantry. One wonders if ever a unit received such a baptism of fire. Advancing, wearing cotton drill, across wet slippery fields interspersed by hedges and ditches entangled with barbed-wire, on went the men in the falling darkness of a chilly late-October day towards an unknown objective, search-lights playing on them, snipers taking their toll. Then followed a three-day battle, without food, in surroundings of wet and mud with friend and foe inextricably mixed. But the gap was closed. The cost: 11 officers and 262 men. There was a loss of another 200 in a stupendous bayonet fight at Festubert a few days later. With the ranks terribly depleted, the Battalion took part in the operations to restore the line in the spring of 1915 following the first use of gas by the enemy. They remained in France until the May, and then went to Egypt to re-make themselves until they were required on the Tigris.

In the battle of the Wadi the Commanding Officer was struck down. To his aid went Sepoy Chattar Singh over a spot as flat as a billiard table and within two hundred yards of the enemy, who kept up a constant fire. Shielding the Colonel with his body, the Sepoy with his entrenching tool slowly, and for hours, made a bullet proof shelter for both.

"King-Emperor's Orderly Officer"
16th Punjab Regiment

At night they were rescued. The full story is too long for these lines, but it concludes with the grant to the valiant Sepoy of the Victoria Cross.

The "Bo-Peeps", as the Battalion were more popularly known to the Army, remained to fight in Mesopotamia until the end of the War. When it was over, of the 'originals' that set out in 1914 only fifteen returned with the unit to India.

The 3rd Battalion, after spending nearly a year in the Suez Canal defences, were at last despatched to France. Before Loos, in their first action, they received in return for many casualties, much glory. Transferred later to Aden, they went on to East Africa in May, 1916.

To this tropical theatre the 3rd Battalion were followed in the December by the 1st from the North-west Frontier. Both had as rough a time of it as many other units did in the more advertised arenas. Not only did the enemy have to be encountered, but also nature. In one action, the 1st ordered to cross a river, were held up by hippopotami, who had to be driven off and despatched by bayonet before the advance could proceed. Both Battalions suffered heavy losses; and after Nyango—the battle which finally drove the Germans to flight—the fighting strength of the 3rd was reduced to a bare hundred. The pluck and tenacity shown by the Indian troops was beyond all praise, as was that of the King's African Rifles and the Royal West African Frontier Force, who all together brought the campaign to a successful conclusion in the face of what seemed to be almost insuperable difficulties.

The 2nd Battalion, after having been drained of trained officers and men for other units, joined the Euphrates force in December, 1915. After nearly three years in Mesopotamia, they went further afield to Salonika, to Turkey, and finally the Black Sea until 1920.

The 10th Battalion saw no active service until the Malakand and Chakdara operations of 1915, thence the Mohmand Blockade of 1916, and the South Waziristan Field Force, 1917. They

embarked in 1918 for the concluding Palestine battles; and afterwards played a prominent part in quelling the internal Egyptian disturbances along the Nile in 1919.

The Afghan intrusion of the same year sent the 3rd Battalion out on active service in the Khyber. They also did a year's duty in Asia Minor, 1921–22, as well as nine month's insurgent pacification in Iraq the year afterwards.

The modern uniform of the Regiment is red with white facings, turban and cumarband blue. When the Battalions had separate existences this was the kit of the 1st and 2nd, that of the others being drab with emerald green facings, except the 4th whose collars and cuffs were of chocolate. The original dress of the Bhopal Contingent was similar to that of the pre-1857 Bengal Native Infantry, drab being adopted in 1865. The official dress of the Punjab Battalions in the beginning was khaki, the present 1st and 2nd Battalions adopting the scarlet of the Sovereign's livery in 1885.

Types of Dress, Indian Officers, 17th Dogra Regiment
Drill. Durbar. Full. Mess. Field.

CHAPTER XLIV

The 17th Dogra Regiment

" Chitral ", " Malakand ", " Punjab Frontier ", " La Bassee, 1914 ", " Festubert, 1914–15 ", " Givenchy, 1914 ", " Neuve Chapelle ", " Aubers ", " France and Flanders, 1914–15 ", " Egypt, 1915 ", " Megiddo ", " Nablus ", " Palestine, 1918 ", " Aden ", " Tigris, 1916 ", " Kut-al-Amara, 1917 ", " Baghdad ", " Mesopotamia, 1915–18 ", " North-west Frontier, India, 1915–17 ", " Afghanistan, 1919 ".

THE enlistment of Dogras into regiments comprising them solely only dates from 1887, when five battalions disbanded on the grounds of economy a quinquennium before were revived in the Army List. Of these five, the new 37th Bengal Infantry became the first all-Dogra unit, and are now the 1st Battalion (Prince of Wales' Own) 17th Dogra Regiment, the subsidiary title together with the right to bear the plume of the Prince of Wales having been bestowed on the 37th Dogras, their title from 1903, just prior to the regimentation of the old Dogra battalions in 1922.

At the end of 1890 the class regimental enlistment of Dogras was still further increased by the conversion of the 38th Bengal Infantry formed from the old "Agra Levy" of low-caste men

during 1857. Trained men were transferred from sixteen other Bengal and Punjab regiments, both infantry and cavalry to form the frame, complete companies coming from "Coke's Rifles" and what is now the 3rd Battalion 14th Punjab Regiment. Recruiting was speedy, and within five months of the orders to re-organize the transformation was completed. Thus came the 2nd Battalion which from point of unbroken formation is actually older than the 1st.

The 3rd Battalion were a product of 1900 as the 41st (Dogra) Regiment of Bengal Infantry, later Dogra Infantry, and in 1903 just the 31st Dogras.

The 10th Battalion—there are only four regular battalions—were raised during the Great War at Jubbulpore in 1917 as the 2nd Battalion of the 41st Dogras.

Dogras, who come from the Punjab Hills, Jammu and Kashmir, make most valuable soldiers. Descended from the old Katoch branch of the lunar race of Rajputs who were defeated by the solars about 1200 B.C., and who later migrated into the Himalayan foothills; by keeping clear of subsequent invasions of the Punjab plains, they have retained to some extent the habits and religion of the old Aryan stock that populated the bulk of Hindustan. Unlike some of the other Rajput tribes of the northern provinces, to them the Koran made no appeal, nor did the fervour of the Khalsa. Some of their kindred tribes did in the end adopt Islam and, since the name is purely geographical, are still in their own country styled Dogras; but in general parlance the description is mostly only used when referring to Dogra Hindus of Rajput origin.

Dogras have courtly manners, great courage, and strong powers of physical endurance. Under Akbar's conciliatory policy and religious tolerance, the tribes became loyal and efficient military feudatories of the Moghul Emperors. The same may be said when those rulers were replaced by the Sikhs. It is their characteristic to be faithful and loyal to any paramount power which offers them opportunities of honour-

able employment. But although enlisted after 1857 into the regular Indian cavalry and infantry, Dogras did not as a body, prior to the Great War, have the opportunity of acquiring the fame that had come the way of some of the other martial classes.

The 1st Battalion formed part of the river column under Brigadier-General Williamson during the Black Mountain Expedition of 1891, the main engagements of which were not by any means serious, the punishment of the offending clans being carried out without much loss. But the affair was, however, noteworthy on account of the great gallantry displayed by the Dogra company of the 4th Sikh Infantry—also in the column—when suddenly and most savagely attacked.

The active service record of the 2nd Battalion began in the Chin Hills of Burma during the winter of 1889–90, just before their departure for India for re-organization, and so the Waziristan operations under General Lockhart four years later was actually their "première" as a purely Dogra corps.

While these expeditions against the Mahsuds were in progress, the Fort of Chitral, then garrisoned by a detachment of Indian regular troops as well as by Dogra soldiers of the Mahrajah of Kashmir, was besieged by a usurper. In the relief force under Sir Robert Low were the 1st Battalion. The Malakand Pass had to be stormed, and on the following day, in the defeat of the Swatis at Khar, the Battalion bore the brunt of the fighting. The pursuit was pushed on and several engagements with tribesmen followed, but in the meantime Chitral had been relieved by another force under Colonel Kelly, and so Sir Robert Low's force withdrew.

In 1897 the whole of the North-west Frontier became very disturbed. As a result of the instigation of fakirs, the tribes in the Tochi Valley broke the peace by the cowardly attack on a small column engaged on a political mission at Maizar. Then followed the outbreak in the Swat, Malakand being attacked by tribesmen in thousands. The Miranzai next came

into the limelight and heavy fighting took place on the Samana. By the August the Afridis, Mohmands, and Orakzais were all in the ring; and the greatest Frontier "show" till then got well under way. Both the 1st and 2nd Battalions were employed, the latter being transported into the Swat arena embussed in ekkas and tum-tums, the first-line transport being provided by the grass mules of "Probyn's Horse". The operations were conducted during intense heat in which the Dogras became renowned for the rapidity and endurance of their marches.

The 3rd Battalion's first duty after formation was the guarding of Boer prisoners-of-war interned in Fort Govindgarh at Amritsar during the South African War of 1899–02. After which a detachment proceeded on active service as mounted infantry to Somaliland. The whole Battalion went overseas to China in 1904, a duty faithfully undertaken without demur, although it meant parting from homes and families, recently victims of the severe Kangra Valley earthquake. It was entirely left to the men to decide as to whether they would like to be replaced by another unit; but none, save two justifiables, demurred. And so at the end of 1905 the 3rd Dogras were to be found in company with troops from France, Germany, Italy, Japan and Russia—the two latter countries were at war elsewhere—garrisoning the coastal terminus of the Great Wall of China where they remained until 1908.

Six years later the 3rd Battalion were again on the sea. But journeying west instead of east to warfare as strange to them as was the land in which it was fought. The Great War started at a most inconvenient time for the Indian regiments selected for France. During the monsoon months troops enjoy their annual vacation. In the case of holiday-making Dogras, their homes were in many cases far remote from the telegraph or the railway. Nevertheless, but less one company who caught measles, the Battalion by the 5th September was on the way to Europe. Arriving on the last day of October to confront Bavarian troops at Festubert at a spot called "The Glory

Hole"; they were greeted, in addition to the normal method of welcome by the foemen, with propaganda addressed to them as Hindus but couched in Moslem terms. The unnecessary shooting of a cow by the enemy eclipsed any form of appeal to Dogra religious views, and they began to teach the Germans that Indian troops were soldiers possessing the unexpected qualities of loyalty and morale.

The Battalion were personally thanked by the British Commander-in-Chief for their services at Neuve Chapelle; and they remained in France altogether for eleven months. Before they left for Mesopotamia, lovers of animals will be pleased to learn, one of the sepoys received the Distinguished Service Medal for his display of gallantry when rescuing the Adjutant's horse from the debris of a house heavily shelled by the enemy.

After sending large reinforcements to the 3rd Battalion in France, the 1st were sent to Mesopotamia where the reverse at Ctesiphon showed that the Turkish armies would have to be more seriously reckoned with if a decision was to be reached as to who was to hold Baghdad. Arriving in the country in the December of 1915 they were quickly in battle; as were the 3rd who arrived a short month later, to earn high tribute for their gallantry in the action at the Wadi.

Both Battalions fought in different Brigades at Sheik Saad, where the medical officer of the 1st—Captain J. A. Sinton, I.M.S.—himself wounded, earned the Victoria Cross for his devoted attention to those stricken in the battle.

At Hanna the senior battalion supported the junior in the desperate attack over the flat, coverless, bullet-swept plain against sunken loop-holed trenches in broad daylight. No troops in the world would have made headway and only twenty-five untouched valiant hearts reached the enemy line.

During the action Lance-Naik Lala of the 3rd Battalion showed a fine example of the Dogra character. He rescued Captain Nicholson of the 1st and Lieut. Lindop of the 3rd, both of whom had fallen wounded, from ground across which

"A Dogra Havildar"

any form of human movement had been considered fatal. He had also dressed the wounds of many men lying in the open in the pouring rain. Exemplary courage and behaviour which resulted in the bestowal of the Victoria Cross.

The heavy casualties incurred by the Battalions could not be replaced—the remnants of the 1st had been brought out of action by a subadar—and so until besieged Kut fell, the two were temporarily amalgamated as "The Dogra Regiment".

The reorganization following on Sir Stanley Maude's appointment to the command in Mesopotamia sent the remains of the 3rd Battalion back to India, while the 1st, made up to strength in the field, carried on with distinction until the end. Among the first units to enter Baghdad, the flag which flew over the Turkish citadel is now one of the Battalion's treasured trophies.

To repel enemy-inspired outbreaks on the North-west Frontier, as well as to train the enormous number of recruits enrolled in the Dogra group, were the primary duties assigned to the 2nd Battalion early in the War. But oversea orders came at last, although they were disappointing, Aden being the destination. Here the strategy was to show the offensive spirit and to keep "Johnny Turk" at a distance; a trying and glamourless undertaking. After four months of this a move was made to the other end of the Arabian Peninsula where the Battalion joined up with the Palestine forces.

The twenty years that followed the Great War were not by any means a term of peace. The 10th Battalion, prevented by the Armistice from joining in the previous fighting, took part in the third Afghan War before Thal; and later in Waziristan during 1920, where they saw their only really serious engagement, before becoming the Training Battalion, with the additional duty of schooling the Indian ranks required for service as drivers and other combatant posts with British infantry units of the Army of India. Much active service on the northern borders, service of a type far exceeding that experienced in the good old days, has also fallen to all the active units; while

the 1st Battalion served in Burma during the rebellion there in 1931–32.

No unusual distinctions in dress make the Dogras outstanding, their uniform following the normal type of Indian line infantry, the facings being the yellow of the old Bengal regiments from which the senior Battalions are descended.

With the creation of the Indian Territorial Force in 1922—an aspect of the Indianization of the Defence Services which provided for the military ambitions of the classes not generally enlisted into the regular army, or for those who wish to prepare voluntarily to serve their country in time of war—the 11th Territorial Battalion was raised as part of the 37th Dogras. Originally centred at Dharamsala—sixty miles from the railway—recruiting was naturally slow; not only from this cause, but from the fact that the newly created force was looked upon as suspiciously like a scheme to obtain soldiers for nothing. Nevertheless at the first annual training there were 250 men, a number doubled in 1924, and from thence the Battalion has far exceeded the expectations of those who were connected with the Dogras' amateurs at the beginning.

All the Battalions are now centred on the Regimental home at Jullundur, the capital of the ancient Katoch Kingdom, the leading descendant rulers of which—the Mahrajahs of Jamnu and Kashmir, and Mandi, themselves Dogra Rajputs—are honorary regimental officers.

To complete the history of the Regiment, mention must be made of the Waziristan operations, 1936–37–38, in which all three active Battalions took part, the 1st being present throughout.

CHAPTER XLV

The 18th Royal Garhwal Rifles

" Punjab Frontier ", " La Bassee, 1914 ", " Armentieres, 1914 ", " Festubert, 1914, '15 ", " Neuve Chapelle ", " Aubers ", " France and Flanders, 1914–15 ", " Macedonia, 1918 ", " Egypt, 1915–16 ", " Khan Baghdadi ", " Sharqat ", " Mesopotamia, 1917–18 ", " Afghanistan, 1919 ".

GARHWALI, in a military sense, implies the Rajput hill races settled in the Himalayas west of Nepal on the Indian side of the United Province's border. The geographical name means the country of forts; and is certainly truly bestowed, for in Garhwal every large hill-top is crowned with the ruins of an ancient citadel.

Originally enlisted into the Gurkha regiments since 1815 on the cessation of the war with Nepal, these hill-men, often mistaken for Gorkhalis, although they are not so stocky, proved themselves of value as a military class; and as the castes and races best suited to a military career came to be more closely studied, proposals were set in train about 1880 to separate them from Gurkhas. But nothing eventuated until seven years

RIFLEMAN, 1ST BATTALION, ROYAL GARHWAL RIFLES, SUMMER FIGHTING ORDER, 1939

later, when orders were issued that the ranks of the 2nd Battalion of the 3rd Gurkhas then in process of formation were to be composed entirely of Garhwalis. An idea of the afterwards Lord Roberts, who had been impressed by the fact that many of the men in the Gurkha regiments who had merited distinction for bravery were Garhwalis.

Organized at Almora, the new unit were transferred to a wooded hill-top—now Lansdowne—and told to construct for themselves a cantonment. An exceedingly difficult job, and one necessitating excessive and laborious fatigues. As a result recruiting was not too good, and by the time the Battalion proceeded on its first active service in the Burmese Chin Hills in 1890, only six companies were Garhwal men, the remaining two being Nepalis. Whilst on this service the idea of a distinct Garhwali battalion developed, and the six Garhwali companies were separated and became the 39th (The Garhwali) Regiment of Bengal Infantry in 1891. The change was not popular. Still what had to be, had to be, and so dropping the badge of the kukri and adopting instead the phoenix—that oriental fabulous bird that periodically rises from its own ashes each time with renewed vigour—the new corps of Garhwalis took over the old number and went on with their job.

The Chins continued their incursions into Burmese territory. Like all warfare on the northern frontiers of that country, there developed affairs of detachments in which the young soldiers of the 39th greatly distinguished themselves by their steadiness and bravery under extremely trying circumstances. The operations were conducted in the difficult country at a pace of about five miles per day. The Chins, armed with, for them, a very effective lethal weapon in the shape of the old Brown-Bess, were very troublesome enemies; experts at concealment and ambuscade, and thus difficult to deal with in the manner best-liked by the soldier. So the Battalion well-earned the praise their apprentice service had brought.

Although wearing green jackets from the beginning, the new

Corps did not receive the official style of Rifles until 1892. For some years they remained at Lansdowne continuing the building of their barracks and creating a parade ground out of the mountain side. They were the last corps to be founded in the regime of the Presidency Armies which became *no more* in 1895, and thus the next service of the 39th was as part of the newly created Indian Army.

That Army only had to wait two years before the North-west Frontier rose with an unexpected vehemence. In the field-force sent into the Malakand were the 39th Garhwalis; and in the night attack on the perimeter camp at Nawagai, on the 20/21st September, 1897, they, with the 2nd Battalion of The Queen's Royal Regiment (West Surrey), withstood the brunt of that savage desperate attack.

Like all Indian infantry regiments, the Corps consisted of one battalion only. In 1901, a second battalion of Garhwalis were raised at Lansdowne originally as the 49th (Garhwal Rifles) Regiment of Bengal Infantry; but a few months later they were ordered to become the 2nd Battalion 39th Garhwal Rifles, the changed title of the original Battalion bestowed a little previously the same year. The Garhwal Rifles, apart from Gurkhas, were thus the first two-battalion regiment of the Indian Infantry.

To really prove their worth the Regiment had to wait nearly thirteen years. And then when their chance did come they were one of the surprises of the Indian Army.

Both Battalions were mobilized in August, 1914, as part of the Garhwal Brigade of the Meerut Division. Both reached the trenches in Northern France at the end of October. Welcomed by a downpour of rain and a heavy shelling they must have thought, were they aware of Western superstition, that there was something in the thirteen business; for their number was a triple of that supposedly unlucky figure, they had landed on the thirteenth, and were now beginning to incur the first of a long casualty list relieving the British 13th Infantry Brigade.

That first winter in France, after the invader had been stayed, was largely one of defence consolidation against further attempts to break down the allied trench line. Towards the end of November Festubert was heavily attacked, and this involved the Garhwalis. In the defence, Naik Darwan Sing Negi of the 1st Battalion won the Victoria Cross. He showed the greatest of gallantry when the Battalion was engaged in retaking and clearing the enemy out of our trenches, and, although wounded in the head in two places, as well as in the arm, was one of the first to push round each successive traverse in the face of severe fire from bombs and rifles at the closest range. Although not the first Indian to earn the highest decoration for remarkable bravery in the presence of the enemy, he was the first ever to receive it from the hands of H.M. The King-Emperor on the field of battle.

Again the Battalions—it was certainly unique in the British system to find the whole Regiment operating side by side—were in the fighting near Givenchy in the Christmas week of 1914. True, the Corps took part in the further unique and unofficial armistice of the first Christmas Day of the Great War. But this gave them no respite and the casualties began again to grow until the battle for Neuve Chapelle in March, 1915. In that engagement the losses reached such a height that both Battalions were amalgamated on the 1st April and became simply "The Garhwal Rifles".

Another Victoria Cross was won on the 10th March, 1915, by Rifleman Gobar Sing Negi of the 2nd Battalion during an attack on the German position; he was one of the bayonet party with bombs who entered their main trench, and was the first man to go round each trench corner, driving back the enemy until they were eventually forced to surrender. But Garhwal did not see the gallant soldier again for he surrendered his own life later in the battle.

It is recorded that during the same action one of the most impressive sights was a dying British officer being carried,

through all the turmoil and confusion, by four Garhwali riflemen with a quietness and tenderness that was astonishing, while the men looked so smart and clean.

The victory at Neuve Chapelle, the first of Sir John French's attempts to break through the German trench system, was quickly followed by the exploiting attack on the Aubers Ridge, an event unluckily which did not bring the hoped-for results. In retaliation for the British success the enemy attacked at Festubert. The Regiment was heavily involved in both actions.

The usual tenacity was displayed later in the year in the attacks in support of the flank of the troops involved in the battle of Loos. In one operation it was impossible to proceed owing to the enemy fire, but nevertheless 2nd Lieutenant Rana Jodha Jang, attached from the Tehri-Garhwal Sappers, valiantly led his company right up to the German wire. It was at Loos that the only Garhwalis to fall prisoners in the Great War were taken, some seventeen who were cut-off having to surrender.

By now the Indian units in France, save the cavalry, had deteriorated to breaking point, due to the heavy casualties and failure of the Indian reserve system; and so at the end of October, 1915, the Regiment was withdrawn. The Garhwal Rifles, who bore a lion's share of the casualties from first to last, did splendid work and, as Sir James Wilcocks says, "They left a name which will be held in high esteem by all who knew them in France, and not least by the Germans". While Lord Kitchener considered their achievements to have placed them in the first rank of the Indian Corps.

And so via Egypt they returned to India where the two Battalions were re-organized and re-fitted; and where in the following summer was raised the 3rd Battalion.

The following year, March, the 2nd Battalion was sent to the Euphrates operations, where they again showed great dash, and at the capture of Ramadi took 2,000 prisoners as well as the Turkish commander on the spot. At Khan Baghdadi they

were in the unique pursuit of the Turk by troops embussed in motor-cars—the precursors of mechanized infantry.

Withdrawn from Mesopotamian soil in the September of 1918 the 2nd Battalion joined the Salonika force whose contemplated operations for a victory there were frustrated by the Armistice. The Turkish Nationalists having refused to sign the peace treaty, further Allied action became necessary, which sent the Battalion into Asia Minor, where Kemalist troops had attacked our posts and cut the communications. Peace was eventually signed and the Battalion arrived back in India in November, 1919.

The 1st Battalion returned to the War in December, 1917, also being sent to Mesopotamia, to take part in the advance on Mosul. On October 28th–29th, 1918, their forced march of thirty-four miles to reinforce the Indian Brigade barring the retreat of the desperate undefeated Turkish army being encountered at Sharqat was a mind-relieving factor to General Cassells, who intimated that all his anxiety for the situation ceased as soon as they arrived to go straight to the front line. After the Armistice the 1st Battalion remained on in now Iraq, taking part in the Kurdish operations and the Arab Rising, eventually returning to India early in 1921.

The 3rd Battalion went on service in the Khyber during the third Afghan War. While the 4th Battalion raised at Dehra Dun in the October of 1918 were the last troops to enter Thal, prior to its investment by a superior and outnumbering Afghan force under the enemy Commander-in-Chief. The garrison, young and inexperienced, had only half rations for three days and no gun ammunition left. A precarious situation which was ably relieved by a force under General Dyer.

As a result of the Afghan War, and propaganda arising from the Great War, the Waziris began to make serious trouble. An unprecedented Frontier campaign followed; unparalleled both for the severity of the fighting in arduous climatic conditions, and the fact that no British troops took part, except a detach-

ment of Mountain Gunners and the Royal Air Force. Demobilization was going on, the Territorial regiments sent to garrison India during the Great War were in process of relief by newly re-organized regular units, and thus unaided, the Indian regiments, some of them young and unseasoned, had to hold the turbulent frontier—a factor which considerably enhanced their prestige. The 4th Battalion were particularly brought to notice for their magnificent spirit and their eagerness to engage hand-to-hand and assert their moral superiority, qualities which reacted throughout the whole force.

The operations did not conclude before another Victoria Cross had been won by the Regiment on the 2nd January, 1920. During the action on the Spin Ghara Ridge a company was holding an advanced covering position which was repeatedly attacked by the Mahsuds in vastly superior numbers. For over four hours ably led by their commander—Lieutenant W. D. Kenny—they maintained their position, repulsing three determined hand-to-hand attacks in which Kenny was repeatedly to the fore, fighting the tribesmen with bomb and bayonet. His gallant leadership undoubtedly saved the situation, and kept intact the flank upon which depended the safety of the troops in the rear. In the subsequent withdrawal this most gallant officer realized that a diversion was necessary if the rear section, which was impeded by the wounded, was to get clear. And so with a handful of his men he turned back to counter-attack the pursuers, to die with the rest of his party fighting to the last; a gallant self-sacrifice which not only enabled the wounded to be withdrawn, but also averted a disaster which would have involved considerable loss of life. The Victoria Cross was not only a reward to Lieutenant Kenny, but also to the little band that backed him up.

The 4th Battalion were replaced in the Waziristan force by the 2nd who served there till 1923.

The post-war organization having decreed that the 4th should become the Training Battalion, they thus became the

10th and proceeded to their permanent station at Lansdowne; where also is based the 11th Territorial Battalion raised in 1922.

The fanatical rebellion, the result of religious-cum-political agitation among the Moplahs, took the 1st Battalion down to the jungles and ghats of Southern India for nine months in 1921. Operations reminiscent of their first service in Burma thirty years before.

For their outstanding services in the Great War and its aftermath the King-Emperor in 1921 conferred upon his youngest infantry corps the style of "Royal". As a result, on their rifle-green uniform a scarlet twisted cord on the right shoulder is worn with a crown as part of the shoulder-title, the black facings remaining.

PRESENTATION OF COLOURS TO THE 4TH BATTALION, 19TH HYDERABAD REGIMENT AT SECUNDERABAD, DECEMBER, 1937. THIS WAS THE FIRST OCCASION UPON WHICH COLOURS WERE RECEIVED BY INDIAN KING'S COMMISSIONED OFFICERS

XIX

HYDERABAD REGIMENT

CHAPTER XLVI

The 19th Hyderabad Regiment

"Nagpore", "Maheidpoor", "Nowah", "Central India", "Burma, 1885-87", "China, 1900", "Megiddo", "Sharon", "Palestine, 1918", "Tigris, 1916", "Khan Baghdadi", "Mesopotamia, 1915-18", "Persia, 1915-18", "North-west Frontier, India, 1914-15, 1916-17", "East Africa, 1914-16", "Afghanistan, 1919".

THE ancestor of His Exalted Highness the Nizam of Hyderabad and Berar, some two centuries ago, threw off the sovereignty of the Moghul Emperor at Delhi, and declared himself an independent ruler: an action which brought him into direct relations with the British and the French, then endeavouring to establish paramountcy in India. Both countries sought the ruler's favours and, eventually, but not without periods of indecision, Britain, in the shape of the East India Company, received his compact—one of H.E.H.'s titles is "Faithful Ally of the British Government".

To cement this affiliation a contingent of Hyderabad troops joined the Company's forces in the third Mysore War; and in the division before Seringapatam commanded by the afterwards great Duke of Wellington, were to be found four battalions of the Nizam which, he records, "behaved well". The

Hyderabad troops formed the nucleus of his charge and the renowned soldier at the age of thirty proudly wrote, "I commanded the Nizam's Army". It was from this force that began the famous Hyderabad Contingent eventually absorbed into the regular Indian Army in 1903, the surviving infantry of which became the 19th Hyderabad Regiment in 1922.

Although these corps cannot claim to have taken part as such in the war that ended the career of Tippu Sultan, the Nizam's Contingent, as it was known until 1854, was for the greater part formed from the Hyderabad corps that did, and the present Regiment, although numerically the junior of the line infantry of India, can claim to be an older fighting corps than most of the regiments that precede them.

The Nawab of Ellichpur—Salabat Khan—a suzerain of the Nizam and the leading feudal lord of Berar, a territory made over to Hyderabad after the Mysore War, had also maintained levies. These, together with the Nizam's troops, rendered further assistance against the Mahrattas in 1803. A war undertaken largely for the protection of the Hyderabad State, but not without an eye on the French still intriguing for power. The war having ended, a state of anarchy began to prevail in Berar due to the raids of freebooters and the truculence of some of the important landlords, and a situation arose with which the seldom, and badly, paid troops of the Nizam and his under-lord were unable to cope.

The necessity for military reforms became glaringly evident, but all proposals ended nowhere until Mr. (afterwards Sir) Henry Russell became Resident at Hyderabad in 1811. Then, through his energy and continual pressure, a move was made, and in 1813 the Russell Brigade—infantry organized on the regular system—came into being. The two battalions of the Brigade are now the 1st and 10th Battalions of the Hyderabad Regiment, their origin being indicated by the subsidiary title "Russell's" still borne. Formerly the 1st and 2nd Infantry of the Hyderabad Contingent into which the Russell Brigade was merged as the military obligations of the Nizam with the paramount government grew, they became the 94th and 95th Russell's Infantry of 1903 and the Great War.

Having reformed the Hyderabad Infantry, Russell turned his attention to Berar; and from two of the regular battalions of Salabat Khan's Ellichpur Brigade raised in 1797 and 1798 descended the 3rd and 5th Infantry of the Hyderabad Contingent, the later 96th and 98th (Indian) Infantry of 1903, and the 2nd and 4th Battalions of to-day—the former still bearing the distinguishing title "Berar".

The two battalions of the Russell Brigade fought side by side in the Mahratta-Pindari War 1817-18. Both were present at Maheidpoor; a victory noted for the steadiness with which the sepoys, although mowed down by the enemy's guns, succeeded in capturing the batteries; and the approach to action across a difficult river by a single ferry in the face of a strongly entrenched position. The battle was decisive and Holkar's entire camp with his guns and military stores fell to the victors, while his power was irretrievably broken. The Pindari leader Cheetoo, who refused to surrender, ended his life by plunging alone into a jungle infested with tigers, where his mangled remains were afterwards found.

Much now remained to be done in the Nizam's Dominions. The country was full of turbulent elements who opposed with force any attempt to establish peaceful rule. So for nearly forty years the Hyderabad troops were constantly engaged until order was finally secured. The greatest exploit, perhaps, of this time was the reduction in 1819, of the fortress of Nowah, held by the chief of the refractory feudatories. A little-known siege, distinguished for the smartness of its execution as well as the brilliancy of achievement; to say nothing of the gallantry displayed by the men of the Russell Brigade in the difficult and costly storm which gave those battalions, as well as the present 2nd Battalion, whose flank companies also took part, an honour peculiar to them.

Colour-Havildar, 95th Russell's Infantry, 1912

The Regiment, since the Great Mutiny affected all Battalions, rendered staunch service in 1857 and the aftermath of the following years in Central India. And there is no doubt that the loyalty of the Hyderabad Contingent effectively stopped the revolt from spreading over the Carnatic and Mysore. The 10th Battalion suppressed the incipient mutiny at Aurangabad, the 2nd were at the defeat of Tantia on the Betwa River—an anniversary still joyfully celebrated—while the 2nd and 4th displayed stout hearts and sharp bayonets in the difficult assault and storm of Jhansi—another day of regimental rejoicing.

For a further twenty-eight years the Contingent remained in Hyderabad cantoning the jungles, restraining law-breakers, and developing their military zeal; hoping, perhaps, that opportunity might arise which would send them to fight outside Hyderabad—a service for which they were at all times liable.

In 1886, King Theebaw's misgovernment of Burma and his financial troubles, which resulted in an absurd and unjust demand on the Bombay Trading Company, followed by the arrest of their employees, led to an ultimatum from India and war. Theebaw was deposed, Burma annexed, and general disarmament of the country ordered. The latter, a fatal mistake, for all Burma swarmed with disbanded soldiery, the rounding up of which was a more difficult proceeding than the original conquest. Many warriors were required from India, and the Hyderabad Contingent were called upon for their share of this toilsome and irritating duty. So across the Bay of Bengal went the 2nd and 10th Battalions to work split into detachments ranging from one hundred and eighty to eighteen men for almost two years.

The 10th Battalion served on the North-west Frontier in 1897, the men receiving medals, but the unit no honours. A process reversed when the 4th Battalion returned after two years absence in Hong Kong during the Boxer Chinese rising that started in 1900.

In 1902, treaty obligations with His Exalted Highness were

reviewed, and as a result the Hyderabad Contingent ceased to be. Formed for the pacification of the Nizam's Dominions, as well as for the unification of India, the force had played a distinguished and conspicuous part during an era now being fast forgotten. The various corps were not, however, all disbanded, and the bulk became units of the then new Indian Army. As such, all achieved an imperishable record in the great conflict which began the following decade.

And so we come to 1914, whence from the October, the 4th Battalion were in constant operations in East Africa for twenty-six months, returning to join the East Persian Cordon until 1920.

The 1st Battalion held the Persian Gulf detached garrisons until November, 1916, and then policed the Frontier against the Mahsuds until sent to Mesopotamia in 1918, where they had much to do in the Arab and Khurdish Rebellions.

The 2nd Battalion, greatly depleted of trained men who had been sent to France as reinforcements, went early to the Persian Gulf and the Tigris line for the whole of the War.

The original expedition in 1914 to protect the Persian oil supplies for the Royal Navy included the 10th Battalion until May, 1915, when they were returned to the Mohmand 'front' until sent to Mesopotamia for a year, where they served on the Lower Euphrates, and later in the 15th Division at Hillah and elsewhere; to proceed thence for a two-year spell in Salonika and on the Black Sea coast, 1918–20.

The 3rd Battalion of the Regiment, since disbanded, was formerly the 4th Infantry Hyderabad Contingent, later the 97th Deccan Infantry; the honours now borne by the Regiment for the victory at Khan Baghdadi, Palestine, and the third Afghan War, were their contribution.

The Hyderabad Infantry for dress and equipment were assimilated to the regular Indian forces, the full-dress having always been red with green facings. White clothing was allowed for hot-weather wear for the first time in 1838.

(*India Office photo*)
"Machine-Gun Instruction," 19th Hyderabad Regiment

The men after Russell's time were chiefly Hindus recruited in the Company's territories; but standards gradually changed, and up to forty per cent Muslims were later admitted. Foreign recruits were afterwards frowned upon, and in 1879, the enlistment of Sikhs from north of Delhi prohibited, a restriction applied in 1895 to all Punjabis. The ranks now comprise Jats, Ahirs, and Kumaoni Rajputs.

From the latter class is recruited a corps which, although a component of the Hyderabad Regiment, is not numbered with them, and is actually a rifle unit—"The 1st Kumaon Rifles." Originally raised at Ranikhet during 1917 as a battalion of the 39th Garhwal Rifles from drafts sent by the existing battalions of that corps and the Burma Military Police, they shortly after formation became the 4th/39th Kumaon Rifles, and in April, 1918, the 1st/50th Kumaon Rifles. As such they fought at Sharon in Palestine where, with the 2nd Guides Infantry, they stormed three lines of Turkish trenches on September 19th, 1918, arriving at their final objective five minutes under the scheduled time. The Kumaoni proved himself a successful soldier, and the Battalion was retained after the war. They were for a year, 1922–23, part of the 9th Jat Regiment. It seems curious that they do not form part of the corps of the Royal Garhwal Rifles with whom they have affinity, military, territorial and ethnological. Being a rifle unit they dress in dark green with black facings, and thus provide another anomalous example of the Indian regimental system.

CHAPTER XLVII

The 1st King George V's Own Gurkha Rifles (The Malaun Regiment)

"Bhurtpore", "Aliwal", "Sobraon", "Afghanistan, 1878–80", "Tirah", "Punjab Frontier", "Givenchy, 1914", "Neuve Chapelle", "Ypres, 1915", "St. Julien", "Festubert, 1915", "Loos", "France and Flanders, 1914–15", "Megiddo", "Sharon", "Palestine, 1918", "Tigris, 1916", "Kut-al-Amara, 1917", "Baghdad", "Mesopotamia, 1916–18", "North-west Frontier, India, 1915, '17", "Afghanistan, 1919".

GURKHAS may be described, not in any derogatory sense, as the Foreign Legion of India, for, although soldiers of the King-Emperor unexampled for their gallantry and loyalty, they are not his subjects, but those of the adjoining Kingdom of Nepal.

Enlisted under special treaties between India and Nepal, these wiry broad-chested little men were first taken into the East India Company's service towards the end of the Nepal

War in 1815, when parties of captured enemy made over their allegiance to General Ochterlony after the surrender of the Nepalese chieftain Amar Singh and the fall of his fortress of Malaun.

Four irregular Nepalese corps were at first authorized; the first two to be known as the 1st and 2nd Nasiri Gurkha Battalions, the name 'nasiri' implying friendly. These two battalions, the second of which was in 1826 absorbed by the 1st, founded the premier regiment of Gurkha riflemen.

The drums of the Regiment bear an emblazonment for the at last successful siege of Bhurtpore by Lord Combermere in 1826. But as only a detachment of the 1st Nasiri Battalion was present, the honour was denied to them for fifty years. It was thus not the first honour borne. These came for the services rendered in the first Sikh War, particularly at Aliwal and Sobraon, where the soldierly qualities of both Gurkha regiments—the 2nd Gurkhas were also *en division*—excited much admiration.

After the Sikh Wars and the annexation of the Punjab, the regular Bengal regiments recruited from Poorbeahs stationed in the new province began to give much trouble, owing to the withdrawal of certain active and foreign service allowances, to which they considered themselves entitled and to which they had been accustomed. Serious acts of insubordination and mutiny occurred. The worst case of all was when the 66th Native Infantry mutinied in Fort Govindgarh at Amritsar in 1849. Sir Charles Napier, the Commander-in-Chief, determined to act at once. He thus ordered the Nasiri Battalion to Amballa, as well as the mutinous corps, and on arrival the latter's *in toto* discharge from the Company's service; while the colours, arms, accoutrements and stores were to be handed over to "the brave and loyal men of the Nusuree Ghoorka Battalion" who were to assume the title of the disbanded corps. The effect of this move was most striking. Mutiny was squashed. And the disloyal sepoys who up till then had defied dismissal on the grounds that new recruits would be prevented from taking their places, now found themselves discharged with ignominy with no possibility of successful united action, since they were confronted with trained soldiers more martial in their spirit, and thoroughly constant in their allegiance. This instance of one unit completely assuming the rôle and title of another by a single stroke is unique in the history of the British arms.

The new 66th marched for Peshawar, and thence commenced the Gurkhas' long record for endurance and bravery among the barren hills of the North-west Frontier. The Regiment spent the next seven years here taking part in wars against the factious clansmen who wished to measure their strength with the new regime then being established.

The Regiment were at Almora in the border hills of Rohilkand when almost all of the Bengal regiments forswore their fealty. Unlike the other three Gurkha corps then extant, the 1st were not concerned in any of the more well-known actions of the time. Nevertheless they had much to do defending Naini-Tal, then full of refugees, keeping Kumaon and Garhwal free from rebels, and clearing Oudh and Rohilkand. At Choorpura on the 10th February, 1858, Lieutenant J. A. Tytler alone attacked the insurgent gunners hand-to-hand—an exemplary act of courage which earned the Victoria Cross.

The Great Mutiny over, the reorganization took the 66th out of the line and gave a separate numbering to the Gurkha corps; so the Regiment became the 1st Gurkha Regiment with the classification Light Infantry. At the same time the Government decided to allot permanent stations in the hills to the Gurkha regiments which were in effect to become their homes, and to which they were to return on the conclusion of such campaigns or emergent duties on which they might be sent. Dharamsala in the Kangra Valley of the Punjab was selected for the 1st and still remains as the Regimental centre.

From this quiet and beautiful spot the 1st Gurkhas have

PIPER, 1ST K.G.V.O. GURKHA RIFLES—1937

many times been called to the wars, the first occasion being in 1868, when the tribes of the Black Mountain in Hazara created considerable disorder which necessitated an expedition. The resistance of the recalcitrants was very feeble, and the expedition is more noted for the wonderful rapidity with which the troops detailed for service were collected, and the endurance shown in the long and rapid marches during great heat—the 1st Gurkhas covered four hundred miles in twenty-three days.

At the end of 1875 when the Malay States were being consolidated, the British Resident in Perak was murdered. Trouble spread, and so the Governor at Singapore called to India for help. The response caused the 1st Gurkhas to be the first Gurkhas to serve overseas. But by the time they arrived the rising had to a great extent been put down and there was little for the force sent from India to accomplish. Nevertheless, detachments of the Regiment were engaged with Malays, and in one fight Captain G. N. Channer received the Victoria Cross, and two sepoys the Order of Merit, for their gallantry in storming alone the pallisade of the enemy stockade at Bukit Putoos.

The services of the Regiment in both phases of the 2nd Afghan War, although very necessary, were irksome and devoid of excitement, being largely escort duty, with the supply of innumerable guards and fatigue parties on the lines of communication, and later in Kabul. Disease took more toll than battle.

The war over, and with the coming of more settled conditions, the new and friendly Amir—Abdur Rahman—visited India. A great review of the army was held in his honour. But the appearance on parade of the 1st Gurkhas caused some consternation to the Viceroy—Lord Dufferin—who, in common with most, thought that all Gurkha regiments were dressed in green. Not so the 1st. In the early days they did wear the rifleman's colour; but on becoming the 66th they assumed the red coats and white facings of that corps, carried their colours, and were in every way assimilated to the ordinary infantry—

a fact of which they were very proud; so that even the hints of the Viceroy that they might care to wear green resiled. The *Lal kurti paltan* had achieved considerable prestige in Nepal, and the men thought that this would vanish if they replaced their clothing by green; while they were reluctant to give up their colours on which the recruits affirmed their obligations to another sovereign. But the matter was again revived the next year, when extended treaty facilities with Nepal enabled second battalions of the Gurkha regiments to be recruited. The new 2nd Battalion was to wear green, and so the red, not without some heartburning, gave way to the green of the Gurkha Brigade. The fact that the 1st Battalion once wore red is commemorated by the facings of the Regiment. The colours, however, continued to be carried for another three years. They were finally abolished from parade in 1891 when all Gurkha corps were styled rifle regiments.

On the adoption of the green uniform a black helmet was prescribed for the officers; and all attempts to get this altered to the more practical white failed, until King George V took umbrage at this most unsuitable head-gear at the Delhi Durbar in 1912. The then King-Emperor was Colonel-in-Chief of the Corps from 1906 when he visited India as Prince of Wales; hence the plumes of the Prince as well as the Royal and Imperial Cypher being borne in the cap badge. The sub-title "The Malaun Regiment", to denote the origin of the Corps, was bestowed in 1903.

The 2nd Battalion first left the Kangra Valley for active service in the autumn of 1888. The Rajah of Sikhim's infringement of his treaty engagements and the attitude of the Thibetans made necessary active operations in that quarter. Despite their useless weapons the enemy were not wanting in courage, and reinforcements had to be sent to Gnatong—hence the 2nd Battalion's employment. The enemy were soon disorganized and completely beaten at a cost of only five casualties.

A different story to the severe hand-to-hand fighting experi-

Types of the 1st Gurkha Rifles,
Winter Full-Dress Uniform—1935

enced by the 1st Battalion, who were unexpectedly attacked without any provocation when escorting the commission endeavouring to delimitate the Afghan boundary of Waziristan in 1894.

No sooner had the 1st Battalion returned to their mountain home, when the 2nd were ordered down for the great Frontier campaigns of 1897–98: they did all that was expected of them, and more.

Then came August, 1914. At once were sent, in the Sirhind Brigade under orders for France, the 1st Battalion. Their brigade being the first to arrive in Egypt, was detached for awhile to defend the Suez Canal, and so they did not reach action in France until the December. And then their story in France, Flanders, Mesopotamia, and finally in Palestine is that of the gallant Lahore Division. As in the Great Mutiny, in Afghanistan, and elsewhere, it was entirely fortuitous that nothing outstandingly spectacular can be used to illuminate the 1st Gurkha's record. But unostentatiously they gave gallant and ungrudging service in foreign lands to a cause of which the men knew but little; ever ready, ever reliable; a comforting thought to any commander who was privileged to have them.

The 2nd Battalion were in Chitral on the outbreak. They afterwards moved to the North-west Frontier for the whole of the War, to see considerable action, as well as suffer much privation, owing to abnormally long spells in great heat, in a country for which the natural physique of the Gurkha is unsuited.

The 2nd Battalion fought in the Khyber actions of the third Afghan War; while a 3rd Battalion raised during 1917 defended Fort Sandeman against the Mahsud attacks of the aftermath.

Normally one battalion of the Regiment remains in Dharmsala, whilst the other is on the Frontier: and it may be said that in all campaigns since 1919 in the north-west of India, including the internal security measures due to the political

situation in which the Gurkha has not the slightest interest, members of the Regiment have played some part.

Sir Hugh Gough, getting on for a century ago, recorded that the Gurkhas armed with the short weapon of their mountains were a terror to the enemy. They have always been so, and will be a terror ever to be reckoned with by those who disturb the King-Emperor's peace, which, with fidelity they have sworn to preserve.

Gurkha Officer with the Truncheon, 2nd Gurkha Rifles

CHAPTER XLVIII

The 2nd King Edward VII'S Own Gurkha Rifles (The Sirmoor Rifles)

"Bhurtpore", "Aliwal", "Sobraon", "Delhi, 1857", "Kabul, 1879", "Kandahar, 1880", "Afghanistan, 1878–80", "Tirah", "Punjab Frontier", "La Bassee, 1914", "Festubert, 1914–15", "Givenchy, 1914", "Neuve Chapelle", "Aubers", "Loos", "France and Flanders, 1914–15", "Egypt, 1915", "Tigris, 1916", "Kut-al-Amara, 1917", "Baghdad", "Mesopotamia, 1916–18", "Persia, 1918", "Baluchistan, 1918", "Afghanistan, 1919".

ALTHOUGH the sterling military worth of the Gurkha had been well proved in India years before, it was not until the Great Mutiny that his fame became world-known. And it fell to the lot of the 2nd Gurkhas to be the principal tribal publicists on account of their outstandingly remarkable feats before Delhi during the summer of 1857.

Raised in 1815 at Nahun in Sirmoor—hence their second title "The Sirmoor Rifles"—the Regiment is the second of the

original four friendly corps formed from Nepalese subjects, who took service with the Indian Government after the first phase of the war brought about by the constant raiding of Indian territory by Gurkha forces.

Sirmoor being early found unsuitable as a cantonment, the Battalion was moved to Dehra Dun, thence to be the permanent Regimental centre. And from there as warriors of the East India Company, the Sirmoor Battalion came down to the plains to join the Grand Army for the Mahratta War of 1817–18 in which they saw but little fighting. So it was not until seven years later when the three companies of the Corps joined in the siege of Bhurtpore, that the first honours for prowess on the field of battle were earned.

The Sirmooris again descended when the Khalsa forces lost restraint in 1845, and by a tremendously forced march, undoubtedly saved Ludhiana. Dressed in green, contrary to the usual practice of corps so garbed, they carried colours. At Aliwal these were nearly shot away, thus indicating the fierceness of the fight, the Regimental colour temporarily falling to the Sikhs, from whom it was recaptured by a small party under a havildar, who cut their way into the enemy lines with most exemplary gallantry. After Sobraon, like the Nasiri corps, the Sirmoor Battalion received much encomium, the Commander-in-Chief feeling obliged to comment upon the determined hardihood and bravery with which the men had invariably met their opponents.

And so to Sunday, the 11th of May, 1857, when the long expected general mutiny of the Bengal regiments began at Meerut. By the 14th idem the Sirmoor Battalion were ordered on service. Taking just what they could on their backs, with two elephants to carry the reserve ammunition, they marched to make history by being the first Indian regiment to level a gun at the mutineers, and the only Indian unit to take part in the opening actions of the siege of Delhi.

Before the Delhi ramparts began the Regiment's long associa-

tion with the King's Royal Rifle Corps of Britain, whose red-faced green coats the 2nd Gurkhas now wear in token of the great days, the two, together with the Guides Infantry, spent holding the main piquet at Hindu Rao's House until the end of the siege. Subject to constant fire for over three months, never once were they relieved from their post, during which time they defeated no less than twenty-six different attacks on what was the key to the besieger's main position.

Even in those times 'red tape' played its irritating part—a recommendation that a superb act of gallantry should be merited by a bestowal of a Victoria Cross being rejected because the report was written in pencil.

After Delhi, the Regiment were employed in Oudh, and altogether they suffered more heavily than any other corps employed in the Mutiny campaign.

Such great services could not go unrecognized, and the Sirmoor Battalion were, after the custom of the time, granted the honour of bearing an extra colour. Later, also as a reward, the Corps became "The Sirmoor Rifle Regiment." An honour which meant giving up the carrying of colours. But Queen Victoria, determined that army usage should not allow the mark of appreciation to disappear, devised a truncheon to be carried in place of the surrendered colours, which was to be honoured in the same way as are the emblems of the line infantry. The truncheon, made of bronze, is six feet in height, and surmounted by three replicas of Gurkha soldiers of the period supporting a crown of silver inscribed "Main Piquet Hindu Rao's House, Delhi 1857". An emblem which is unique in the Armies of the British Commonwealth, if not of the world. Upon it all recruits now swear their allegiance to the soveriegn of another country with whom their own king has been in close and faithful alliance for over a century.

The Regiment were employed in the Mohmand expedition of 1854, the Black Mountain of 1868, and on the North East Frontier in 1871–72. In the latter campaign against the Chins

and Lushais, before Lal Gnoora's stockaded village, Major D. Macintyre's gallantry earned for him the Victoria Cross.

Further honours came their way when King Edward VII, as Prince of Wales, became the Colonel, an association which gave them as emblazonments that monarch's imperial cypher, and earlier his principality's plumes. As "The 2nd Prince of Wales' Gurkha Rifles" they were the first of their kin to visit Europe, being the only Gurkhas selected for the Indian garrisons in the Mediterranean during the Russo-Turkish War in 1878.

The Corps had just returned from what was a pleasant, but for them by no means unexciting duty—they were nearly shipwrecked in the Red Sea—when the Afghan War started. Sent up to the Khyber, they joined in the columns which entered the Bazar and Bara backwaters to deal with the tribesmen who persistently interfered with the army's communications. In the second phase of the War, selected from the Kabul garrison, they marched to the relief of Kandahar. In the final battle their joint charge with the Gordon Highlanders upon the Afghan artillery decided the day. One excited little Gurkha reaching the guns first, sprang on one, and loudly proclaimed amidst the din, his Regiment's ownership; and to show there should be no possible doubt about it, promptly pushed his cap down the muzzle.

The 2nd Battalion of the Regiment came into existence in 1886, their first service being in the expedition against the primitive tribes in Northern Burma in 1888–90.

The storm of the Dargai heights made world-famous by the exploit of the Gordon Highlanders was also the outstanding feature of the 2nd Gurkhas' part in the Frontier outbreaks of 1897. An adventure which cemented the friendship between the two regiments, begun at Kandahar not quite twenty years before.

During these border wars, various attempts were made to introduce the turban as a headdress for Gurkhas, the small Kilmarnock cap with its khaki cover and neck shade being useless in great heat. But Gurkhas dislike *pagris*, for in their social order it is not the dress of the worker; and so later, when the 1st Battalion was sent upon blockade duties in Wazirstan in 1901, the now familiar Kashmir 'slouch' hat was tried out in the Battalion, with results that have made it the fashionable battle wear of all Gurkha regiments since.

The contribution to the Allied cause in the Great War began with the 2nd Battalion's departure for France with the Meerut Division in September, 1914. Thence, when the Indian infantry left for other theatres, the Suez Canal and Egypt, to return to India in 1916 for duty on both N.E. and N.W. Frontiers, Afghanistan, and the Mahsud 'pacification' that followed.

The 1st Battalion—which had been on active service against the bow and arrow armed Abors in Upper Assam during 1911–12—did not leave India for overseas until February, 1916, when the situation in Mesopotamia called for urgent reinforcements. The Battalion fought in all the principal actions which eventually led to the fall of Baghdad—a place the little hill-men, after inspection, were surprised should have been worth fighting for. Later the Battalion went on to the coast of the Caspian Sea with the idea of keeping revolutionary Russians out of Persia. From whence, after undergoing much hardship and fighting many small actions, they were withdrawn in 1921.

"The 2nd King Edward VII's Own Gurkha Rifles" are always a source of interest to Caledonians, for their "pill-box" cap is adorned with a diced border similar to that worn by most Scottish regiments on their headdresses. It was, of course, introduced by a Scotch Colonel, but whether as just mere fancy—a good deal of liberty with uniform was allowed in the old days—or to mark the memory of some long ago particular association with a Highland regiment is now unfortunately obscure.

RIFLEMAN, 3RD GURKHA (RIFLE) REGIMENT, 1897

CHAPTER XLIX

The 3rd Queen Alexandra's Own Gurkha Rifles

"Delhi, 1857", "Ahmed Khel", "Afghanistan, 1878–80", "Burma, 1885–87", "Chitral", "Tirah", "Punjab Frontier", "La Bassee, 1914", "Armentieres, 1914", "Festubert, 1914–15", "Givenchy, 1914", "Neuve Chapelle", "Aubers", "France and Flanders, 1914–15", "Egypt, 1915–16", "Gaza", "El Mughar", "Nebi Samwil", "Jerusalem", "Tell-Asur", "Megiddo", "Sharon", "Palestine, 1917–18", "Sharqat", "Mesopotamia, 1917–18", "Afghanistan, 1919".

TWO Gurkhas of this Regiment, the junior of the original corps founded in 1815, represent so far the ranks of the Gurkha Brigade in the most gallant order of the wearers of the Victoria Cross.

Rifleman Kulbir Thapa, the first Nepali to be so honoured, won his during the battle of Loos on the 25th September, 1915. Himself wounded, he remained for a day and a night with a badly wounded British soldier behind the first line of the German trenches. Continually urged to save himself he would

not do so, and in the mist of the following morning brought his protégé through the enemy wire to comparative safety. Then, leaving him, he went back for two of his own countrymen, bringing them in one after the other. Finally ending his adventure by again going back for his British comrade to carry him most of the way in broad daylight under enemy fire.

The second Cross was earned in even more daring circumstances in Palestine on the 10th April, 1918. The official report, not usually given to seemingly extravagant language, described the award as being for most conspicuous bravery, resource in action under adverse conditions, and utter contempt for danger.

During an attack, Rifleman Karanbahadur Rana with a few other men, succeeded under intense enemy fire in creeping forward with a Lewis gun in order to subdue an enemy machine-gun which had caused severe casualties during previous attempts to put it out of action. The leader of Karanbahadur's party on opening fire was at once shot dead. Without any hesitation our hero pushed the dead man off the gun, and in spite of the bombs thrown at him and the heavy shooting from the flanks, he opened fire and knocked out the machine-gun crew. Karanbahadur then silenced the fire of the enemy in front of him, showing the greatest coolness in removing defects which on two occasions prevented the gun from working. Throughout the day he did magnificent work, and when a withdrawal was ordered, assisted with covering fire until the enemy were close upon him. And although the official report does not mention it, he also rescued his Company Commander, himself a V.C. This gallant son of Nepal received the Victoria Cross at the hands of His Majesty the King-Emperor at Buckingham Palace on the conclusion of the special peace march through London by representative detachments of India's Army on the 2nd August, 1919.

The foundation members of the corps were styled "The Kumaon Battalion"; they were not all pure Gurkhas, but largely men from Nepalese feudatory districts, such as Garhwal and that from which they took their name, who had lost heart in the Gurkha cause after the first phase of the Nepal War. Descendants of these originals born and bred in the regimental home at Almora, known as "line boys", later formed a goodish proportion of the ranks.

For forty-one years the Regiment policed the Nepal border, and not until the Great Mutiny did the soldier's hope eventuate. By then the Regiment had moved down to Rawalpindi from whence in June, 1857, they started off for Delhi in the interest of peace and good order.

Unable to move with the same rapidity as the Guides had done the previous month, since they were charged with the escort of a three-mile long artillery and ammunition train, to say nothing of large sums of treasure, their march, nevertheless, is worthy of comparison with other remarkable troop movements of those times as it called for the highest discipline, endurance and courage in very great heat for which the hillman is most unsuited.

The Corps reached Delhi at "the crack of dawn" on August 1st to go straight into a heavy action. After the storm of the City, in which they were detailed to assault the Kashmir Gate following its epic demolition by the Bengal Sappers and Miners, the Corps moved down into Oudh where their operations finally terminated in February, 1859.

Except for the uninteresting dull campaign against the Bhutanese in 1865, but nevertheless one of great hardship and much sickness, the Regiment were not required to leave the Himalayas until the unreasonable hostility of the Amir of Afghanistan to India—to whom he owed his throne—led to war. In the first phase (1878–79) the 3rd Gurkhas were sent into the country via Quetta. The rough nature of the ground broke down the gun-bullocks, so the Gurkhas had to do the pulling. And it was only due to their stout hearts and strong arms that the "forty-pounders" reached Kandahar. The troops there were eventually ordered back to India, but—as

we say to-day—owing to a deterioration of the situation they were to come via Kabul. They did, but not before a fanatical host of Afghanis endeavoured to bar their path at Ahmed Khel. These *ghazis* very nearly succeeded, and would have done so had not the 3rd Gurkhas formed rallying squares and checked for a time the reckless rush of the murderous religious maniacs who could not even be stayed by case-shot from the field artillery.

As will be noted from the battle honours the Regiment took part in the third Burmese War, their theatre of action being in the Shan States.

Unlike those of the senior regiments of Gurkhas the 2nd Battalion was not raised until 1891, difficulties having arisen through a desire of the then Commander-in-Chief—the afterwards Lord Roberts—to have the new Battalion comprised entirely of Garhwalis; and as we have seen previously, the Royal Garhwal Rifles started life as the 2nd Battalion of the 3rd Gurkhas. And also unlike the senior regiments, the 3rd Gurkhas have a different centre for each Battalion; Almora for the 1st, and Lansdowne for the 2nd: difficulties over sanitation having caused the separation of the Regimental home in 1894.

A diversity of tasks come the way of India's soldiers. One example is the reconnaissance made along her borders in 1889. At this time the authorities were becoming somewhat worried by the constant "intelligence" that the Thibetans were encroaching into Indian territory, and constructing a fortification on the wrong side of the border. Relations with the Lhassa government at the time being not too cordial, Simla ordered a composite double-company of both Battalions to ascertain and report. They did so; and in so doing crossed the Himalayas at a height of 18,000 feet—probably the greatest altitude ever reached by an organized body of disciplined troops whilst on the march, and that, too, a march of the greatest severity. No fortifications were found, however, only a camping-ground wall

erected to give shelter on the trade-route against the prevailing strong and merciless wind.

The 1st Battalion served in the Tirah and adjunct operations of 1897. And in Waziristan, 1901–2, where the regimental scouts put in some very good work, and which, incidentally, as operations were a very severe test on the men.

Another severe ordeal was that set by Lord Kitchener to test the efficiency of the regiments of the Army in India in 1905. When the final result came through, the great man was at Almora inspecting the Battalion. They had come out second. The now 5th Battalion of the Baluch Regiment had won. Without decrying the efforts of the winners, and as consolation to the runners-up, one of Lord K.'s few indiscretions may be recorded. After breakfast, not dinner, came the utterance:—"Having now seen both battalions, had the result been left to me the cup would have been the Gurkhas'".

Football as a sport among the Indian Army came in via the Gurkhas. The 3rd played it over fifty years ago, but the chronicles of the time indicate that the game was somewhat capricious, with indefinite rules which varied according to the fancy of the senior officer playing, while the sides consisted of any number with the ball oval or round, according to whichever was handy.

In 1907 King Edward VII decided to confer a mark of appreciation on the Regiment and made them "The 3rd Queen's Own Gurkha Rifles". His Queen, however, disliked the title as being too impersonal, there having been so many queens. And so the Corps became "Queen Alexandra's Own". That gracious Danish princess, whose cypher now forms their badge, took a very great personal interest in her Regiment. Even to the extent of intercession when another of those many efforts to put Gurkhas into *pagris* was attempted.

The limits of the old-world Great War journeyings of the 2nd Battalion have been indicated in the story of the V.C.s.

To Palestine also proceeded in 1918 a war-raised 3rd Battalion. In this theatre a volunteer detachment from both Battalions served under Lawrence of Arabia. They did sterling work mounted on camels, of which they became skilled riders in strong contrast to the effort to turn Gurkhas into mounted infantry in the pre-war days.

In the original planning of possible expeditionary forces from India in the event of major war, it was ruled that both battalions of a Gurkha Regiment were not to proceed overseas, since owing to the limitations placed upon recruitment in Nepal the replacement of wastage might be of some difficulty. So the 1st Battalion did not proceed abroad until the end of 1917. Nevertheless, they had four adventurous years in Mesopotamia.

Nepal, however, did not become a close preserve for the recruiting officer and some 200,000 Nepalese subjects served the King-Emperor between 1914 and 1919, some 20,000 of which were fated never to return. In addition several of the crack regiments of the Nepalese Army served in Indian garrisons on internal security duties during that time. Thus was made possible the raising of the 4th Battalion who represented the Regiment in the Afghan War of 1919.

Many conundrums are asked, many enigmas stated, concerning soldiers. But of those old warriors and others who delight in doggerel histories, few could answer the question as to which regiment in the Great War stopped in the heat of a serious action to cheer the exploit of one solitary little man. It was the Rifle Brigade, whose black facings and uniform the 3rd Gurkhas copy; and the incident took place at Neuve Chapelle in 1915, when Rifleman Gane Gurung of the 2nd Battalion proudly marched eight fat German prisoners, whom he had cowed into submission, towards the advancing British line. He had outstripped his comrades attacking from another direction, and the sight of triumphant joy on the Gurkha's face was too much for the British green-jackets, and created an incident quite unique in history. Never before, to quote Sir James Willcocks, "had the contemplation of the deed of a single individual Gurkha been heartily cheered by a British regiment in action".

CHAPTER L

The 4th Prince of Wales' Own Gurkha Rifles

"Ali Masjid", "Kabul, 1879", "Kandahar, 1880", "Afghanistan, 1878–80", "Waziristan, 1895", "Chitral", "Tirah", "Punjab Frontier", "China, 1900", "Givenchy, 1914", "Neuve Chapelle", "Ypres, 1915", "St. Julien", "Aubers", "Festubert, 1915", "France and Flanders, 1914–15", "Gallipoli, 1915", "Egypt, 1916", "Tigris, 1916", "Kut-al-Amara, 1917", "Baghdad", "Mesopotamia, 1916–18", "North-west Frontier, India, 1917", "Baluchistan, 1918", "Afghanistan, 1919".

THE 4th Gurkhas must not be confused, as they sometimes have been, with the hill corps which, through temporary insubordination at Jutogh at the beginning of the "troubles" in 1857, scared a good deal of Simla into the jungles and delayed the departure of the Commander-in-Chief for the scenes of action. That corps was the New Nasiri Battalion raised in replacement of the original when the latter, now the 1st Gurkhas, became the 66th regular infantry of Bengal in 1850. The trouble, let it be said, was quite minor and the New Nasirs later did good work, keeping open the Punjab-Delhi road until

they were dispersed among other corps when the new era of Indian government began in 1861.

With the object of holding the Kumaon Hills, the authorities ordered to be raised at Pithoragarh during the same summer, another corps to be known as the Extra Gurkha Regiment, the skeleton being found by the present 1st Gurkhas. Clothed in *dusuti* lined with blanket cloth, and armed with old muzzle loaders the new corps' first action was at Choorpura. Musketry was primitive, the men firing ten rounds a year, four hits in ten being considered very good shooting. The first proper uniforms to be issued in 1858 were khaki, which was replaced by the more familiar green in the following year.

The transference of the Indian forces to the Crown brought the Extra Regiment into the Bengal Army proper as the 19th of the infantry, but this a few months later gave place to the present basic title and number—the style "Prince of Wales' Own" coming as an honour sixty-three years later in 1924.

The Regiment were but two years soldiers of the Empress when they were sent to join the Ambeyla Expedition in the winter of 1863; a campaign which was to be renowned for the severity of the fighting for many decades, the commander of the gallant 9,000 employed—Sir Neville Chamberlain—being himself wounded in action. At this time the Regiment had a British Sergeant-Major who with many of the Gurkha ranks distinguished themselves in the attack on the Conical Hill.

Further active service was seen against the Black Mountain tribes in 1868, and on the other side of India in the Chittagong column during the Lushai operations of 1871–72.

The peregrinations of the Regiment in the second Afghan War covered a wide field: first with Sam Browne at Ali Masjid; then on the Khyber line and into the Bazar Valley; and later in the operations based upon Jalalabad. In the second round they were brought up to Kabul and took a brilliant share in the fighting around the Afghan capital. They finally marched

the 320 miles to Kandahar—a memorable feat of 10,000 fighters and 8,000 followers which not only required good leadership, but the most thorough co-operation and effort from general to the youngest bugler. Lord Roberts, who achieved world fame through this epic, considered, however, that his advance on Kabul via the Kurram Valley in the early winter of 1879 to have been a more difficult and dangerous military operation. But the march to Kandahar captured the public imagination, due no doubt to the presence of important pressmen who accompanied the column.

The work of the 2nd Gurkhas was still by no means over, for they were sent into Baluchistan to chastise the truculent Marris who had been interfering with the Army's convoys during the war; operations which entailed great hardship in the year 1880.

The Regimental story must now be divided, since the 2nd Battalion came into existence at Bakloh, the Regimental home near Dalhousie, in 1886. Their first service was on the eastern borders with the Chin-Lushai force in 1889–90, and again in the Chin Hills in the next winter. They also had a sharp fight with rebels during the Manipur outbreak in 1891.

Both Battalions were at work on the North-west Frontier during 1895; the 1st in the Wana column of the Mahsud punitive force, the credit for which in the official story of the Bengal Army is wrongly given to another regiment; the 2nd in the expedition sent to relieve the Fort of Chitral, they taking part in the fight which forced the strongly-held Malakand Pass and in the following actions of the flying column. The same component of the Regiment were very active sharers in the actions of Gaselee's Brigade during the Tirah campaign of 1897, several Orders of Merit being won for conspicuous gallantry by the men.

Also under General Gaselee, the 1st Battalion sailed to join the forces of the nations then assembling in China, as a result of the activities of an ancient Celestial society known as the

Boxers, which had been for some time carrying on an agitation against foreigners. Their past efforts had been suppressed by the Manchu Emperors, but like most associations of the type the society was not exterminated. Unfortunate missionary interference in affairs political led to a recrudescence of subversive activities, and the Boxers set out to destroy the foreigner and all his works, objectively that their own countrymen might pursue their own ways and evolve their own salvation. All very praiseworthy in a way, but as history time and again has shown, the patriots rapidly developed into fanatical assassins. The situation got beyond the control of the Chinese authorities to such an extent that the principal world powers had to dispatch their own troops to put down the disturbances, and protect the lives and property of their subjects. And so America, Australia, Britain, France, Germany, India, Italy, Japan and Russia all sent armed forces which took part in naval and military operations in the North of China and the Yangtse Valley between June and December, 1900. Thus the 1st Battalion have served farther east of Khatmandu than any of their brother Gurkha regiments.

During the Somaliland campaign of 1903 Captain Walker of the Regiment won the Victoria Cross for outstanding gallantry when serving with the Bikaner Camel Corps.

The 1st Battalion were the Regimental representatives on the Western Front during the Great War. They were, however, unlucky in being in the Sirhind Brigade which, owing to the situation in Egypt and on the Suez Canal, was detached from the Lahore Division to hold the fort until Territorial troops could be dispatched from Britain. Although fate had made them too late to earn the 1914 Star, they were not too late for the 1914 fighting, the Battalion having been relieved in time to reach the Givenchy trenches early in the December.

Before Givenchy the most severe hand-to-hand fighting took place after the enemy had mined our trenches. The Battalion was forced back—no amount of bravery could have prevailed

against the tremendous odds, and to remain would have meant annihilation without the attainment of any good purpose.

The Battalion showed their mettle at Neuve Chapelle, the anniversary of which is now a day of celebration; they very modestly disclaim, however, any share in the after battle of Aubers for which they received the honour only because they happened to be part of a higher formation the other units of which participated.

When the French troops broke before the use of gas as a weapon of war in what later became known as the second battle of Ypres, the 1st/4th Gurkhas were literally flung into a defensive line which simply did not exist. They excelled themselves, and many deeds of daring, gallantry and devotion were performed in that battle. But the fountain of honour trickled slowly in the early days of the War, and in the heat of the fighting responsible officers are too harassed to record the individual deeds they see, or else are killed before they can report them. The bravest deeds are often performed unseen, and there is no doubt that many Victoria Crosses were earned in those critical days when men went into action with the only protection against asphyxiation being a damp rag.

The call for Gurkhas now came from the Dardanelles. Sir John French could only spare one battalion, so he sent the 1st/4th. Made up to strength by welcome drafts from the 2nd Gurkhas, the Burma and the Assam Military Police, they reached the Gallipoli Peninsula on the 7th September, 1915. They were too late to be of use in turning the tide of misfortune—evacuation had almost been decided upon—and so they took no part in any great action in that theatre.

After a spell in Egypt the 1st Battalion returned to the North-west Frontier theatre in 1917, and participated in another Marri punitive expedition in 1918. The next year they were with the brigade that stormed the Afghan fort of Spin Baldak in return for the unprovoked attack on India.

The 2nd Battalion went to the Tigris in time for the last

but unsuccessful attempt to relieve besieged Kut-al-Amara which unhappily fell on the 29th April, 1915. During the ensuing summer little could be done, but when General Maude began his series of masterly operations which finally drove the Turks from Kut, the 2nd/4th were well in the van, so much so that at the attack at the Dahra Bend they were allowed to carry on alone with the result that a "bag" of nearly four hundred prisoners, including three regimental commanders, was sent back to the rear. Their Brigade was the first into Baghdad, where the Battalion were left to keep order—Baghdad Day is their anniversary holiday. Later operations were those around the Khalis Canal, the advance up the Adhain, and the third action at Jabal Hamrin. This finished their Mesopotamian career and they were despatched overseas to the Balkans, from there to the Caucasus, thence to Trans-Caspia, finally ending up in Constantinople.

Space forbids the details of the operations of the Regiment in keeping the Borders in the recent era, but it can be truly said that in them the most world-travelled regiment of the Gurkha Legion again enhanced their reputation.

RIFLEMAN
5TH GURKHA REGIMENT, P.F.F.—1890

CHAPTER LI

The 5th Royal Gurkha Rifles (Frontier Force)

" Peiwar Kotal ", " Charasiah ", " Kabul, 1879 ", " Kandahar, 1880 ", " Afghanistan, 1878–80 ", " Punjab Frontier ", " Helles ", " Krithia ", " Suvla ", " Sari Bair ", " Gallipoli, 1915 ", " Suez Canal ", " Egypt, 1915–16 ", " Khan Baghdadi ", " Mesopotamia, 1916–18 ", " North-west Frontier, India, 1917 ", " Afghanistan, 1919 ".

THE development of scouting as a military science is generally attributed to Lord Baden-Powell, who is said to have developed the idea during the Matabele campaigns in Rhodesia. But the principles were age-old to frontiersmen the world over, and would thus be natural to the Gurkha. And since the 5th Gurkhas were originally formed for frontier protection, it would be innate that the evolution of the modern military pathfinder should have come to the Indian Army through them. "B-P's" idea became the basis of a world youth movement, though few to-day realize that the much ridiculed in the early days nether garments now universally worn by the Boy Scout were introduced to circles martial and youthful by the Regimental Scouts of the 5th Gurkha Rifles during the Tirah campaign in 1897–98.

Not for many years, however, did shorts receive official approval as an article of uniform dress. And the story is told

that a certain Commander-in-Chief after the Great War gave serious consideration to abolishing from the soldier's wardrobe what he considered most unmilitary garments.

Cradled at Abbotabad to hold the Hazara frontier, the basis of the Regiment was the 25th Punjab Infantry (The Hazara Goorkha Battalion) founded by transferring hillmen from the existing regiments of the Frontier irregular units. This was in 1858. Their baptism as soldiers came in 1860 when 3,000 Mahsuds tried a fifteen-minute cold-steel dawn scrimmage with the defenders of the Palesina camp. A terrible hand-to-hand business, which resulted in heavy casualties to both sides before the tribesmen were put about. The following year they became the 5th Gurkha Regiment.

During their first major campaign—the 2nd Afghan War—, the 5th brought themselves greatly in the eye of the afterwards Lord Roberts, with whose column in the Kurram Valley the Regiment operated. Side by side with the now Seaforth Highlanders they fought in many brilliant actions; particularly in the turning movement which shook the entrenched Afghans from the Peiwar Kotal. A dawn attack which brought the Victoria Cross to Captain John Cook for his signal gallantry in leading the charge under terrific fire. Seven Orders of Merit were won by the men in this, and the following rear-guard fight through the Mangiar defile, when the steadiness of the Regiment warded off continued vigorous attacks for five hours without losing a single load of baggage. The same standard and comradeship with the Highlanders was kept up in the second round of the war to the great impression of "Bobs Bahadur" who, to show his esteem when later raised to the peerage, chose for the heraldic supporters of his coat of arms two hillmen—a Seaforth Highlander and a 5th Gurkha.

The renowned soldier afterwards became Commander-in-Chief. As a result of his policy of eliminating from the ranks the not so martial classes and their substitution by more war-like men so that the Indian Army could be as perfect a fighting

machine as possible for the money available, second battalions for the then existing Gurkha regiments were raised in 1886. This policy, the commencement of which is often wrongly attributed to Lord Kitchener, is still carried on to-day; and whatever political value the enlistment of classes to whom a military career is not inherent may be, it is entirely overshadowed by the economic necessity of having the best possible army at the lowest possible price.

The field apprenticeship of the 2nd Battalion was served in the Hazara Black Mountain Expedition of 1891. The same year a detachment of the 1st was operating among mountain snows and glaciers on a part of the Frontier not usually the scene of warlike operations. In the far northern corner of Kashmir were two truculent tribes, the Hunzas and the Nagirs. Both were feign to recognize the authority of the Kashmir Durbar and were making overtures to Russia and China. Usually at daggers drawn, this time they made common cause. An opportunity was wanted to try out the new Kashmir Imperial Service Troops. Here it was. Nevertheless, a smattering of regular Indian Army soldiers was considered necessary, hence the Gurkhas. The tribes were subdued after their almost inaccessible positions had been valiantly stormed. And the somewhat unique campaign records closed with the reward to two officers of the 5th Gurkhas—Lieutenants Guy Boisragon and J. Manners-Smith—of the Victoria Cross. The latter climbed 1,500 feet of precipitous cliff surmounted by fortifications, it being due to his skilful leading that the position was captured.

During the Frontier stormy years of 1897–98, only a wing of the 1st Battalion saw service in their proper sphere. But the two Battalions supplied the first organized scouting body ever used in an order of battle in India. Scouting had been developed in a previous expedition, and during the Tirah campaign the Gurkha Scouts fairly came into their own. Born cragsmen, crack shots armed with special rifles, well-accustomed

to guerrilla tactics, they completely beat the tribesmen at their own game. So much so, that before the campaign ended a whole battalion had been formed to which had been added men from the 1st and 3rd Gurkhas. The experiment was repeated in the Afghan War of 1919 with detachments from all the Gurkha Regiments.

The 2nd Battalion enjoyed themselves, it seems, during Sir James Willcocks' "week-end" campaign of 1908. The big General Sahib was somewhat taken aback by the presence of a smart little rifleman who approached with the request that an important letter to the wife should go with the General's. The free field post-office services did not lend themselves to appreciation by the little man. It did not use stamps—there seemed to be a catch in it—and the Gurk' was not going to be caught. Greatly amused, Sir James took the letter.

The Suez Canal and Gallipoli honours come from the 1st Battalion. Sent early to Egypt they shared in the attack on the Turks approaching the Suez Canal at El Kubri. Then, after Sir Ian Hamilton's urgent requests to Kitchener for Gurkhas had at last received attention, they went on to the Dardanelles. All the world knows what the Gurkha Brigade did on the Gallipoli Peninsula. There the 1st Battalion of the 5th remained fighting game until the evacuation. They had the post of honour in the finale of that disappointing but glorious episode, an officer and twenty-five men of the 1st/5th being the last to leave at Anzac.

One little incident on the Peninsula is worth an outstanding place. Naik Dhan Singh Gurung, in charge of a covering party in an action that went agley, was captured by the Turks. Seizing his opportunity, he wrested himself free and jumped over the cliff, and in luck reached the sea beach. Out of luck, he was again re-taken. But that did not disturb him; he again broke loose and plunged into the sea. Weighed down by his clothing and equipment and under a heavy rifle fire, out into the blue he swam; and then turning parallel to the coast continued on until he landed within the confines of the British lines and duly reported for duty.

After the close-down in the Dardanelles the 1st/5th returned to India to the Mohmand Blockade. And after a spell there refurbishing, to the Tigris force in 1917, where they joined in brigade with the 2nd Battalion, who had been serving in that theatre since the year before.

A 3rd Battalion existed from 1916 to 1921. This took part in the third Afghan War, and in the Arab rising in Iraq during 1920.

The services of the Gurkhas as a whole in the Great War were rewarded by the 5th Regiment being created "Royal" by the King-Emperor in 1921. And as distinction in their dress they wear a red shoulder cord. Great honour is attached to this cord: a recruit receives it personally from the hands of the Commanding Officer the day he is sworn in, while no man undergoing a sentence for any military crime is allowed to wear it. And so is *esprit de corps* created.

There have been many wars on the northern borders of India since that war which was to end war ceased. The Regiment has been well represented in all. The most notable outstanding achievement being the defence of the piquet at Damdil during the Waziristan operations of 1937, by a Lance-Naik and eight men of the 2nd Battalion against forty murderously intent fanatics. The four Gurkha survivors of the bloody fight were all wounded, while their commander was unable to stand. An incident that has few equals, and amply proves the century-old reputation that Gurkhas in bravery are excelled by none.

EARLY TYPE OF NEPALESE RECRUITED INTO THE GURKHA REGIMENTS

CHAPTER LII

The 6th Gurkha Rifles

"Burma, 1885–87", "Helles", "Krithia", "Suvla", "Sari Bair", "Gallipoli, 1915", "Suez Canal", "Egypt, 1915–16", "Khan Baghdadi", "Mesopotamia, 1916–18", "Persia, 1918", "North-west Frontier, India, 1915", "Afghanistan, 1919".

NOW centred at Abbotabad in the North-west Frontier Province, for over eighty years the sixth regiment of Gurkhas served on the eastern borders, being first embodied in Orissa as the Cuttack Legion to police the hinterland, where the tyranny of the petty officials had brought about open rebellion among the aboriginal tribes. That was in 1817. Six years later this irregular body moved into Northern Bengal, receiving a fresh title "The Rangpur Light Infantry". And again after five years yet another, "The Assam Light Infantry".

The early years were all spent in consolidating the East India Company's Bengal Presidency; and until after the first Burma War none but plainsmen were, it seems, enlisted. Then, in 1828, on the decision to permanently garrison Assam, Nepalese were enrolled into two of the twelve companies of the Regiment for the first time. This proportion gradually increased until 1886 when none but Gorkhalis and kindred races of the Himalayas were recruited.

Although not surrounded by the glamour of the North-west Frontier, work on the North-east was none the less strenuous and, moreover, distinctly unhealthy. The fighting and the terrain were as unromantic as the enemy—head-hunters with poisoned weapons who either played tip-and-run screened by dense bush, or else defended almost unassailable stockades in thick forest where sickness often took a greater toll than enemy action. Altogether warfare of a most unpleasant type. And from the time the Corps moved into Assam hardly a year passed without them being engaged in some activity suppressing predatory Nagas, Daphlas, Abors, Chins, Lushais, Manipuris—to say nothing of the operations of the 2nd Burmese War—until, as the 42nd Gurkha Rifle Regiment of Bengal Infantry, the Corps were transferred to the other Border in 1899. The number 42 was the allotment of 1861, the present title evolving in 1903.

The 2nd Battalion came into being the following year, succeeding the then disbanded 65th Carnatic Light Infantry. Though taking over that regiment's funds and property, the new unit did not, as in the case of the Punjabi regiments created in almost the same way a little earlier, take over the old unit's honours—"Carnatic", "Sholinghur", "Mysore", "Pegu". Similarly it is worth noting that the ancestral Rangpur Light Infantry from which the 1st Battalion has direct descent earned "Assam" for the 1st Burma War in which that unit played a most distinguished part, but this finds no place in the present day roll.

When came 1914, both Battalions at Abbotabad had little thoughts that either would enter the struggle, but as the days went on prospects grew and by mid-October the 1st Battalion was mobilized. Turkey had entered the war. The Suez Canal would require defences. And so over the sea went the 1st/6th Gurkha Rifles on the greatest campaign of their career. The voyage to the land of the Pharaohs was certainly something new, but what the stocky riflemen would not understand was the necessity for fresh-water conservancy, since there was *pani* in abundance. And a pint of salt straight from Neptune's tankard had to be administered to the unconvinced before water wastage was ended.

The Battalion fired their first shots in anger for twenty-four years before Kantara on the 26th January, 1915, when the Turks attempted to seize the Suez Canal.

Came April, 1915. Sir Ian Hamilton arrived and inspected some of the men—the proposed military expedition to force the Dardanelles and thus open a way to Russia and create a diversion from the stalemate reached in France, was by now common property. The Royal Navy had tried alone and failed, for without a military force in support no progress could be effected. Thus a landing on Gallipoli, hitherto considered an impossible operation, was to be made. Sir Ian was determined it should succeed. An old campaigner of the Indian Frontier he knew the Gurkha. And he wanted Gurkhas to support the Australian and New Zealand troops selected to make the most difficult landing of all. He asked Lord Kitchener, the War Secretary, for a brigade of them:—"Ghurkas . . . they would be at their brilliant best . . . each little Gurk' might be worth his weight in gold." Lord K. did not reply, and Sir Ian did not get his Gurkhas until after the original landing, and then only the 1st/6th. They arrived on the 30th April, five days too late.

But they quickly made up for lost time. The right of the Turkish line in the Cape Helles sector rested upon a three-hundred-foot-high sea-cliff prominence converted by the enemy into a powerful bastion which the most gallant attacks of two Irish regiments had failed to overcome. Three days later the 6th Gurkhas were ordered to take the feature. They did. And that is why on all mariner's charts to-day you will find "Gurkha Bluff" inscribed to the landmark—a tribute to the heroism of India's Army.

The ferocity of the fire fight in the later attempts by Turk

to turn out Gurk' is evidenced by the fact that at one time the rifles became too hot to hold, the protective wood-work charred, and cool weapons had to be passed up from the rear.

And now the saddest story of all in that disappointing but glorious epic in the history of the forces of the British Empire. A great plan was devised. The Turks were to be driven back on to Constantinople by an overwhelming break-out from Anzac to which area the 6th Gurkhas were transferred. There were also other features, such as the landing at Suvla Bay to provide a winter base, and other tactical moves too lengthy to mention here. The break-out happened. After three nights and two days of fighting the final ridge was reached, to end in the most bitter hand-to-hand close encounter, perhaps of the War. The Turks gave up. The key to the Dardanelles was in the Battalion's hands. Below them was the retreating Turkish Army. Every minute was precious. Now was the chance of the supporting battalions. But they had lost their way in the darkness. By mistake the fleet bombarded the crest line, and, temporarily, panic set in—it always does when men are shot-up by their own. The Turk, brave fighter saw it, rallied, and the position so courageously won passed out of the Gurkhas' hands. What might have been!

Trench warfare now set in, with all its discomforts added to by a most unusually severe blizzard, until the evacuation.

So back to India—to Peshawar. In spite of all their achievements, no one troubled to meet them, none sent martial music to cheer them to their camp.

The 1st Battalion having arrived, the 2nd departed for a long spell on the lower Euphrates, moving up to the higher waters for the action which became the victory of Khan Baghdadi. Then to Salonika, where in true Gurkha style the newly issued equipment was dyed the rifle black, things which certainly do matter in regiments with self-respect. More adventures on the Black Sea, at Batum, in Georgia, Armenia, and finally, Caucasia.

Whilst all this peregrination to keep off the Bolsheviks and settle affairs with Turkey was going on, the 1st Battalion—which had landed for lines of communication duty on the Tigris in February, 1918, were marching through Persia the 600 miles to the Russian front. Both Battalions arrived within a few miles of each other on the shores of the Caspian Sea. But they never met; and when no longer required returned their own ways to India; the 1st Battalion to tarry awhile during the Kurdish Revolt. What time the temporary 3rd Battalion (1917–21) were adding Regimental honours for their work in Waziristan and the 3rd Afghan War.

The Battalions of the Regiment during the War all bore considerable burdens in strange places, even if, except in the Dardanelles, they did not see much actual fighting. But they all did with merit what was required of them then, as they have done since, without the compensation of flood-lit display.

Types of Limbu and Rai Nepalese enlisted into the 7th and 10th Gurkha Rifles

CHAPTER LIII

The 7th Gurkha Rifles

"Suez Canal", "Egypt, 1915", "Megiddo", "Sharon", "Palestine, 1918", "Shaiba", "Kut-al-Amara, 1915–17", "Ctesiphon", "Defence of Kut-al-Amara", "Baghdad", "Sharqat", "Mesopotamia, 1915–18", "Afghanistan, 1919".

THIS Corps obtains seventh place in "India's Foreign Legion" through fortuity rather than seniority. Originally founded at Thayetmyo in 1902 as the 8th Gurkha Rifles, the following year they became the 2nd Battalion of the 10th. Five years later, in 1907, when the Nepal Durbar permitted the number of Gurkha corps in India's service to be increased, a reshuffling of units became necessary so that the ranks of the two battalions of a regiment should comprise the same classes. To achieve this the 2nd/10th were divided and the two wings recruited up to form the two battalions of a new 7th Gurkha Rifles composed of Limbu and Rai tribesmen from Eastern Nepal. As the original 7th recruited a different class of Nepali, they became the 2nd Battalion of the 8th Gurkhas.

The origin of the Limbu and Rai Nepalese tribes is obscure. They are generally considered to be the aboriginal inhabitants of Nepal, who, when Hinduism became the general religion of the country accepted that creed, their own religion having been of a primitive Buddhistic type. In the earlier days when they

were freely enlisted into all the Gurkha corps, they developed a reputation for being quarrelsome—probably owing to their indifference to orthodoxy—in consequence their recruitment gradually ceased.

The two tribes are pastoral peasantry closely resembling the Magar and Gurung Gurkhas in appearance, but generally more stocky in build, shorter, and possessing a more fair complexion. Since they have been recruited into the two regiments which comprise them solely—the other is the 10th Gurkhas—they have proved themselves successful soldiers, equal in all respects to the other classes enlisted as Gurkha Riflemen.

The Regimental baptism of fire took place, not amidst the barren rocks of the North-west Frontier, nor in the fever-laden jungles of Assam; but many many miles from their native Himalayas, when the Turkish Army under Djemal Psha attempted unsuccessfully to seize the Suez Canal in January and February, 1915. An honour which fell to the 2nd Battalion.

The main threat to this vital artery having been successfully disposed of, it became essential that no loophole be left whereby the Turk could mine the sea approaches. The collection of an enemy force at Tor—a quarantine station for Mecca pilgrims on the Gulf of Suez—gave inkling of such enemy intentions. To deal with this menace half the 2nd/7th were embarked and sent to dispossess the enemy. They did, the amphibious operation being carried out with a speed which, while costing them only one man, caused considerable loss to the enemy and the capture of the whole garrison. In fine, the storm of Tor was a most laudable object lesson in the art of surprise; while it proved that in efficiency and daring the youngest Gurkha corps were not going to be in any way inferior to their elder brethren.

After Tor, the Battalion were transported to the Mesopotamian Delta just in time for the *soldiers' battle* of Shaiba—so called because the victory was attributable more to the splendid fighting qualities of the regimental officers and men, rather than to the direction of the higher command. Now came the opera-

tions in Arabistan which culminated, after nearly a month's advance and fighting over marshes and muddy creeks in overwhelmingly oppressive heat and clouds of malarial mosquitoes, in the capture of Nasiriyeh. It is not therefore surprising that under such hardships and conditions the Battalion strength fell from 900 to a third thereof. Nevertheless they were now sent up to join Townshend's force in the advance up the Tigris to Ctesiphon. And there, when the strategic retreat began, 300 of the 2nd/7th Gurkhas with 100 of the present 4th/14th Punjab Regiment held up the whole of an enemy division. The repeated attacks failed to dislodge the little band, and despite that the enemy worked all round them, none succeeded in getting anywhere near. Then, together, Nepali and Punjabi arose from their small stronghold—now known as Gurkha Mound—and finally drove back the remnants of the thousands of riflemen of the Turkish 35th Division. A feat which brought forth tributes both then and later from enemy soldiers and archivists.

Then followed twelve days of march without sufficient food or water and often deprived of sleep—a retreat executed in such a soldierly way as to be historic. Next, a stand to starvation point. Lastly, surrender. And finally—the bitterest fate of all for the soldier—honourable captivity, bravely borne in circumstances that must for ever be a blot upon the Turkish reputation, well-knowing that they, the only Gurkhas in the besieged force, had done their best without discredit to themselves, their army, or the record of their countrymen. It is easy to be wise after the event, but we all know now that the great disaster of Kut-al-Amara should never have happened.

The 2nd Battalion was afterwards resurgated from drafts from both Battalions and the Assam Military Police then in the country to take part in Maude's successful advance, eventually reaching Samarra. Thence in March, 1918, the 2nd were sent to join Allenby in Palestine.

The 1st Battalion which had sent many drafts to the various theatres of war, meanwhile came out from India and joined

in the further advance from Samarra taking part in the capture of Mosul. They remained in the country until 1920, seeing a somewhat strenuous year's active service in Kurdistan.

A 3rd Battalion formed in 1918 saw service in the Baluchistan area in the third Afghan War before disbandment as a war formation. The 2nd Battalion also served in the Kohat area.

In the post-war period both Battalions spent most of their time holding the Baluchistan frontier. Quetta the capital being now considered unofficially the "home town" of the Regiment.

CHAPTER LIV

The 8th Gurkha Rifles

" Burma, 1885–87 ", " La Bassee, 1914 ", " Festubert, 1914, '15 ", " Givenchy, 1914 ", " Neuve Chapelle ", " Aubers ", " France and Flanders, 1914–15 ", " Egypt, 1915–16 ", " Megiddo ", " Sharon ", " Palestine, 1918 ", " Tigris, 1916 ", " Kut-al-Amara, 1917 ", " Baghdad ", " Mesopotamia, 1916–17 ", " Afghanistan, 1919 ".

IF all the regiments of India bore a territorial title, then the 8th Gurkha Rifles could surely be described as the Assam Regiment. For both Battalions started their careers at Gauhati and Sylhet in 1824 and 1835, while the Regimental centre is in the Assam capital, Shillong. As a matter of fact the 2nd Battalion did actually bear the name of the province from the raising until 1885.

Founded as part of a civil force which eventually developed into a frontier protection corps, their composition was mixed both as to class and capacity, mounted men at one time forming part of the establishment, as well as an artillery section which latter the 2nd Battalion did not cease to employ until 1905.

Both Battalions fought in the Bhutan War of 1864–65, the 2nd engaging in severe action and having to fight their way

back to the plains heavily harassed by the enemy, and short of water, rations and ammunition.

For the greater part of the nineteenth century the two Battalions kept the North-east Frontier, during which period twenty-five minor campaigns came into their picture, as well as the Great Mutiny operations. In the expedition against the troublesome primitive Nagas in 1879 Lieutenant R. K. Ridgeway of the 2nd won a Victoria Cross in circumstances of great gallantry, during the attack on the stronghold of Konoma.

The first honours to be carried, however, were for the guerilla aftermath in Burma, following its annexation to India.

The revolutionary tactics which developed in the internal affairs of Manipur during 1891 involved the whole of the Regiment. The orderly retirement of a small party of thirty-five of the 2nd Battalion, and their subsequent behaviour after joining forces with Lieutenant Grant and a detachment of fifty of the now 3rd Battalion/1st Punjab Regiment who endeavoured to relieve the Manipur mission was the outstanding feature. At Thobal, Grant's small force was confronted by a rabble of several thousand armed Manipuris who kept up for some days an almost ceaseless heavy fire on their entrenchments. The fierce attacks of the enemy were only kept back by the disciplined rifle fire of the Gurkhas, as the other infantrymen were raw recruits armed only with Sniders. Later, under orders, the force withdrew. Lieutenant Grant, who received the V.C., reported that the conduct of the Jemadar of the Ghurkhas was perfect, which brought him as reward the Order of British India; while every man of the mixed force, for the combined magnificent gallantry, received the Order of Merit. Thobal Day, April 6th, is thus an anniversary proudly kept up by the Battalion.

Gurkhas were in a small number recruited from the beginning, and in 1832 complete companies from the 1st and 2nd Gurkhas were transferred to the 1st Battalion. From 1886 only Gorkhalis were enlisted in either Battalion, and that year saw the

incorporation of Gurkha in the then Regimental titles. The two units had earlier been brought into the Bengal Army in 1861 as the 43rd and 44th Native Infantry, remaining numbered as such until 1903, then, on the separation of the Gurkha corps, the 43rd became the 7th Gurkha Rifles and the 44th the 8th.

Shortly after this the Government decided to increase the number of Gurkha battalions and muster out certain Madras units. The Nepal Durbar were willing to allow increased recruitment, but not of Magar and Gurung tribesmen of which the then 7th and 8th Gurkhas were composed. Thus, since it was decreed that in future all Gurkha corps were to be two battalion establishments, one to support the other in war, second battalions to the two corps under review, which were not of the same class of Gorkhalis were impracticable. So the senior corps, the 7th and the old 43rd, became the 2nd Battalion of the 8th—the old 44th.

The next service of the Regiment was unique. For some time in the late eighteen-nineties the unneighbourliness of Thibet had been causing anxiety to Simla. The Lhasa government had been flirting with Russia and there were signs that a secret Treaty with Moscow was in course of negotiation; which was contrary to Thibet's treaty with India. In 1903 Russian arms began to appear in the country, Indian territory was encroached upon, and protective measures became essential. A political officer was sent to the frontier to treat and discuss. In his escort was the 1st Battalion.

After waiting four months no responsible negociants for the other side appeared and so an advance to Gyantse was in the end lothfully ordered. At first the march was peaceful, but upon nearing the objective some opposition was encountered until Gyantse was finally occupied. Later the forces became beleagured and some serious fighting took place. In one sortie one company of the Battalion scaled a cliff 2,000 feet high in blinding snow, the action itself being fought at an altitude of 18,000 feet above the sea—probably the highest ever fought

at such a height. On the 6th July, 1904 the Jong or Fort of Gyantse had to be stormed. In this the Battalion excelled, Lieutenant J. D. Grant winning a Victoria Cross for scaling the sheer rock after he had once been thrown down and wounded; an inspiring action which resulted in the successful issue of the assault.

The force in the end had to go on to the "Forbidden City" of Lhasa and here peaceful relations were established which have remained ever since.

The campaign is noteworthy for the severity of the hardships undergone, and that, for the small expedition of only 1,600 men, a special medal was struck, with an additional clasp for those who fought at Gyantse. The Thibetans proved grim foemen, and disdained to run, but solemnly formed-up and marched off whenever the fight became too hot for them. The services of the Battalion were specially commended, and several honours for gallantry were awarded to the Gurkha ranks.

The bow and arrow armed Abor tribes of border Assam having massacred a civilian party in 1911, a punitive expedition followed which included the 1st Battalion, upon whom fell the brunt of the fighting during operations lasting seven months of the following cold weather. The paths were quite unfit for use by laden coolies, thus the progress of the force was exceedingly slow. Nevertheless retribution was effected, and the murderers brought to trial. The most important result, however, was the freeing of the Lakimpur districts north of the Brahmaputra from the dominance of the Abor clans.

Two years afterwards the 2nd Battalion were in the thick in France. Owing to their wearing slouch hats, unlike the rest of the Indian corps sent to Western Europe, they did not receive quite the same vociferous welcome as did the other regiments—they were thought to be Boers or Chinese—until the followers came into view when they, wearing turbans, were quickly recognized as Indians and received all the cheers.

The Battalion's first experience of trench warfare nearly ended in disaster, since they were made in an emergency to occupy earth-works too deep for the men to see to shoot, and as a result they had to give way sullenly to an overwhelming enemy on the 30th October, 1914. An incident which increased the determination and spirit which characterized the Battalion's actions until the attack on the flank of the Loos operations on the 25th September, 1915, when they more than got their own back with the addition of many prisoners; but support failed, and the action cost them all but 150 men. Sent to Egypt to refit, the Battalion afterwards returned to India in 1916 to be engaged for the next four years in arduous and meritorious service on the North-west Frontier.

The 1st Battalion went overseas to the war in the endeavour to prevent Kut from falling. They remained in the "Garden of Eden" or thereabout until after Baghdad fell, when they were sent to the Palestine front for the rest of the War.

Political preachings caused a lot of trouble during the aftermath of the Great War; and India had more than her share. One such outbreak was the rebellion of the Moplahs in Malabar. The country was difficult for ordinary troops to achieve any measure of success, and so hill-troops were sent down to the south—incidentally for the first time—among them the 2nd Battalion. The rounding-up of the rebels was an extremely difficult business. It was done; but not without loss and suffering to both parties. One incident stands out. At Pandikkad on the 14th November, 1921, two hundred of the Battalion encamped for the night. Two thousand fanatical desperadoes thought they saw a wonderful opportunity. But Malabar is a long way from the Himalayas and the fame of the Gurkha had not spread to the jungles of the southern coastal foot-hills. The determined rebels attacked at dawn with a ferocity that broke down the protective earth-work. The fight which followed cost the rebels 234 in killed alone with wounded too numerous to mention, as against the Gurkhas' 4 killed and

36 wounded. It was a hand-to-hand encounter of the first order well and bravely done.

Both Battalions were heavily involved in the Quetta Earthquake of 1935, officers and men earning several decorations—including the Albert Medal (the peace Victoria Cross)—for the gallantry displayed in the rescue work.

CHAPTER LV

The 9th Gurkha Rifles

"Bhurtpore", "Sobraon", "Afghanistan, 1879–80", "Punjab Frontier", "La Bassee, 1914", "Armentieres, 1914", "Festubert, 1914, '15", "Givenchy, 1914", "Neuve Chapelle", "Aubers", "Loos", "France and Flanders, 1914–15", "Tigris, 1916", "Kut-al-Amara, 1917", "Baghdad", "Mesopotamia, 1916–18", "Afghanistan, 1919".

OF the serially numbered ten Gurkha regiments the 9th is the senior and takes precedence over all the others. Curious—since they were not raised until 1823, while the first Gurkha regiments of India were established in 1815. The original corps were not, however, considered part of the regular army until some decades later, whereas the 9th Gurkhas, who did not commence their regimental life as Gurkhas for many years after their nativity, were a regular corps from 1823. Thus the old rules of 'caste' allow them the privilege of stepping past in rifle time, before the superiorly numbered regiments of the "Indian Foreign Legion" whose records as Gurkhas certainly entitle them to the position which to the uninitiated they would seem to have. No disparagement of the 9th is intended, the point is made to illustrate the inconsistency with

PRIVATE
9TH BENGAL INFANTRY—1890

which the units of the Indian Army are numbered, and that the number has no reference to the place of precedence when parading together.

The 9th Gurkhas have their foundation in the Mainpuri Levy which became the 63rd Bengal Native Infantry in 1824. Some companies took part in the first Burma War, but the Corps as a whole did not see service until the siege of Bhurtpore, the Jat rulers of which had for years shown marked defiance to the East India Company. Lord Lake had tried to subdue the fortress during the Mahratta War, but his several brilliant and costly efforts failed. In January, 1826, Lord Combermere succeeded and put an end to the increasing feeling in the suzerain states that the Company's troops were effete, and British rule about to end.

The 63rd B.N.I. acquitted themselves well at the Battle of Sobraon in 1846, when they captured a Sikh Standard which is still in an excellent state of preservation. But during the Great Mutiny their loyalty seems to have been doubtful, and they were disarmed for some years; no untoward incident occurred, and the Regiment were retained in the re-organization of 1861, being renumbered the 9th. The composition was altered, and one company of the eight became Gurkhas or other hillmen, those taken being principally the latter. The Corps were then the only infantry, other than the then five Gurkha regiments proper, allowed to enlist this class.

With this new composition the Regiment took part in the Bhutanese campaign of 1864–66, as well as in the 2nd Afghan War, but they saw little, if any, fighting. They had some severe work in the Chin Lushai expeditions of 1888–90. During the last ten years of this era no Gurkhas were in the composition.

In 1893 sixteen mixed class units of the Bengal Army were converted into single class regiments. The 9th were one so selected. So far Khas Gurkhas had not been enlisted for service in the Indian Army. The new 9th were to be composed of this class.

The origin of the Khas Gurkha is uncertain. They are not of aboriginal Nepalese stock, nor do they descend from the Mongols, as do the Magars and Gurungs enlisted into the first six regiments and the eighth. They also differ in appearance from the other Gurkha clans, being slighter in build and aquiline rather then celestial in appearance. The general acceptance is that the class is Aryan stock of Rajput origin. The Khas certainly formed a large portion of the King of Gurkha's army when he sallied forth from his tiny mountain state to conquer the Nepal Valley in 1768. The leaders of the Nepalese Army in 1815 were also principally Khas. So their previous history denotes them as having very good martial qualities. Being more strictly Hindu than the other classes enlisted, who are more inclined to Buddhism, the difficulties of caste probably made the Khas less appreciated as a soldier on foreign service. Thakurs, the class having the highest social standing in Nepal, were also enlisted. The Regiment to-day recruits from these classes only, being the sole one so to do. From their behaviour since the 9th became a purely Gurkha corps, both at home and in foreign fields—particularly in France where circumstances for high-caste Hindu soldiers were as difficult as they could be—the Khas have shown themselves to be in no way inferior to their brother clans to whom socially they consider themselves superior.

Thus arose the 9th (Gurkha Rifle) Regiment of Bengal Infantry, dressed in green, facings black, with all the other traditional differences that denote the rifle corps. Three years later came the new Gurkha corps' first active service—in the Mohmand country and the Tirah, 1897–98.

A second battalion raised in replacement of the 11th Coorg Regiment of the old Madras Army came on to the establishment at the end of 1904. The Regimental centre was thence situated at Dehra Dun.

So as part of the Dehra Dun Brigade the 1st Battalion found themselves at Marseilles by October, 1914, ready in spirit and

Gurkha Officer
9th Gurkha Rifles—circa 1910

courage to fight the Kaiser's hosts, but unready in equipment and clothing for the severe test they were to undergo, and from which they emerged sadly depleted with a turnover in personnel equal to their fighting strength fourteen months later.

War brings privations and shortages to all classes, male and female. It is not recorded doubtless in the French economic archives what the ladies of Marseilles had to say concerning the corner in safety-pins, made by the Gurkha battalions to pin-up beneath their drill shorts the long woollen nether-garments provided to warm their stocky bodies by a benevolent ordnance department, pending the issue of more suitable battle-clothing.

The following winter found the 1st Battalion at Basra, on the way to which they underwent, for most people, the nerve-racking experience of having their ship attacked by an enemy submarine.

From now on Mesopotamia was to be the theatre of Regimental exploit in the War. The 1st Battalion had the stage for the battles up to the fall of Kut-al-Amara, until relieved in September, 1916, by the 2nd Battalion from India, where the 1st returned.

The first action before an enemy of the 2nd Battalion, then composed of almost boys enlisted since the war had started, was at the Shumran Bend on the Tigris during the pursuit of the Turk to Baghdad. The operation entailed the crossing of a four-mile wide flooded river in small boats in the early dawn, swept by fire the whole way, and then the ejectment of the enemy from a strongly entrenched position. It was done. And it was as gallant an exploit as any of the noted amphibious operations of the War, and one which brought to the Regiment a Victoria Cross awarded to Major G. C. Wheeler, for his intrepid leading from which mainly resulted the Battalion's success.

Before their entry into the theatre of war the 2nd Battalion had formed the Viceroy's escort when he visited Basra during 1914–15. And they had the unique honour of providing the personal escort of the Commander-in-Chief of the forces in Mesopotamia during their two and a half years' stay in the country.

The third Afghan War of 1919 not only employed the 1st and 2nd Battalions, but also the "war-child" 3rd. The senior took part in the only serious fighting of the war, while the junior defended Thal.

In the twenty years that followed the two components of the Regiment have been employed in Waziristan, Malabar, the Khyber, Chitral, Chittagong and Bengal; experiences novel, exciting, and sometimes unpleasant.

During the siege of Thal by the Afghans in 1919 the fort flag halyards were shot away and down came the flag. This would never do. So one of the riflemen of the 3rd Battalion under close and accurate sniping climbed the flagstaff and nailed the Union Jack to the mast. And so it remained throughout the siege. He received no special medal, the Regimental archives do not even record his name. But the action of this solitary small legionary typifies the old Gurkha proverb: —"It is better to die than be coward."

RIFLEMAN
10TH GURKHA RIFLES—1937

CHAPTER LVI

The 10th Gurkha Rifles

"Helles", "Krithia", "Suvla", "Sari Bair", "Gallipoli, 1915", "Suez Canal", "Egypt, 1915", "Sharqat", "Mesopotamia, 1916–18", "Afghanistan, 1919".

ONE of the causes which led to the foundation of the 10th Gurkhas was the intention of a dissolute potentate to establish in Mandalay a French bank, together with a detachment of soldiers of that Republic to guard it. This, together with many other unfriendly actions towards India, led to his dethronement and the Viceroy assuming the government of his territory until its recent establishment as a separate entity within the British Commonwealth. King Theebaw, the concerned ruler, was deported. Then followed extreme lawlessness; much blood flowed, and many good bodies sickened before the military proper could bring about control. Order eventually evolved, and to keep it military police battalions were raised—one of them to hold the Kubo Valley, itself a peaceful district in the extreme west of Burma, but one liable to tribal raids from the eastern border hills. This unit, the Kubo Valley Military Police Battalion of May, 1887, comprised of volunteer ranks from the regular Indian Army, principally Gorkhalis, were the forefathers of the present Regiment. They

appear to have been a well-behaved lot in contradistinction to some of the others, and their conduct as police towards the "locals" most satisfactory.

Burma was then in the Madras military command, and in order to create a permanent force in Burma, certain Madras infantry corps were to be localised in the country and the men of the Carnatic mustered out. A move which turned the Kubo Valley Police into soldiers as the 10th Madras Infantry. They were, however, to be uniformed and equipped in like manner to the regular Gurkha corps of the Bengal Army. Later, the title became the 10th Regiment (1st Burma Gurkha Rifles) Madras Infantry, under which they fought, for the first time as a regiment, in the Chin-Lushai subjection which began in 1890.

Ever since the annexation of Upper Burma the Chins, Lushais, and other primitive savage aboriginals on the eastern frontier had been a source of trial. A number of punitive expeditions had been effected, but none had succeeded in stopping the depredatory incursions into the controlled territory. The establishment of a proper administration appearing to be the only course open, the country was occupied; while the only effective way of controlling rebellion was the constant patrol by small detachments. In this arduous work of hunting an elusive enemy, often on short rations and with only the barest necessities, the Regiment remained until 1894.

In 1903 the Corps entered the Indian Foreign Legion and became the 10th Gurkha Rifles, with instructions to recruit entirely in Eastern Nepal. Thus ever since the Regiment has been composed of Limbu and Rai Gurkhas. Unlike the Madras units that were converted into Punjabis at the same time the battle honours earned by their parent were not conferred upon them.

The present 2nd Battalion eventuated in 1908 by a division of the 1st. Thus the years immediate to 1914 were ones of preparation for both. The Regiment were then based upon

Maymo, and many of the grass rides cut through the jungle which are one of the amenities of that salubrious Burma station were the result of their sojourn.

The junior had the honour of being the pioneer in war, they departing for Egypt in October, 1914, to enter combat with the Turk in the pontoon attack on the Suez Canal. The first attempt failed, but not abandoning hope a final attempt was made near the Tossum signal station. Here, on the eastern bank of the Canal, was a large detachment of the 2nd/10th greatly outnumbered by the attacking force. But the Gurkhas in holding their ground showed such a determined front that the enemy's operation was brought to a standstill.

The meeting of Sir Ian Hamilton's urgent wish for Gurkhas landed the 2nd Battalion in Gallipoli. It is sufficient here to say that, like those of the other Gurkha corps engaged, their deeds could not have been excelled; and to record that it is generally accepted that at Suvla Bay one party of the Battalion penetrated further inland than any other troops. Egregious generalship failed them, and their effort in that battle could not affect the tide of misfortune.

Sadly diminished, the 2nd Battalion returned via Mesopotamia to Burma during July, 1916, in relief of the 1st Battalion, then under orders for Basra. The 1st had not by any means been inactive, however, for during the earlier months of the War they had been on service putting down attempts at insurrection among the Kachins and other Shan tribes on the Sino-Burmese border.

The 1st Battalion arrived in the eastern theatre during the preparatory period for the avengement of Kut and the advance to Baghdad. One of the essentials was the building of a railway from Basra to Nasiriyah and up the Euphrates. In fine, the eastern end of the long contemplated rail-link from Berlin to the Persian Gulf. To protect the construction was the part allotted to the Battalion, and this disappointing duty kept them out of the fighting until almost the end of the campaign.

They were, however, privileged to be in at the death, when after ten days of strenuous engagement and pursuit the Turk laid down his arms. By this time the Battalion comprised considerable numbers of volunteers from the Burma and Assam Military Police.

The 1st did not immediately return after the victory. The Arabs and Kurds soon began to listen to the teachings of the agitator, and had no wish to follow the dogma propounded by the politicians in a Paris suburb making, as they thought, world tranquillity for ever. The result was that many brave and tired men were kept behind to lose many comrades before the country became settled. The 1st/10th were depleted down to two companies, and so, to make an effective fighting unit, two companies of the 116th Mahrattas, then in similar straits, were co-joined. This composite unit, "The 126th Gurattas", carried out many gallant and successful fights. For four and a half years the 1st/10th Gurka Rifles remained in the old Turkish empire in Asia campaigning the length and breadth of the land. Deprived by fortune of the earlier victorious fighting they fully made up for it in the end, bestowing upon themselves their unofficial motto, "Per Arida ad Astra".

The only other honour carried by the Regiment, other than those of the Great War, is "Afghanistan 1919", earned by the 2nd Battalion who were thus the first to represent the Regiment in any affair on the North-west Frontier, where nowadays one or other is generally to be found serving. Burma, however, is now no longer their home, the post-war requirements having given them a permanent centre with the similar class regiment of the Gurkha Brigade at Shillong.

H.H. The Prime Minister of Nepal, the temporal ruler of the country, is Honorary Colonel of all the Gurkha Regiments, as since the treaty of Segowli, Nepal has always made the King-Emperor's cause her own. During the Great War 200,000 Nepalese subjects served overseas in India's service, some 20,000 never to return to their Himalayan homes. Despite its

sparse manpower, the whole of Nepal was thrown open to recruitment, which enabled an 11th Regiment to be formed, as well as an additional battalion for the existing regiments for frontier and internal service in India. Cheerful, hardy, most disciplinable and courageous in the extreme, these proud and virile men, bound only by the legionary's oath to the service of a foreign land, represent the quintessence of the fighting man. As such the Gurkha has no equal.

COMMISSARIAT OFFICER—1870

CHAPTER LVII

The Royal Indian Army Service Corps

IN all ages the operations of armies have been influenced by the necessity of providing and distributing food, forage, munitions and stores for men and horses. In some cases, such as on the Indian frontiers, transport and supply difficulties have often far outweighed all problems of country or enemy, and have in effect controlled the campaign.

As time progressed these supplies have become more and more varied as weapons developed in complexity, power and accuracy of workmanship. In proportion, the branches of an Army which are charged with duties of "supply and transport" have become specialized as regards recruiting, training and organization.

The Royal Indian Army Service Corps perform these duties in India, and are thus divided into two main branches; transport, supply. In fact, prior to 1923, the designation was the Supply and Transport Corps. The origins of the Corps were the Army Commissariat Departments of 1810 formed for the purpose of organizing supplies to the Presidency Armies. Separate transport branches were raised in 1883 and amal-

R.I.A.S.C. Mule Transport Carts in a Frontier Defile

gamated with the Commissariat Departments in 1887, the three Departments being put on an all-Indian basis as the Commissariat-Transport Department in 1889. The title "Royal" was bestowed on the occasion of the Silver Jubilee of H.M. King George V.

The supply branch undertakes the purchase and distribution of food and forage for all personnel and animals of the Army, as well as the purchase, storage and distribution of petrol and lubrication for mechanical vehicles.

Prior to the Great War the duties of the supply service were much less onerous, as Indian troops and the Indian cavalry, save three regiments, were fed—horse and man—under unit arrangements. On the other hand certain duties in respect of clothing for British troops and other miscellaneous articles, are now dealt with by the Ordnance Corps.

In order to carry out these duties the Corps maintains butcheries and bakeries, store depots, as well as a military food laboratory to scientifically examine samples of edibles suitable for the Indian soldier who, compared to pre-war days when he fed himself, is now rationed upon a scientifically constructed diet both in quantity and quality, and he has no temptation to economize as was previously the case.

The transport service comprises all forms of transport required to convey the stores and baggage of a military force operating in the field. And without trained and efficient transport service, an Army cannot be fought to its best advantage. It is therefore necessary to maintain a highly developed transport organization which receives its training in peace on the carriage of military stores in cantonments and on manœuvres, for which civilian transport would otherwise have to be hired. Wheeled transport is the more economical, but in view of the terrain of the frontiers, pack-animal transport in large numbers is required.

The pack-animal drivers often come under close enemy fire, and it is recorded that one of the most courageous acts ever seen on a battle-field was an unknown mule-driver, himself

(India Office photo)

Royal Indian Army Service Corps

A feature of the R.I.A.S.C. is their mule-carrying lorries depicted above. These are for quick conveyance of pack mules to the foot-hills of the Frontiers beyond which mechanical transport cannot go.

(India Office photo)

ROYAL INDIAN ARMY SERVICE CORPS
LORRIES AND DRIVERS LINED UP FOR INSPECTION

hit, holding on to his struggling and wounded mules under very heavy rifle fire during the operations in Mesopotamia.

Before the Great War, India had no military mechanical transport. Temporary units were formed during the War and did yeoman service in the Frontier and Afghan operations. After the War the organization was in the hands of the Royal Army Service Corps of the British Army. But since 1928 the whole has been part of the Indian Army proper and, with the exception of the officers generally, all the ranks are now Indian, while a large number of Indian non-combatants are employed as artificers and followers.

The uniform of the Corps is blue with white facings; and Field-Marshal H.R.H. The Duke of Connaught is Colonel-in-Chief.

CHAPTER LVIII

The Medical Services

THE military health departments of India's Army are four:

"The Indian Medical Service"
"The Indian Medical Department"
"The Indian Military Nursing Service" and
"The Indian Hospital Corps".

The first, which is more well-known than the others, comprises solely officer personnel, all of high qualification, who in the main, after spending their early military life with an Indian regiment, pass later to the control of the civil government as administrative medical and health officers, or as tutors and professors in the field of Indian medical education; and thus passing form a reserve of superior military medical staff available in the event of war. A system which is now said to have outlived its usefulness, and one the subject of much criticism in the economic field. It is probable that the future will bring a change, and that before long the Army's needs will be met by a Corps limited in its responsibilities entirely to the military sphere.

The I.M.S. has its foundation in the Bengal, Madras, and

OFFICER, INDIAN MEDICAL SERVICE, 1939

Bombay Medical Services constituted by the East India Company as long ago as 1764, to embody all the surgeons ashore and afloat then in their service. These gradually evolved into the military-cum-civil system of to-day, the three being put on an all India basis with the present nomenclature in 1897.

The Indian Medical Department comprises a class of medical men all of whom with few exceptions are statutory natives of India. The Assistant Surgeons are largely utilized upon duties in connection with the British wing of the Army in India; while the Sub-Assistant Surgeons are employed practically entirely with Indian troops.

The Indian Military Nursing Service is recruited from ladies possessing nursing qualifications and with experience of Indian civil hospitals, whose duties are to supervise and teach the nursing orderlies serving in the Indian Military Station Hospitals. The Service is an innovation since the Great War. Prior to this, no nursing arrangements for the Indian sepoy existed other than that rendered by regimental ranks not in the least skilled in a science that is as necessary to health restoration as is the doctor.

The Indian Hospital Corps is an amalgamation of the Army Hospital Corps and the Army Bearer Corps existing prior to 1914, who did such yeomen service to the stricken in the Great War. The Corps is organized into sections comprising clerks, storekeepers, male nurses, ambulance men, and others, including menial servants and domestics whose duties entirely relate to the preservation of, or restoration to, the health of the sick and wounded.

Four officers of the Indian Medical Service have received the Victoria Cross. Surgeon John Crimmin was the first in 1889 for gallantry in the Burmese War. Drawing his sword he fought hand-to-hand with Karen tribesmen over the body of a wounded soldier to whom he was attending. A good case of the Red Cross rule which permits medical personnel to be armed for the protection of their charges.

The second was Surgeon-Captain H. E. Whitchurch for going to the rescue, under extremely heavy hostile fire, of an officer dangerously wounded in the sortie from the Fort of Chitral during the siege in March, 1895.

Then, Captain J. A. Sinton for most conspicuous bravery and devotion to duty during the Battle of Sheik-Saad in Mesopotamia, January, 1916. Although shot through both arms and the side, he refused to go to the hospital and remained as long as daylight lasted attending to the wounded lying in areas heavily shot-swept by the Turks. In the three previous actions, Captain Sinton had displayed the utmost courage. He was Regimental Medical Officer of the now 1st Battalion 17th Dogra Regiment.

The fourth, Captain H. J. Andrews on the 22nd October, 1919, when as the Senior Medical Officer of the Khajuri Post in Waziristan, on hearing that men had been wounded, he took out an Aid-Post under very heavy fire, establishing such in a position that while it afforded some protection to the casualties, it did not to himself. He attended most devotedly to the stricken, all Indian soldiers, and showed the utmost disregard of danger then, as well as later, when collecting them under the very noses of the tribesmen, and placing them in an ambulance van. He was killed when about to step into the ambulance on the completion of his task. The tribesmen of course consider the Geneva rules quaint.

Captain Andrews was for over thirty years an officer of the Salvation Army, and a pioneer of their medical work in India. His hospital at Moradabad, was taken by the military authorities as a model, and his system of hospital administration as applied to the handling of Indian troops now largely forms the basis of the Indian Military Hospital of to-day.

The first member of any Medical Service of India to win the coveted decoration was, however, a young Hospital Apprentice of the Bengal Sub-Medical Department—Arthur Fitzgibbon. During the capture of the North Taku Forts in China on the

21st August, 1860, he displayed great coolness and courage in proceeding in circumstances of the greatest danger, to attend a wounded *dhooli*-bearer and others; an action during which he himself was severely wounded.

In the early days, surgeons supplied their own instruments, medicines and dressings, receiving therefor an annual grant. As was natural, the grant was looked upon as a source of profit, with most unpleasant results for the wounded man. At the Battle of Maheidpore, 1817, of the 612 wounded, 200 died for want of proper medical treatment. In the field hospitals there was scarcely any dressing plaster for the officers and none for the men; nor was there a single set of amputating instruments other than those belonging to individual surgeons, and these were so blunt as to be useless. So thus two out of every three wounded men died in hospital.

Even up to the Great War the hospital arrangements for the Indian soldier were extremely primitive. There were none of the fine station hospitals that exist to-day. Casualties were treated in their own regimental hospitals, accommodation varying according to the ideas of the unit, often the whims of the C.O. The supply of drugs and surgical necessities was meagre, and at times insufficient. The patient brought his own clothing, lived largely on his own food. Nursing was provided by untrained sepoys who did their best, assisted by a few menials. The Regimental Medical Officer had little scope to exercise his professional skill and found very little opportunity for increasing his general vocational knowledge.

During the Great War, particularly in Mesopotamia, the Medical Services of India went definitely agley. And much ill-founded criticism was abroad. There may have been faulty administration, and the break-down is not denied. But what is not generally realized is that the heavy casualties which the Indian Army suffered in the Mesopotamia theatre were entirely unexpected, and in volume unlike anything before experienced. Medical equipment was rudimentary, and efforts to obtain even

NURSING SISTER, INDIAN MILITARY NURSING SERVICE

a modicum of modern apparatus and methods of transport were short-circuited for want of available cash.

Nevertheless, handicapped as they were, the I.M.S. performed some very fine feats. After Ctesiphon and during the retreat to Kut, not a single wounded man was left behind—at the former battle there were 500 wounded. And the individual devotion of the officers to the stricken humanity of both sides there, as well as in other theatres, or in enemy captivity, was all, and more, that the great traditions of those who devote themselves to the healing of the sick demanded. One officer—Captain Som Dutt—after rendering aid in France to a badly-wounded German Colonel, was rewarded by the latter with his Iron Cross—the German equivalent of the V.C. Hardly a single regiment in its archives fails to pay high tribute to their medical men.

The tribulation of Mesopotamia can never happen again. To-day all is changed, both for peace and war. There are now eighty-one military hospitals for Indian troops, as well as eighteen Indian wings to hospitals where British troops are also treated. All are provided with adequate staffs, modern accommodation, and the latest equipment; hospital clothing and bedding are supplied; should he need it, the sick soldier receives such special diet as he may be ordered. In fact the treatment of the Indian soldier when in hospital, compares favourably to that provided by any army in the world. True this costs almost a crore and a half of rupees—double that of 1914. But is it not worth it? Consider that the Medical Service of to-day concerns itself with every phase of the soldier's life; not merely to get him well when he is ill, but to prevent him ever getting so. The hygiene of cantonments, the quarters, his clothing, the accoutrements he has to wear, the food he is given to eat—all are now the close concern of the medical officers. In no sphere of Indian Army organization has such improvements been effected since the Great War. The doctor's responsibilities to-day are heavy.

The "Via Dolorosa" of the wounded man has become a thing of the past. Then, he made his way to the rear as best he could, any vehicle returning from the line was good enough for his carriage. The *dhooli* and the bullock-drawn tonga have given place to the swift-moving articulated motor-ambulance, warmed in winter, cooled in summer. And since 1937, the ambulance aeroplane is available for severe and urgent cases.

The followers—most important persons in the Indian Army—are now treated on exactly the same footing as the combatants.

The uniform of the Medical Services is blue with black velvet facings for the I.M.S. and I.M.D. That of the Indian Nursing Service white with navy-blue facings. Combatant rank was bestowed on I.M.S. officers in 1898; and the Senior Assistant Surgeons of the I.M.D. also hold similar rank as Lieutenants, Captains and Majors, the Juniors being Warrant Officers. The Sub-Assistant Surgeons rank as Viceroy's Commissioned Officers.

The I.M.S. was the first of the services to be Indianized, and many distinguished Indian medical men have passed through its ranks. So too have many famous Europeans. Sir Ronald Ross, the discoverer of the malarial parasite, was one; were it not for his work the Panama Canal would have been an impossibility. The Service can also claim the credit for the system of water purification now in use in most modern cities, including London. And the conception of mass inoculation against epidemic disease is also theirs. In the building of modern India the Military Medical Services have set an example difficult to excel. The standards of sanitation and public health in the military garrisons and cantonments are model; their example must have widespread influence in a most effective way on the future social well-being of the populace. Time will appraise the value of their work; history will, when true perspective becomes undimmed, award its honours.

ARMY REMOUNT DEPARTMENT INDIA

CHAPTER LIX

The Remount and Veterinary Services

ONE of the greatest problems the Army in India has had to face from the time of its foundation, has been the provision of horses suitable for cavalry and artillery which could meet the ever-increasing demand the development of military science required from the mounted arms. The indigenous breeds, by reason of their smallness, poor physique and consequent liability to ailment, were early found unsuitable. Quantity was also lacking, and on more than one occasion the cavalry had to fight dismounted because no horses were available to replace the wastage of war.

The first attempts at breeding types of horses suitable for Indian military needs was in 1793, when the East India Company commenced a stud in each Presidency. These experiments were unsuccessful, principally because modern methods of breeding were unknown, and veterinary practice was then in the early stages of infancy. In India there were no veterinary surgeons as we know them to-day at all. And in England only two; and those graduates of the famous French school at Lyons. Animal ill-health was dealt with—when it was dealt with at all—by rule of thumb. And the first orders on the subject relative to animals suspected to be suffering from infectious

VETERINARY OFFICER, 1860

disease stated that they should be removed from the cantonment and inspected by "persons who may be judged capable of distinguishing the disorders of horses;"—this was in the year 1793.

As for the Company's horse-breeding operations, they went from bad to worse through ignorance and mismanagement. The progeny were freaks, and so the Madras and Bombay Studs were closed and written off as costly failures. That established in Bengal was about to follow suit when the Company engaged William Moorcroft, who had been joint head of what is now the Royal College of Veterinary Surgeons, to superintend the breeding operations in Bengal.

Moorcroft arrived in the country about 1808, and from then on there was a complete change. He succeeded in producing at Pusa in Bihar, far superior animals to those brought into the equine kingdom by his predecessors; and at the same time made great progress in the suppression of contagious disease, demonstrating what western veterinary science could effect in this direction.

Moorcroft's remounts were sent to the cavalry regiments by batches, and each officer was allowed to select one: but at a cost to his pocket of Rs.800 which, considering the purchasing power of the rupee in those days, made Pusa-bred horses almost an extravagance.

Moorcroft, after much experiment, held the idea that neither the English nor the Arab horse was suitable as a sire for producing animals best suited to India's needs. Having heard of the Turkoman breeds in Central Asia, he journeyed there in 1812 and again in 1819. On his travels he discovered the source of the Beas and the Chenab. And also made a treaty with the Government of Ladak by which the trade of Central Asia was thrown open to Hindustan. He never returned to India and his end is unknown; some say that he died of fever beyond the Pamirs, others that he was assassinated in Bokhara in 1825. On his last journey he rested at Lahore, and the small house

which he occupied still stands in the famous Shalimar Gardens. Many Punjabis from the great horse-breeding areas of the north, must have read the small tablet that Mahrajaha Ranjit Singh ordered to be affixed thereto. Moorcroft, however, is just described as a traveller, and thus the reader would not know that he was in effect the first Director of Remount and Veterinary Services in India.

In 1821 the first Army Veterinary Officer was appointed and posted to the Governor-General's Body-Guard with instructions to train Assistant Apothecaries of the Medical Department as veterinary surgeons. Later, more qualified men were sent out from England, and by 1826 the Indian Veterinary Department, now entirely a civil department, was founded.

Contrary to what we find to-day there was much friction in the early days between the officers of the Stud Department and the newly appointed veterinary officers; as a result no very good progress was made, but nevertheless the output began to improve. The supply, however, was never sufficient to meet the demand, the country-bred horses purchased through the commissariat agents were the principal source of supply, with the addition of regular seasonal importations of stock from England and South Africa.

The veterinary officer, like the medical, contracted to supply his own medicines, the rate allowed being two annas a month for each horse. The system was that of the time, and history records that large personal profits resulted.

The veterinary officers of course accompanied their regiments in the field, and the Service has been ubiquitous in all the campaigns of the Indian Army since its formation. At Khushab, during the Persian War of 1857, in the remarkable charge of the 3rd Bombay Light Cavalry against an infantry square, it was the veterinary officer of the regiment who captured the enemy standard. In the Great Mutiny many members of the Service met their end—one fighting gallantly in the historic defence of Lucknow.

Although in earlier days India was a good deal behindhand in matters military as compared to the armies of Europe, she can claim to be the first ever to institute in a British Empire campaign the provision of field hospitals for sick animals. This was during the war in Abyssinia in 1868; an expedition organized entirely from India. The arrangements, compared with those of to-day, were certainly crude; but they were successful. And at the conclusion of the campaign, which depended so much upon the fitness of the animals, the work of the Veterinary Service was mentioned for the first time by a Commander-in-Chief in his field-despatches.

After Abyssinia, in 1875, the present remount depot at Saharanpur came into being as a reserve depot for the reception and treatment of temporarily useless horses.

Prior to the Great War, the Army Remount Department had developed into an organization for the purchasing, rearing, and issue of horses required for British regiments in India, as well as for the three regular or non-*sillidar* Indian cavalry corps. The rest of the Indian cavalry being on the *sillidar* basis, mounting was done under regimental arrangements. The Department also supervised horse, mule, and donkey breeding in certain selected areas in Northern India. The duties were thus extremely limited in scope, and in general only confined to the issue of animals.

The Great War experience and the break-down in remounting the *sillidar* cavalry necessitated change. And change there has been. The Army Remount Department is now responsible for the supply of all horses, camels, mules, draught bullocks, etc., for all services and all units; the enumeration throughout India of all animals available for transport use in the event of war; and the general efficiency of all animals of the entire army. Moreover, the development of horse-breeding operations with a view to making the country as self-supporting as possible in the provision of horses and mules for defence requirements is also their particular work. Were it not for the efforts of the

Officer, Indian Army Veterinary Corps, 1939

Remount Department, the horse and mule breeding industry which is now one of national importance, might well have become extinct in India.

The Veterinary Service in 1914 was solely responsible for the veterinary care of the animals required by British units. Indian units looked after themselves, except when contagious disease broke out. Other than officers, all personnel was borrowed from combatant units. Units themselves had their own hospitals; and their own ideas as to treatment.

Again there has been vast change. There is now an Indian Army Veterinary Corps organized upon a scientific basis, which contains all the personnel required for the veterinary service of the animals of the military. Divided control has been abolished, and proper station hospitals established; as well as a complete mobilization cadre to provide for the field requirements of animal casualties, from ambulance sections to convalescent hospitals. Indian Commissioned Veterinary Officers have been appointed—a thing unheard of before.

The first veterinary treatment to animals of the army was rendered by the shoeing-smiths of the unit. In each cavalry squadron there were two important personages—the trumpeter and the farrier. Both had distinctive dress. But whereas that of the trumpeter was gaudy, that of the farrier was the reverse; but it had one peculiar feature, the coat was always of different colour to that of the fighting ranks. As all British cavalry corps in the early days—with one exception—wore red, the farrier's coat was blue. He also carried an axe for the better despatch of hopeless cases. The blue-coated farriers of the King-Emperor's Life Guards who, carrying ceremonial axes, ride in rear of all state processions of His Majesty when in London, are the present-day signs of this old custom.

When the Veterinary Service was first started, the officers wore blue also, but in later years, until about 1883, the uniform of the regiment to which they were attached was more usual.

Later, on the institution of the I.A.V.C. proper, blue uniform with maroon facings became the dress order.

Veterinary considerations sometimes play a very large part in the strategy of a campaign, an instance being during the Great War in East Africa, where a number of veterinary officers from India served. The belts of tsetse fly which are peculiar features of that country and which preclude the passage of any form of equine or bovine life had a great bearing on the plan of the operations, and probably threw greater responsibilities on the veterinary staff than in any other theatre of the War.

CHAPTER LX

The Indian Army Ordnance Corps

THE Dictionary defines the word "ordnance" as either firearms of large calibre, or else that which is ordained. The word is thus very apt when used in regard to the corps to which it is applied, as the Ordnance Corps is charged with the duty of supply and distribution to the fighting and ancillary units of the army not only of lethal weapons of all kinds, but that which it is ordained the soldier shall wear, carry, and use. In fact, just as the medical and veterinary services keep the army in health, the service corps provides its food and transport, so in general the ordnance service supplies the parts which enable a great complexity of healthy, well-fed humans and animals to function with machine-like utility always ready to operate in defence of the national safety.

From the earliest times the duty of the supply of ordnance rested with the artillery, and thus the foundations of the Ordnance Corps started as an offshoot from the gunners under the title of the Military Train and Magazine Establishment. Each of the three Presidency armies had their own. About 1796 the title Ordnance Departments evolved, the three being amalgamated into the Indian Ordnance Department in 1884.

OFFICER, BENGAL ORDNANCE DEPARTMENT—1856

From the Indian Ordnance Department in 1922 developed the Indian Army Ordnance Corps, the result of the division of labour which was found desirable after the vast experience of the Great War. The former, now almost entirely staffed by civilian technical experts and Indian artisan and clerical personnel, deals with the manufacture of weapons, vehicles, military equipment, clothing and all the other munitions of war, and is organized entirely upon commercial industrial lines.

The functions of the Indian Army Ordnance Corps have been described in outline, and may be repeated as the storage and distribution to the troops of military arms and equipment, stores, clothing and boots, the manufacture of which has been arranged for by the Indian Ordnance Department; as well as their repair, maintenance, and efficiency in garrison or in the field.

The I.A.O.C. staffs the arsenals and depots placed at suitable points all over India, each of which is equipped with up-to-date engineering shops where all normal repairs can be carried out from the re-building of an internal combustion engine to the minutest repairs to the most delicate optical and electrical technical instruments. In these establishments the bulk of the personnel is Indian, and their expertly supervised work in this sphere is comparable to the best of any nation.

The I.A.O.C. is thus in effect a behind-the-line institution working with none of the excitement of the fighting man, nor does it display honours or other emblems as the result of martial prowess. But it has won more Albert Medals for bravery displayed in circumstances of the greatest danger in peace, than any corps in the British Empire. And five members of the Corps, or rather of its forebears, have won the Victoria Cross for deeds of gallantry in the face of the enemy.

It is unnecessary to dwell upon the Great Mutiny and its causes, but all can admire brave men. And there were certainly none braver than those nine resolute members of the Bengal Ordnance Establishment, who, on the 11th May, 1857, defended

the Delhi Magazine against hordes of mutineers until, when all hope of succour had gone, they set fire to the stocks of ammunition and created an explosion which destroyed thousands of the enemy and in which they themselves expected to perish. Four, however, survived—the commander, Lieutenant George Willoughby (later to fall in action), Lieutenants George Forrest and W. Raynor and Conductor J. Buckley. The three who eventually lived received the decoration instituted by the first Queen-Empress.

Conductor James Miller, of the Bengal Ordnance Department, received the Cross for his action on the 28th October, 1857, when he went to the assistance of a wounded officer, and carried him out of action at great personal risk, himself being wounded whilst doing so.

Privates Samuel Morley and James Murphy of the 2nd Battalion of the Military Train—the field ordnance establishment—on the 15th April, 1858, at Azimghur earned theirs for saving the life of Lieutenant Hamilton of the 3rd Sikh Cavalry, fighting over his body until others came to their aid.

The earliest uniform for the Ordnance staff laid down in 1787 was green with red facings; this two years later was changed to blue with facings of the same colour, the lace being white. The supervising officials wore cocked hats with black feathers. Uniform, by order of the Governor-General, had to be worn at all times, and the cumbersome cloth frock-coat with lace frills was about as suitable wear for work in the Fort William cannon foundry during the Bengal hot weather as would be the full-dress robes and wig of a High Court Judge. But they stuck it, and what is more wore swords and periwigs or long unshorn hair in the doing.

The badge of the Corps is that of the old Board of His Majesty's Ordnance instituted in 1414—three field pieces and three cannon balls upon a shield with the motto "Sua Tel Tonati"—Science has wrested the weapons from nature. Both the meaning and the origin of the motto have been hotly

contested: some attribute it to Cromwell, others to Milton. It is the badge of all the Ordnance Corps of the Empire.

Since it is the duty of the Ordnance Corps to supply these things, here may be recorded the important historical changes in material in the Indian Army.

Colours were first issued to infantry about 1764. Each company carried a set which were the colour of the men's facings and bore in the centre the device of the subadar—a dagger, a sabre, or a crescent. Those of the grenadier companies had the Union Jack in the upper canton. Company colours only lasted for nine years, after which the present system of two per battalion came into being.

Standards for the cavalry on the basis of one per squadron, were presented in 1818. The first or royal standard was blue and bore the arms of the sovereign; the second or the Company's carried that corporation's arms and was of crimson; the third or regimental, the colour of the facings with the corps number in the centre. Standards, save in certain exceptions where awarded as honours, were abolished from parade or the field in 1864.

Bazaar tom-toms and trumpets were forbidden in 1773, in which year drums and fifes were issued. Thirty-five years later, two bugle horns and five whistles were distributed to the light infantry companies.

Havildars of infantry carried pikes from 1802 to 1831, when fusils took their place.

The cavalry fire-arm was unregulated by the Ordnance Department until 1864, when the Victoria Carbine carried slung across the back became 'de rigueur'. In order to secure rapid uniformity since in the majority of the cavalry regiments each man owned his own, the new weapon was given out free to all ranks who gave up fire-arms in exchange, and half-price to others.

The muskets of the dismounted arm remained generally the smooth bore "Brown Bess" until the introduction of the

(*India Office photo*)

A Trade School of the Indian Army

Enfield Rifle between 1861–71—this was the arm, the attempts at earlier introduction of which had been one of the many causes of the trouble of 1857. The Snider followed the Enfield, 1871, the Martini-Henry, 1892, to be replaced by the Lee-Metford and Lee-Enfield in 1905.

Ten rounds of ammunition for musketry practice was an innovation of 1778. The targets were of painted canvas and it was ordained that a butt of earth was to be erected behind to preserve the fired balls, which had to be returned to the local depot for remarking.

The cavalry sword was obtained under regimental arrangements until and during the Great War, and so long as such was of a good cutting type, the pattern was left to the Colonels.

Regulations regarding the lance were a feature of 1864; these stipulated that the shaft was to be of bamboo with a bayonet-shaped head between ten and eleven-and-a-half feet long and not to weigh more than four pounds.

Light field-pieces, called galloper guns, the founders of the Horse Artillery, were the fashion of 1801. And in 1827 all field-ordnance was ordered to be horse-drawn; such cannon of this type as could not be so removed was to be utilized as fixed armament.

Camp equipage came in 1787, but the soldier received no shelters for himself, only eleven bell-tents per battalion for the arms being permitted. The officers were allowed seven marquees; the sergeants three smaller tents.

Equipment varied greatly in pattern in the different regiments of the three Presidency armies, but widely speaking, the first issues were of black leather. This was changed to the buff pipe-clayed white in 1805. Buff was replaced by brown in 1879. Haversacks of canvas in place of the unwieldy knapsack were products of 1845.

The first infantry dress was a coatee issued free, each man purchasing his own nether garments until 1817, when trousers

were issued for the first time. Cumarbands were regulated to a width of six inches in 1809. The coatee changed to a zouave tunic with slashed cuffs in 1869. Puttees came in 1877, replacing white gaiters in active service order.

The cavalry wore French grey from 1811, this colour being retained by the three regular regiments for over a century. The yellow coat for Skinner's Horse was approved the same year; but irregular cavalry and the later *silladar* regiments wore very much what they liked until all were regularized in 1922. Breeches of the mounted men were of brown leather, imitated in the irregular corps by garments of cotton *dusuti* dyed *multani mutti*; 1851 saw these replaced by cloth overalls. The head-dress of the light cavalry regiments was then a fur busby. Jack-boots even with the top of the knee came in in 1868.

The head-dresss of the infantry in the beginnngs undoubtedly followed the style of the indigenous *pagri*. Later, when desired to assimilate the Indian line to that of the British in order to deceive the enemy, it developed into a heavy European style shako built on an iron frame and later of bamboo.

The more comfortable Kilmarnock cap for field wear, now worn in abbreviated form at a jaunty angle by the hillmen recruited rifle corps, developed about 1840–44; when first introduced the cap was larger and placed evenly on the head; it was most popular and had very hard wearing qualities.

Indian officers at first wore the coatee with tight pantaloons and half boots. Rank was denoted by rows of gold beads, two rows for subadars, one for jemadars; these were substituted by crossed swords and a single sword in 1856, the stars of to-day being adopted thirty years later.

Khaki was introduced as working dress in 1861. Three years later it was displaced by white, and its general re-introduction deferred until 1881. Supplies of this and all uniform other than full-dress was, until the Great War, effected under regimental arrangements.

(*India Office photo*)

"LEARNING TO BECOME MECHANICS"

In the beginning the East India Company had little interest in the local manufacture of munition supplies for its troops, preferring their importation; since it was the policy then to keep the country ignorant of the theory and manufacture of artillery and similar weapons. Later all this changed, and by the time of the Mutiny, the Company had established factories to supply gunpowder, brass cannon, gun-vehicles, bullets, and other martial stores. The manufacture of small fire-arms, however, was not attempted. The earliest factors of gunpowder were the Company's surgeons.

To-day India has modern and well-equipped factories all capable of expansion in time of war. Guns and shells come from Cossipore; rifles, automatic guns and other small arms from Ishapore; gun-carriages from Jubbulpore; clothing from Shahjahanpur; cordite from Aruvankadu in the Nilgiri Hills, and ammunition from Kirkee; harness and saddlery from Cawnpore. The products of these establishments are all made from indigenous materials. There are still some munitions of war which India does not yet manufacture, but it is the policy of the Government to make the country year by year increasingly self-reliant and self-productive in the matter of armaments as her means and resources will allow; and the bulk of the less technical stores for the forces are now supplied entirely by the Indian trade.

The Ordnance factories employ many thousands of Indian workmen, while Indian personnel of high social status are being trained by apprenticeship in the engineering and allied professions to fit them for the higher responsible posts in the supervising sphere. The standard required is high. A lowering would be fatal to the country; for the Ordnance Department must ensure that India's fighting men are equipped with material able to stand the heavy wear and tear of war but produced at the lowest cost to the tax-payer. An army may still march on its stomach, but a nation now fights with its factories, the control of which is normally the duty of the Master-General of the Ordnance. To assist him in the distribution of their products to the consumers, and to give service after sale—to put the matter in commercialized form—is the duty of a Corps seldom in the public eye, but whose functions are the very essential to success in war of any form.

CHAPTER LXI

The Military Engineer Services

BEFORE 1914 the units of the Indian Army themselves built and maintained their own barracks or hutments—generally the latter constructed of mud—receiving from the Government an allowance for their upkeep. Nowadays they are provided with large modern well-lit airy barracks, and are most comfortably housed. The duty of their construction and maintenance, as well as that of all engineering military works of a civil character, falls upon the Military Engineeer Services.

The history of the Service goes back to the eighteenth century, when military engineering requirements preponderated over those of the other departments of the developing civil government. As conditions became more settled there came into being a Public Works Department under the control of the "Military Board". This Department was manned by the Indian Corps of Engineers, whose character was entirely military. Civil works, however, gradually began to assume a greater importance, especially during the construction of the Grand Trunk Road from Calcutta to Delhi, and the Ganges Canal from Hardwar to Cawnpore. Civilian engineers began to be employed in large numbers; at the same time provincial governments began to express dissatisfaction with the military control over civil works executed on their behalf. Thus in 1851 the Public Works Department was made a civil department, no separate organization being considered necessary for military works, it being thought more economical for the same department to execute both military and civil requirements. After the evolution of control directly by the Crown, there was a

boom in the construction of civil works, and the Public Works Department was greatly expanded.

In course of time the Army developed the same complaints as the civil government had done previously. It found that it suffered from the loss of control over military works—these were not expanded as wished—and barracks fell into disrepair. Complaints which resulted in 1871 in the control of large military works, as well as works in all the cantonments of Bengal being placed in charge of a Military Works Branch of the Public Works Department. But it was not until 1881 that the control of this branch became vested in the Army. Later, in 1889, the Military Works Department took over all military works in India, and in 1899 became entirely military in character, the system of the Public Works Department being abolished.

The extended requirements consequent upon the provision of barracks for Indian troops already referred to, led to a reorganization of the Department, and on the 4th December, 1923, the present title of "The Military Engineering Services" evolved.

The Service is divided into three branches: buildings and roads, electrical and mechanical, and stores. The units are divided into: power stations, ice factories, water supplies, road roller groups, workshops, engineering stores, furniture depots—with a principal supply depot for engineer stores at Lahore, and a large workshop at Rorkhi.

The Military Engineer Services, whose uniform is red with blue facings, control all military works in India and the Persian Gulf, except in the case of a few small outlying military stations which are still in charge of the P.W.D.; they also control all the works for the Royal Air Force and the Royal Indian Navy, and certain civil works in the North-West Frontier Province and Baluchistan.

CHAPTER LXII

The Military Farms Department

FORMING part of the supply service of the Indian Army, and controlled by the Quarter-Master General, who also controls the R.I.A.S.C., is the Military Farms Department, wearers of a red uniform with blue facings. This also originated as a branch of the Commissariat Department on the 7th June, 1884, becoming a separate department in 1912. It maintains the Grass and Dairy Farms which, though primarily directed to ensuring in peace and war an adequate supply of fodder for Army animals and pure dairy products for military personnel, have actually attained a wider range of public utility. The scientific methods of cultivation employed by the Grass Farms Department have been of direct benefit, by the way of example, to the Indian cultivator; and the Okara Farm in the Punjab in particular may be quoted as a model institution. Military Dairies have stimulated the demand for pure milk and butter and are doing much to bring into existence a new Indian industry, which is now trying to supplant its progenitor. Many useful experiments have been carried out with imported breeds of cattle and their crossing with Indian stock. These have been of direct value to Indian agriculture, and many of the calves bred on the Farms are given away to local farmers, who use them for breeding, with positive benefit to their own herds. Thus does the Army contribute to the economic improvement of the country.

CHAPTER LXIII

The Indian Army Corps of Clerks

THE "Service Corps", unlike its counterpart in other armies, does not undertake the clerical duties for the army headquarter staffs, this function in India being carried out by the Indian Army Corps of Clerks, comprised of British and Indian wings. This branch of the service, too, has accompanied any portion of the Indian Army when campaigning, and has behind-the-scenes responsibilities without which the brains of the army could not function. Military clerks are posted in the field to all formations of a brigade and above, and although the fighting man is apt to scoff at the soldier behind the typewriter, it should be remembered that the accurate speedy production of a commander's operation order, often effected under the nerve-racking experience of enemy bombardment, is as potent a factor to success as anything else in the military scheme.

(India Office photo)

OFFICERS OF THE JODHPUR STATE FORCES AT FIELD EXERCISE IN NORTHERN INDIA

The people of Jodhpur—a Rajput State in Central India—have great military traditions.

CHAPTER LXIV

The Indian States Forces

IN the years following upon the conclusion of the Great War a Royal Commission appointed to suggest measures of political development in British India, previewing the ground to be covered, realized that the scope of their enquiry must be widened, since no conspectus of modern India could be entire unless the territories of the Princes—almost one half of geographical Hindustan—were included in the picture. Thus any review of the forces of India would be incomplete without a reference to the armies maintained by those territorial divisions of the country known as the States.

For very many decades the independent rulers, as well as the states feudatory to British India, maintained large armies systemized to some extent upon European lines and commanded or organized in many cases, by European officers; some, just soldiers of adventure, others, lent by their governments for the purpose. The names of De Boigne of Gwalior, Raymond of Hyderabad, Avitabile of the Sikhs, Perron, Gardner, Skinner, are all familiar to the students of Indian history of the eighteenth and nineteenth centuries. Some of these troops were led against the British, others fought by their side; and in the forces of the Principalities to-day are to be found regiments bearing honours for Laswaree, Bhurtpore, Kabul—1837, Moodkee, Mooltan, Delhi—1857, Lucknow, Afghanistan—1878–79, etc. In fact almost every Indian campaign since the East India Company's government was established has seen a contingent of some State or other serving alongside the troops of the paramount power. It was the

excellent services rendered by the Pubjab States, as well as those of Rajputana and others, which enabled the insurgent soldiery to see the error of their ways during the mad months of 1857–58.

Before proceeding it seems wise to point out that the armies of the States must not be confused with the various contingents, such as the Hyderabad Contingent, which were officered by the supreme government and which were only State forces in so far as the particular State paid for them, or ceded territory so that the Indian government might do so.

It was the Penjdeh incident of 1885 which brought home to the Ruling Princes the consequences to them of a Russian penetration of India with a view to securing the supremacy of government. The war which seemed likely then, happily was averted, but it emphasized the desirability of organizing the forces of the Princes so that they could be utilized in the defence schemes of the country.

Thoughts in this direction became definite when His Exalted Highness the Nizam offered a considerable sum from the revenues of his Hyderabad Dominions as a contribution towards the defence of the N.W. Frontier on the occasion of Queen Victoria's Jubilee in 1887. This, as would be natural, led many other rulers to follow suit. The Government of India, however, decided that assistance by the States against foreign aggression could be better afforded in man power. And so there evolved the scheme by which certain units of the small armies of the Princes could be earmarked for the defence of the Indian Empire, and trained to a standard which would make them efficient to stand in the battle-line with the regular Indian troops of the Queen-Empress.

Thus arose "The Imperial Service Troops" as distinct from the feudal forces of the State.

By 1889 the movement was well under way; and units of cavalry, infantry, engineers, transport corps—to say nothing of the unique Camel Corps of Bikaner—all officered by Indian

(*India Office photo*)

SAPPER
MALERKOTLA SAPPERS AND MINERS, 1937

SOWAR
H.H. THE HOLKAR OF INDORE'S ESCORT, 1937

gentlemen and financed at the expense of the State, had come into being. Training was supervised by officers lent from the regular Indian Army, while schools of instruction for the Imperial Service Troops were started.

Kashmir, whose frontier marches with that of Russia, undertook to maintain two batteries of mountain artillery. And it was the Imperial Service Troops of the Maharajah of that State who first earned glory and showed that the experiment had more than the germ of success in the difficult Hunza-Nagar campaign of 1891, and later in the gallant defence of the Chitral Fort in 1895.

In the grand Frontier War of 1897, units of the Imperial Service Troops of Gwalior, Jaipur, Jind, Kashmir, Malerkotla, Nabha, Sirmoor and Patiala all took part; as did the present Kapurthala Jagatjit Infantry, a detachment of which greatly distinguished themselves and gallantly perished to a man when cut-off and overwhelmed by savage and merciless tribesmen in the Kermana defile.

Alwar, Bikaner, Jodhpur, and Malerkotla sent detachments to serve with the International force in China during the Boxer rebellion of 1901—the Maharajah of Bikaner and Maharajah Sir Pratab Singh of Jodhpur themselves also taking the field. The German Field-Marshal—the Graf von Waldersee—who commanded the force, expressed himself agreeably surprised at the steadiness and soldierly bearing of the Imperial Service Troops. In one of the actions a Dufadar from Jodhpur won the Order of Merit—then the sepoy V.C.—for gallantry in the field.

In 1902, Indian troops were engaged in Somaliland, and on hearing of the urgent need for mounted troops, the Maharajah of Bikaner offered a strong contingent of the Bikaner Camel Corps for service on the African continent. They proved of great value in scouting, establishing posts, to say nothing of desert action in one of which the Rathori sowars full upheld the Rajput traditions.

By 1914, the Imperial Service Troops of the various States had reached an approximate total of four companies of engineers, two mountain batteries, nearly fifteen regiments of cavalry, three camel corps, thirteen infantry battalions, and seven transport corps. When war came that year, the Princes characteristically displaying their unswerving loyalty to the King-Emperor, immediately offered the whole of the resources of their States. Of the 22,000 Imperial Service Troops then existing, some 18,000 served overseas. And throughout the four odd years of the War they were maintained in the field at the expense of their Rulers. The majority of the units were absent from India for over four years, some of them longer. Those few units too small to be employed overseas, rendered useful service in India on internal security duties, and in the training of remounts. Many of the corps earned renown, some reputations which can be envied; many of the individuals received honours for gallantry and devotion to duty.

An Imperial Service Cavalry Brigade, consisting of the Hyderabad, Mysore, and Patiala Lancers, with detachments from several other States, together with the famous Camelry of Bikaner—again in action commanded by their ruler—served in Egypt and the Suez Canal defences from October, 1914, onwards.

The Jodhpur Lancers, reinforced by those of Alwar under the command of H.H. Maharajah Sir Pratab Singh, served in France with the expeditionary force sent to Europe at the beginning. With this force also went detachments of Sappers from Tehri-Garhwal, Faridkot, and Malerkotla, as well as the Transport Corps of Bharatpur, Gwalior, and Indore.

To Egypt went infantry from Alwar, Gwalior, and Patiala, the latter sending two companies as reinforcements to Gallipoli, where they sustained heavy casualties in the fighting on the Peninsula. All were later very actively employed in Palestine in the victories so complete as to be unique in the history of

warfare. The Mysore and the Jodhpur Lancers also joined these operations, the latter making a brilliant charge which captured Haifa, believed to be the only occasion when a fortified town has ever been captured by cavalry at the gallop.

To East Africa went units from Bharatpur, Gwalior, Jind, Kapurthala, Kashmir, and Rampur. All served with excellence in many actions, some of them severe. The Faridkot Sappers, in the same theatre, were also singled out for distinction.

The campaign in Mesopotamia employed the sapper corps, who went in detachments to France and who were afterwards reformed into complete units. Here from the beginning, the Sirmoor Sappers did splendid work, earning a record difficult to surpass. Defenders of Kut, they suffered considerable hardship in captivity.

The Transport Corps of Bahawalpur, Bharatpur, Gwalior, Indore, Jaipur, Khairpur, and Mysore, all did yeomen service—some of them enduring severe casualties. Their work was often carried out under trying conditions of rain, cold, and excessive heat; and no praise can be too high for them. Seldom has an unarmed transport company actively assisted in the winning of a battle. A claim which can easily be made by the Jaipur Corps sent to Mesopotamia; at Shaiba the dust caused by the rapid advance of the company to remove the wounded from the field was mistaken for charging cavalry and caused panic to prevail in part of the Turkish line.

Although the standard of training of the Imperial Service Troops was high, their establishments differed from those of the regular army, and when serving together in the field, this dissimilarity proved a source of difficulty. So after the War, when the whole of India's military requirements were under review, a committee of the Princes, together with experts from the central Government of India, evolved the present scheme.

Thus we now have "The Indian States Forces" whereby 'all

(*India Office photo*)

PATIALA STATE INFANTRY

units of the Princes are included in the scheme of efficient training and armament from which the States will be able to offer such as they can spare for service with His Majesty's Forces in the defence of the Commonwealth as and when the occasion arises. This scheme does not include the irregular forces still maintained by some rulers on an almost medieval and feudatcry basis.

The Indian States Forces number nearly 50,000 men, more than double the number that existed during the Great War. Forty-nine States are in the scheme, and their armies vary in strength from the equivalent of the division maintained by Hyderabad, Gwalior, and Kashmir; down to one platoon. In most cases the troops are well barracked, well clothed and equipped, and fairly paid; there are, however, great variations, but many of the States are now rapidly falling into line with the prevailing conditions in the Indian Army proper.

The distribution by arms is approximately three companies of engineers; two horse artillery batteries, three mountain batteries; one camel battery; sixteen regiments of cavalry; nearly forty-six battalions of infantry; and twenty-two troops of transport. A number of the units are held ready for immediate employment on mobilization.

The training of the force is carried out by military advisers who are specially selected for the duty, the whole being directed by a specially-selected General Officer known as the Military-Adviser-in-Chief. This officer is not under the control of the Defence Department, but works in close touch with it, as the matter of the forces of Indian India falls within the scope of the Political Department of the Governor-General.

Officers and men of the Indian States Forces are now sent to the same schools of instruction as are those of the Indian Army, and the standard now expected is very high. The various units carry out joint manœuvres with the troops of the Indian Army and assimulate themselves in every way as part of the defence forces of the coming Federal India. All the Rulers take

intense pride and personal interest in the welfare of their men and the efficiency of their armies. But it must not be forgotten that the military aid which they lend to the Government of India is entirely at their discretion. In any just cause of His Majesty the King-Emperor, such aid will never be withheld.

(*India Office photo*)

MEN OF THE ROYAL ENGINEERS, AUXILIARY FORCE, INDIA, UNDERGOING TRAINING IN BRIDGING

CHAPTER LXV

The Auxiliary Force, India

THE Auxiliary Force, India, more popularly known as the A.F.I., is the present day name for the old Indian Volunteers raised from Europeans and persons of mixed descent some sixty years ago.

A volunteer force was first formed in the days of the Great Mutiny, the Madras Volunteer Guard, now the Madras Guards, having been raised in 1857, other infantry corps being constituted some three to four years later. Mounted volunteer corps came into being in the next decade, artillery in 1879, the Madras Artillery Volunteers being the doyen. The great railway companies of India commenced the formation of infantry units from their own employees in 1869, the earliest being the East Indian Railway Volunteer Corps.

These volunteer regiments were organized on the same lines as regular units of the British Army, the personnel being trained with the special object of local security. Each corps had an adjutant from the regular army, otherwise the units comprised entirely volunteers, both officers and men. No definite obligations were imposed as to the amount of training to be undergone, and the efficiency standard varied greatly in the different units.

Generally speaking the volunteer system before 1914 was not very efficient, and with the commencement of the European war, steps had to be taken to improve matters. The units of the force were utilized in many cases to relieve regular troops of ordinary routine Garrison duty, but as the statutes governing the use of the volunteers limited their activity to the localities

in which they were raised, this and other reasons such as the needs of the volunteers' civil employment, could not allow the force to be employed to its capacity or as a whole.

Public opinion in the European sphere demanded some form of compulsory service, and as a result the Indian Defence Force Act was passed in 1917. By this Act all the units of the Indian Volunteers became units of the Indian Defence Force, and service therein was made compulsory generally for all European British subjects in India between the ages of 18 and 41; at the same time certain units were allowed to accept as volunteers certain classes of Indian British subjects.

The Indian Defence Force remained as such until 1920, when a new force was created styled the Auxiliary Force, India, and a renumbering and renaming of the volunteer units transferred to it was effected. The liability of the volunteer for training and service was clearly defined, but again service was to be local.

The Auxiliary Force comprises all branches of the service, the organization being based upon the corresponding arm of the British Army in India. The composition of units, however, is not all of one arm, and the sub-units may be of different branches of the fighting forces. Their role is that of local defence, and the local military authority has the power of calling out the volunteers for local service in the case of emergency. Training is carried on throughout the year. Pay is given for each day's training, and a bonus on completion of the requisite period at annual camp. Enrolment is for an indefinite period, but discharge can be claimed after four years' service or on attaining the age of forty-five years.

The future of the Force is obscure, and it is more than likely that the future will see a gradual depletion in the number of units and their eventual absorption into the Indian Territorial Force which is now becoming an important feature of India's Army, membership of which carries with it a liability for something more than local or home defence.

BRIEF HISTORIES OF THE UNITS OF THE AUXILIARY FORCE, INDIA

CAVALRY

The Bihar Light Horse

Formed as the Bihar Mounted Rifle Corps on 8th December, 1862. Designation changed to the Bihar Light Horse on 29th February, 1884.

Uniform—Khaki.
Facings—White.
Badge—A wild boar.
Motto—"Necaspera terrent."
Honorary Colonel—The Governor of Bihar.

The Calcutta Light Horse

Formed as the Calcutta Volunteer Lancers on 22nd August, 1872. Reconstituted as the Calcutta Mounted Volunteer Rifles on 7th October, 1881. Designated the Calcutta Mounted Rifles, 1st October, 1886, and the Calcutta Light Horse on 20th May, 1887. Absorbed on 1st November, 1901, the Central Bengal Light Horse raised on the 27th June, 1884.

Uniform—Blue.
Facings—White.
Badge—An eight-pointed star with crown, "C.L.H." and motto in scroll.
Motto—"Defence not Defiance."
Honorary Colonel—The Viceroy of India.

The Calcutta Light Horse has had the honour of escorting all the members of the Royal family who have visited the second city of the British Empire.

The Surma Valley Light Horse

Formed as the Sylhet Volunteer Rifle Corps on the 22nd October, 1880. Became the Cachar and Sylhet Mounted Rifles on the 6th April, 1883. Present designation 13th August, 1886.

Uniform—Khaki.
Motto—"Non Sibi Sed Patriae".
Honorary Colonel—The Governor of Assam.

The Assam Valley Light Horse

The Lakhimpur Volunteer Rifle Corps, raised on the 3rd November, 1882, became the Lakhimpur Mounted Rifles on the 27th July, 1888. The Sibsagar Mounted Rifles, formed on the 29th February, 1884, became the Sibsagar Mounted Infantry on the 30th April, 1886, and resumed their original designation on the 15th February, 1889. The Darrang Mounted Infantry, formed on the 3rd June, 1887, became the Darrang Mounted Rifles on the 15th February, 1889. These units, with the Nowgong Mounted Rifles, raised on the 27th July, 1888, and the Gauhati Rifles, formed on the 13th November, 1885, were amalgamated as the Assam Valley Mounted Rifles on 6th November, 1891. Present designation 25th September, 1896. The Shillong Volunteer Rifles, formed on the 11th August, 1882, were absorbed on the 1st May, 1906.

Uniform—Blue.
Facings—White.
Badge—A.V.L.H. Monogram.
Motto—"Semper Paratus".
Honorary Colonel—The Governor of Assam.

The Regiment had ths honour of sending a detachment on service with the Abor Expedition on the North-east Frontier of India in 1911-12.

The United Provinces Horse (Southern Regiment)

The Allahabad Troop Light Horse was formed on the 12th September, 1884, from the Allahabad Volunteer Rifle Corps and became the Allahabad Light Horse in 1890. The Cawnpore Light Horse, formed on 5th February, 1886, from the Cawnpore Volunteer Rifle Corps; the Ghazipur Light Horse, formed on 7th January, 1887 from the Ghazipur Volunteer Rifle Corps; the Gorakhpur Light Horse, formed on 20th May, 1887, and the Oudh Light Horse, formed on 10th June, 1887, from the Oudh Volunteer Rifle Corps, were amalgamated as the United Provinces Light Horse on the 1st April, 1904. Designation changed to 1st United Provinces Horse, 1st April, 1909, and to 7th (Southern Regiment) United Provinces Horse on 1st April, 1917. Present designation 1st October, 1920.

Uniform—Khaki.
Badge—A swordsman with a drawn sword on a Prancing Charger. Below it inscribed "Relief of Lucknow, 1857" and "United Provinces Horse."
Honorary Colonel—The Governor of the United Provinces.

The Northern Bengal Mounted Rifles

Formed on 6th August, 1873, as the Northern Bengal Volunteer Rifle Corps. Absorbed the Darjeeling Volunteer Rifle Corps on 5th August, 1881. Reorganized on 15th February, 1889, under the present designation.

Uniform—Scarlet.
Facings—White.
Badge—The Bengal Tiger.
Motto—"Fideliter".

The Punjab Light Horse

Raised on the 12th May, 1893.
Uniform—Blue.

Facings—White.
Badge—An eight-pointed star surmounted by a crown.
Motto—"De Bon Voulir Servir le Roy".
Honorary Colonel—The Governor of the Punjab.

The Southern Provinces Mounted Rifles

Formed on the 12th February, 1904.
Uniform—Blue.
Facings—White.
Badge—A bugle.
Honorary Colonel—The Governor of Madras.

The Chota Nagpur Regiment

Formed as the Chota Nagpur Mounted Rifles on 7th August, 1891. Became the Chota Nagpur Light Horse on 2nd March, 1910. Present designation, 1st April, 1917.

Uniform—Khaki.
Facings—White.
Badge—A horse shoe surmounted by a crown encircling a cobra in the act of striking.
Motto—"I strike to kill".
Honorary Colonel—The Governor of Bihar.

The Bombay Light Patrol

Formed from the Bombay Light Horse and the Bombay Light Motor Patrols of the Bombay Contingent. Constituted the Bombay Light Patrol, 1933.

Badge—An eight-pointed star surmounted by a crown. In the centre of the star the Bombay Coat of Arms encircled by a band inscribed with motto. Below a scroll inscribed "Bombay Light Patrol".
Motto—"Urbs Prima in Indis".

ROYAL ARTILLERY (A.F.,I.)

The Bengal Artillery

Constituted on the 18th December, 1925, by an amalgamation of the I. (Calcutta) Field Brigade and the V. (Cossipore) Field Brigade.

The Calcutta Brigade were founded as the Calcutta Naval Artillery Volunteers on 21st September, 1883, and became the (Calcutta Port Defence) Group Garrison Artillery on 1st April, 1917. Reconstituted 1st October, 1920, as I. (Calcutta Port Defence) Field Brigade.

The Cossipore Brigade started life as the Cossipore Artillery Volunteers on 1st February, 1884. Reorganized as the Cossipore Artillery Volunteers, 7th September, 1888. Became I. (Cossipore) Brigade Mobile Artillery, 1st April, 1917. Reconstituted 1st October, 1920, as the IV. (Cossipore) Field Brigade.

The Bengal Artillery were the first artillery in India to be tractor drawn.

The uniform is blue with scarlet facings. Badge as for Royal Artillery.

No. 3 (Madras) Field Battery. "The Duke's Own".

Raised on 1st January, 1879, from the Madras Volunteer Guards and designated the Madras Artillery Volunteers. Became 2nd (Madras) Group Garrison Artillery on 1st April, 1917. Constituted as II. (Madras) Field Brigade "The Duke's Own", 1st October, 1920, the subsidiary title being conferred by H.R.H. The Duke of Connaught on his visiting India to inaugurate the "Montfort" reforms in the Indian Government of that year. Reorganised as No. 3 (Madras) Field Battery "The Duke's Own", 1933.

No. 10 (Bombay) Battery

Formed as the Bombay Volunteer Artillery on 6th June, 1887; became the 4th (Bombay) Group Garrison Artillery on 1st April, 1917, and V. (Bombay) Field Brigade on 1st October, 1920. Received present title, 13th April, 1933.

No. 13 (Lucknow) Field Battery

Formed as the Oudh Volunteer Rifle Corps on 10th August, 1865. Reorganised as the Lucknow Volunteer Rifle Corps on 10th January, 1872. Became the Oudh Volunteer Rifle Corps on 6th June, 1884, and again the Lucknow Volunteer Rifles, 24th July, 1903. Reconstituted as the 5th (Lucknow) Group Garrison Artillery on 1st April, 1917; became VI. (Lucknow) Field Brigade on 1st October, 1920. Present designation, 1933.

No. 15 (Kirkee) Field Battery,

Constituted as F. (Kirkee) Battery Mobile Artillery on 1st April, 1917. Present designation, 1st October, 1920.

No. 17 (Agra) Field Battery

Formed as the Agra Volunteer Rifle Corps on 9th August, 1876. Became the 19th Agra Company on 1st April, 1917. Present designation, 1st October, 1920.

ROYAL ENGINEERS (A.F.,I)

No. 1 (Calcutta) Fortress Company

Raised 1st April, 1902, from the Calcutta Port Defence Volunteers. Became No. 1 Electrical Engineer Company, 1st

April, 1917, and No. 1 (Calcutta) Field Company, 1st October, 1920. Present designation, 1st April, 1933.

No. 3 (Bombay) Fortress Company

Raised 1st April, 1902, from the Bombay Volunteer Artillery. Became No. 3 Electrical Engineer Company, 1st April, 1917. No. 3 (Bombay) Field Company, 1st October, 1920. Reconstituted as No. 3 (Bombay) Fortress Company, 1933. Present designation, 13th April, 1933.

No. 4 (Karachi) Fortress Company

Raised 1st April, 1902, from the Karachi Artillery Volunteers. Became No. 4 Electrical Engineer Company, 1st April, 1917. No. 4 (Karachi) Field Company, 1st October, 1920. No. 4 (Karachi) Electrical and Mechanical Company, 1st May, 1925. No. 4 (Searchlight) Company, 10th February, 1928. Present designation, 1st April, 1933.

ROYAL CORPS OF SIGNALS (A.F.,I)

No. 1 (Madras) Signal Company

Raised from the Madras Artillery Volunteers and became No. 5 (Madras) Field Company, 1st October, 1920. Reconstituted with present designation, 23rd June, 1928.

INFANTRY

The East Indian Railway Regiment

Raised 17th July, 1869, as the East Indian Railway Volunteer Rifle Corps. The Sibpore College Volunteer Rifle Corps, formed 19th August, 1881, incorporated on 20th May, 1890. St. Michael's School Cadet Corps, raised 15th April, 1892, amalgamated as the 7th East Indian Railway Battalion on 1st April,

1917. Became the 7th East Indian Railway Corps, 8th November, 1918. Present designation, 1st October, 1920. Amalgamated in 1926 with the Oudh and Rohilkhand Railway Battalion.

Uniform—Khaki.
Facings—White.
Badge—The crest of the East Indian Railway Company surmounted by a crown and encircled by a laurel wreath bearing the words "East Indian Railway Regiment Auxiliary Force", and the motto.
Motto—"Strong without Rage".
Honorary Colonel—The Viceroy of India.
The Regiment has two battalions.

The Eastern Bengal Railway Battalion

Raised 1st March, 1873, as the Eastern Bengal State Railway Volunteer Rifle Corps. The Northern Bengal State Railway Volunteer Rifle Corps, raised 24th January, 1879, were amalgamated in 1882 as the Eastern Bengal Railway Volunteer Rifle Corps. Resumed original designation on 21st November, 1884. Became the 12th Eastern Bengal Railway Battalion, 1st April, 1917. Present designation, 1st October, 1920.

Uniform—Khaki.
Facings—White.
Badge—A tiger's head encircled by a laurel wreath and surmounted by a crown; above the crown a scroll inscribed with the motto, below a scroll inscribed "Eastern Bengal Railway Battalion".
Motto—"Semper Paratus."

The Great Indian Peninsula Railway Regiment

Raised 29th December, 1875, as the Great Indian Peninsula Railway Volunteer Corps. The Midland Railway Volunteer Corps, raised 22nd August, 1890, was amalgamated as the 2nd Battalion on 5th August, 1908. Became the 13th Great

Indian Peninsula Railway Battalion, 1st April, 1917. Present designation, 1st October, 1920.

Uniform—Khaki.
Facings—Scarlet.
Badge—Stringed bugle surmounted by a crown. Below a scroll inscribed "G.I.P. Rly. Regt."
Honorary Colonel—The Governor of Bombay.
The Regiment has two battalions.

The Bombay, Baroda and Central India Railway Regiment

Raised 3rd August, 1877, as the Bombay, Baroda and Central India Railway Volunteer Corps. The Ghadeshi Volunteer Rifle Corps, formed 23rd December, 1885, were amalgamated in 1886. The Rajputana-Malwa Volunteer Rifle Corps, raised 8th August, 1882, were amalgamated as 2nd Battalion in 1887. Became the 17th Bombay, Baroda and Central India Railway Battalion on 1st April, 1917. Present designation, 1st October, 1920.

Uniform—Khaki.
Facings—White.
Badge—Crossed bayonets and flags and crown with railway crest superimposed.
The Regiment has two battalions.

The Bengal and North Western Railway Battalion

Formed as the Tirhoot State Railway Volunteer Rifle Corps on 17th January, 1879. The Ghazipur Volunteer Rifle Corps, raised 14th June, 1879, became the Ghazipur Volunteer Rifle Battalion on 11th November, 1881, resumed former designation in 1885, and amalgamated as the Bengal and North-Western Railway Volunteer Corps on 17th June, 1892. Became 22nd Bengal and North-Western Railway Battalion, 1st April, 1917. Present designation, 1st October, 1920.

Uniform—Khaki.
Facings—White.
Badge—St. Andrew's Cross in a wreath of thistle.

The North-Western Railway Battalion

Formed as the 3rd or Sind, Punjab and Indus Valley Railways Volunteer Rifle Corps on 5th March, 1886. Designation changed 1888 to 3rd Punjab (North-Western Railway) Volunteer Rifle Corps. Became the North-Western Railway Volunteer Rifles on 27th July, 1892, and the 24th North-Western Railway Battalion, 1st April, 1917. Became the North-Western Railway Regiment, 1st October, 1920. Present designation, 1st April, 1933.

Uniform—Khaki.
Badge—N.W. Railway.
This Corps frequently supports the regular forces during trouble on the N.W. Frontier by manning armoured trains.

The South Indian Railway Battalion

Formed 5th August, 1884, as the South Indian Railway Volunteer Rifle Corps. Became the 29th South Indian Railway Battalion, 1st April, 1917. Present designation, 1st October, 1920.

Uniform—Khaki.

The Madras and Southern Mahratta Railway Rifles

1st Battalion formed as the Madras Railway Volunteers on 21st August, 1885. 2nd Battalion formed as the Southern Mahratta Railway Rifle Corps on 20th October, 1886. The Bellary Volunteer Rifle Corps, raised 24th July, 1885, were amalgamated, 28th January, 1898. Reconstituted the Madras and Southern Mahratta Railway Rifles, 4th June, 1910. Became the 32nd Madras and Southern Mahratta Railway Rifles, 1st April, 1917. Present designation, 1st October, 1920.

Uniform—Khaki.
Facings—White.
Badge—An eight pointed star surmounted by a crown with a bugle encircled by a garter in the centre.
Motto—"One and all".
The regiment has two battalions.

The Bengal Nagpur Railway Battalion

Raised as the Bengal Nagpur Railway Volunteer Rifle Corps on 15th June, 1888. Amalgamated 18th November, 1898, with the Orissa Volunteer Rifles. Reorganised into a Corps of two Battalions, 1903. Became the 36th Bengal Nagpur Railway Regiment, 1st April, 1917. The Bengal Nagpur Railway Regiment 1st October, 1920. Present designation, 13th April, 1933.

Uniform—Khaki.

The Assam Bengal Railway Battalion

Formed as the Assam-Bengal Railway Volunteer Rifles on 23rd January, 1901. Became the 42nd Assam Bengal Railway Battalion, 1st April, 1917. Present designation, 1st October, 1920.

Uniform—Khaki.
Facings—Green.
Badge—Bengal tiger encircled by laurel wreath, ensigned by a royal crown.
Motto—"In Omnia Paratus".

The Madras Guards

Raised on 2nd July, 1857, as the Madras Volunteer Guards. Became the 1st Madras Guards on 1st April, 1917. Present designation, 1st October, 1920.

Uniform—Blue.
Facings—Scarlet.
Badge—The Royal Arms within a garter—over it a crown with motto underneath.
Motto—"Ready aye Ready".
Honorary Colonel—The Governor of Madras.

The Nagpur Rifles

Formed 29th December, 1860. Absorbed the Berar Volunteer Rifle Corps raised 21st February, 1879, on 1st October, 1904. Became the 2nd Nagpur Rifles on 1st April, 1917. The Nagpur Rifles on 1st October, 1926. The Nagpur Regiment on 24th December, 1927. Present designation on 12th June, 1937.

Uniform—Khaki.
Badge—The Light Infantry Bugle and Cord.
Honorary Colonel—The Governor of the United Provinces

The Punjab Rifles

Raised as the 1st Punjab Volunteer Rifle Corps on 1st April, 1861. Absorbed on 15th July, 1871, the Punjab Light Horse (Volunteers) raised on 27th March, 1867. Became the 3rd Punjab Rifles on 1st April, 1917. Present designation 1st October, 1920.

Uniform—Khaki.
Facings—Scarlet.
Badge—The Punjab Coat of Arms.
Motto—"Crescate Fluviis".
Honorary Colonel—The Governor of the Punjab.

The Simla Rifles

Raised on 2nd May, 1861, as the 2nd Punjab or Simla Volunteer Rifle Corps. Became the Simla Volunteer Rifles in 1904

OFFICER, THE PUNJAB RIFLES, A.F.,I.—1937

and the 4th Simla Rifles on 15th April, 1917. Present designation, 1st October, 1920.

Uniform—Khaki.
Facings—Rifle Green.
Badge—The Monogram "S.R." surmounted by a crown, pendent therefrom a bugle horn stringed. "The Simla Rifles, A.F.I." on a scroll beneath.
Motto—"In Defence".
Honorary Colonel—The Viceroy.

The Calcutta and Presidency Battalion

The Calcutta Battalion—formed on 3rd February, 1863, as the Calcutta Volunteer Rifle Corps. Reconstituted as the 1st Battalion, Calcutta Volunteer Rifles, on 24th March, 1898. Became the 5th Calcutta Battalion on 1st April, 1917, and the Calcutta Battalion on 1st October, 1920.

The Calcutta Presidency Battalion—raised as the Presidency Volunteer Reserve Battalion on 30th March, 1888. Became the Presidency Volunteer Rifle Battalion on 3rd July, 1891; and the Presidency Battalion, Calcutta Volunteer Rifles, on 24th March, 1898; the 37th Calcutta Presidency Battalion, 1st April, 1917. The Calcutta Presidency Battalion, 1st October, 1920. Amalgamated, with present designation, 23rd July, 1926.

Uniform—Khaki.
Facings—White.
Badge—Arms of Calcutta surmounted on crossed rifles.
Motto—"Per Ardua Stabilis Esto".

The Bangalore Battalion

Raised as the Bangalore Rifle Volunteers on 21st December, 1868. The Coorg and Mysore Rifles became a separate unit on 21st November, 1884, but were amalgamated as the 6th Bangalore, Coorg and Mysore Battalion on 1st April, 1917. Present designation, 1st October, 1920.

Uniform—Rifle Green.
Facings—Scarlet.
Motto—"Defence not Defiance".

The Allahabad Rifles

Formed 11th January, 1871; became the 8th Allahabad Rifles, 1st April, 1917. Present designation, 1st October, 1920.

Uniform—Khaki.
Honorary Colonel—The Governor of the United Provinces.

The Dehra Dun Contingent

Formed 24th July, 1871, as the Mussoorie Volunteer Rifle Corps. The Thomason College Volunteer Rifle Corps, formed 19th August, 1872, and the Mussoorie Volunteer Reserve Corps, formed 13th August, 1889, were incorporated on 4th March, 1901. Became the 9th Mussoorie Battalion on 1st April, 1917; the Mussoorie Battalion, 1st October, 1920. Amalgamated with the Dehra Dun Detachment, U.P. Horse (N.), and the Meerut Detachment, No. 5 Coy., M.G. Corps, and reconstituted as the Dehra Dun Contingent 10th July, 1925.

Uniform—Khaki.
Badge—Himalayan Bear with the Motto.
Motto—"Strike Fast".

The Bareilly Contingent

Formed as the Naini Tal Volunteer Rifle Corps on 26th July, 1871. Became the 10th (Naini Tal) Company on 1st April, 1917; and the Naini Tal Rifles, 1st October, 1920. Amalgamated with the Bareilly Detachment U.P. Horse (N) and reconstituted as the Bareilly Contingent, 18th December, 1925.

Uniform—Khaki.
Badge—Crossed rifles upon a laurel wreath surmounted with a crown with the word "Bareilly" upon a scroll below.
Honorary Colonel—The Governor of the United Provinces.

The Bombay Battalion.

Raised as the Bombay Volunteer Rifle Corps on 15th August, 1877. Became the 16th Bombay Battalion 1st April, 1917. Present designation, 1st October, 1920.

Uniform—Khaki.
Badge—Crown on bugle.
Honorary Colonel—The Governor of Bombay.

The Cawnpore Rifles

Formed as the Cawnpore Volunteer Rifle Corps on 16th August, 1877. Became the 16th Cawnpore Rifles on 1st April, 1917. Present designation, 1st October, 1920.

Uniform—Khaki.
Badge—The Cawnpore Angel at the Well.
Motto—"Defence not defiance".
Honorary Colonel—The Governor of the United Provinces.

The Nilgiri Malabar Battalion

Raised as the Nilgiri Volunteer Rifles on 29th October, 1878. The Coimbatore Volunteer Corps, formed 7th August, 1885, incorporated on 4th January, 1892. The Malabar Volunteer Rifles, formed 14th August, 1885 from the Calicut and Tellichery Volunteer Corps, were amalgamated as the 20th Nilgiri Malabar Battalion on 1st April, 1917. Present designation, 1st October 1920.

Uniform—Khaki.
Badge—Stag's head encircled by a garter inscribed "Auxiliary Force India", and surmounted by a crown. The whole encircled by a coconut palm wreath and below a scroll inscribed "Nilgiri Malabar Batt".
Honorary Colonel—The Governor of Madras.

The Sind Rifles

Raised as the Sind Volunteer Rifle Corps on 11th November, 1879. Became the 23rd Sind Battalion, 1st April, 1917. Present designation, 1st October, 1920.

Uniform—Khaki.
Honorary Colonel—The Governor of Sind.

The Hyderabad Rifles

Formed as the Hyderabad Volunteer Rifle Corps, 7th July, 1882. Became 26th Hyderabad Rifles, 1st April, 1917. Present designation, 1st October, 1920.

Uniform—Khaki.

The Eastern Bengal Company

Raised as the Dacca Volunteer Rifle Corps on 14th November, 1884. The Chittagong Volunteer Rifle Corps, formed 24th July, 1885, amalgamated as the Eastern Bengal Volunteer Rifles on 6th November, 1900. Became 41st Eastern Bengal Company, 1st April, 1917. Present designation, 1st October, 1920.

Uniform—Khaki.
Facings—Yellow.
Badge—A crocodile.
Motto—"Vis Et Tenax".

The East Coast Battalion

Raised as the Godavari Rifle Volunteers on 9th June, 1885, and amalgamated with the Vizagapatam Rifle Volunteers raised 10th October, 1885, to form the East Coast Rifle Volunteers on 14th October, 1890. Became the East Coast Volunteer Rifles on 14th October, 1903, and the 38th East Coast Battalion on 1st April, 1917. Present designation, 1st October, 1920.

Uniform—Khaki.
Badge—Dolphin surmounted by Tudor crown.
Motto—"Manu Forte".
Honorary Colonel—The Governor of Madras.

The Poona Rifles

Raised as the Poona Volunteer Rifles on 4th January, 1887. Became the 35th Poona Battalion, 1st April, 1917. Present designation, 1st October, 1920.

Uniform—Khaki.
Honorary Colonel—The Governor of Bombay.

The Kolar Gold Fields Battalion

Formed from a detachment of the Bangalore Rifle Volunteers on 26th June, 1903. Became the 43rd Kolar Gold Fields Battalion 1st April, 1917. Present designation, 1st October, 1920.

Uniform—Khaki.
Badge—Crossed pickaxe and hammer in a circle superscribed "The Kolar Gold Fields Bn." The whole surmounted by a crown.

The Calcutta Scottish

Formed as the Calcutta Scottish Volunteers on 1st August, 1911. Became the 44th Calcutta Scottish 1st April, 1917. Present designation, 1st October, 1920.

Uniform—Scarlet.
Facings—Yellow.
Badge—The Cross of St. Andrew with the Coat of Arms of Calcutta superimposed; in the upper angle of the cross, a crown.
Tartan—Hunting Stewart.
Glengarry—Green and white diced border; green pom-pom.
Motto—"Per Ardua Stablis Esto".

The Delhi Contingent

Formed 4th February, 1921, from detachments of the Punjab Light Horse and the Punjab Rifles.

Uniform—Khaki.

The Coorg and Mysore Company

Formed from the Bangalore Coorg and Mysore Battalion as No. 1 (Coorg and Mysore) Coy. and reorganized with the present designation in 1933.

Uniform—Khaki.
Facings—Green.
Badge—Bison Head.

The Lucknow Rifles

Constituted as No. 7 (Lucknow) Company Machine Gun Corps, 1st October, 1920. Reconstituted as the Lucknow Rifles, 1933. Present designation, 13th April, 1933.

Badge—The Lucknow Residency with "Defence of Lucknow, 1857".

The Yercaud Company

Formed from No. 13 (Yercaud) Company, Machine Gun Corps. Constituted as an Infantry unit, 1933. Present designation, 13th April, 1933.

The Bhusawal Company

Constituted 1st August, 1936.

MACHINE GUN CORPS (A.F.,I.)

No. 2 (Karachi) Company

Formed as the Karachi Artillery Volunteers on 25th October, 1892. Reorganized as the 2nd (Karachi) Brigade Mobile Artillery on 1st April, 1917. Reconstituted, 1st October, 1920.

Uniform—Khaki.
Badge—A camel surrounded by a garter within the rising sun, surmounted by a crown, superscribed "Karachi Corps Auxiliary Force".
Honorary Colonel—The Governor of Sind.

No. 5 (Agra) Company

Formed as No. 5 (Agra) Company, Machine Gun Corps, 1st October, 1920.

Uniform—Khaki.

The Bangalore Armoured Car Company

Formed from the No. 14 (Bangalore) Field Battery, Royal Artillery, 1933. Present designation, 13th April, 1933.

Uniform—Khaki.

CHAPTER LXVI

The War and Meritorious Service Medals issued to the Army of India

THE Honourable East India Company were the first government in the world to originate the practice of giving a medal to all ranks to commemorate a successful campaign, and many medals were issued by the Directors during the course of their long rule. The following notes briefly describe the principal war and meritorious service medals issued to Indian soldiers.

The Army of India Medal.
The first form of General Service Medal in India was that of the Army-in-India medal issued by the Company with the approval of Queen Victoria on April 14th, 1851, and given to the survivors of campaigns engaged in wars in India for which no special medal had been issued between 1799 and 1826. The ribbon is pale-blue corded silk. Twenty-one bars were struck with the medal to commemorate the following actions—ALLIGHUR, DELHI, ASSAYE, ASSEERGHUR, LASWARREE, ARGAUM, GAWILGHUR, DEFENCE OF DELHI, DEIG, CAPTURE OF DEIG, NEPAUL, KIRKEE, POONA, KIRKEE AND POONA, SEETABULDEE, NAGPORE, SEETABULDEE AND NAGPORE, MAHEIDPOOR, CORYGAUM, AVA, BHURTPOOR, the bars being issued to those who took part in the respective battles.

India General Service Medal 1854.
This was sanctioned by Queen Victoria on January 23rd, 1854, primarily for reward to those who had taken part in the

"BRAVE AS A LION"
SUBADAR-MAJOR MAULADAD KHAN, C.I.E., I.O.M. ENLISTED IN 1847 AND RETIRED IN 1890. A REMARKABLE TYPE OF THE OLD STYLE INDIAN OFFICER OF THE PUNJAB

Burma War of 1852-53. To obviate the necessity for giving medals for each campaign, and to avoid multiplying the number which might adorn a warrior's chest, it was later styled the India General Service Medal and was given for twenty-three different small wars between 1852 and 1893; a separate clasp being given for each campaign. The ribbon is dark crimson with two dark blue stripes. The following clasps or bars were issued—PEGU, PERSIA, NORTH-WEST FRONTIER, UMBEYLA, BHOOTAN, LOOSHAI, PERAK, JOWAKI, BURMA 1885-87, BURMA 1887-89, NAGA 1879-80, SIKKIM 1888, HAZARA 1888 CHIN-LUSHAI 1889-90, SAMANA 1891, HAZARA 1891, BURMA 1889-92, LUSHAI 1889-92, NORTH-EAST FRONTIER 1891, HUNZA 1891, CHIN HILLS 1892-93, KACHIN HILLS 1892-93.

The Great Mutiny 1857-58.

The medal for the campaigns of the Great Mutiny was the last medal to be issued by the East India Company. This ribbon is scarlet and white stripes arranged alternately, three of white, two of red. The clasps awarded were—DELHI, DEFENCE OF LUCKNOW, RELIEF OF LUCKNOW, LUCKNOW, CENTRAL INDIA. No man was able to earn all five clasps, and only a few received four, while the infantry soldiers did not earn more than two.

The India General Service Medal 1895.

This was issued in 1895 after the Chitral campaign. The ribbon is red with two green stripes the same width as the red spacing between them. Clasps issued were—DEFENCE OF CHITRAL, RELIEF OF CHITRAL, PUNJAB FRONTIER 1897-98, MALAKAND 1897, SAMANA 1897, TIRAH 1897-98.

The India General Service Medal 1903.

This has the same ribbon and design as the 1895 medal except that it bears the head of King Edward VII. One clasp only was issued—WAZIRISTAN 1901-2.

The India General Service Medal 1908.

First issued in 1910 to mark the North-west Frontier campaign of 1908, with a clasp NORTH-WEST FRONTIER 1908. This medal bore the head of King Edward VII. The ribbon is green with dark blue edges. The medal was re-issued for subsequent campaigns bearing the head of King George V with the following clasps—those who had received the earlier medal receiving clasps only—ABOR 1911-12, AFGHANISTAN 1919, MAHSUD 1919-20, WAZIRISTAN 1919-21, MALABAR 1921-22, WAZIRISTAN 1921-24, MOHMAND 1933, NORTH-WEST FRONTIER 1930-31. BURMA 1930-32, NORTH-WEST FRONTIER 1935.

The India General Service Medal 1936.

A new medal for the border campaigns in India of 1936-37 was struck in 1938, the initial clasp being NORTH-WEST FRONTIER 1936-37. The ribbon is khaki in the centre, flanked on each side by a narrow red stripe and edged with broader green stripes.

The following medals were issued to commemorate special campaigns conducted in or from India outside British territory:—

China 1842.

Issued in January, 1843. Ribbon—Red with broad yellow edges.

First Jellalabad.

Issued April 30th, 1842. Ribbon—Shaded rainbow crimson to yellow and blue. Given for the Defence of Jellalabad.

Second Jellalabad.

Issued in March, 1845, to replace by exchange the crudely designed First Jellalabad Medal, and has the same ribbon. Few recipients, however, changed their medals.

Ghuznee, Candahar, and Cabul 1842, Meanee, Hyderabad.

Medals of more or less the same design, bearing on the reverse the name or names of these actions or campaigns, were issued to those who took part. The ribbon of the medals is the rainbow type.

Gwalior.

Six-pointed bronze stars bearing the words PUNNIAR OR MAHRAJPOOR 1843 were issued for this campaign, the ribbon being the rainbow type.

Sikh Wars.

The Sutlej Medal. A medal bearing the words "Army of the Sutlej" was issued to those who took part in the first Sikh War, 1845-46. The first engagement the recipient took part in was engraved upon the medal, subsequent actions being commemorated by clasps. Thus a man who took part in all four actions had MOODKEE on the medal, and clasps FEROZESHUHUR, ALIWAL, SOBRAON. The ribbon is dark blue, edged with dark crimson. This was the first Indian medal to be issued with clasps.

The Punjab Medal. This is similar to the Sutlej medal in design and was struck for the campaign of 1848-49; it bears the words "To the Army of the Punjaub". The clasps issued were MOOLTAN, CHILIANWALA, GOOJERAT. The ribbon is dark blue with bright-yellow stripes at the edges.

China 1857-60.

This medal and ribbon were the same as for the 1842 campaign, the clasps being CANTON 1857, TAKU FORTS 1858, TAKU FORTS 1860, PEKIN 1860; a clasp FATSHAN 1857 was issued to sailors and marines only.

Abyssinia 1864.

A medal was granted for this campaign in 1869, the ribbon being red with broad white borders.

Afghanistan 1878-80.

Six bars were issued with this medal granted on March 19th, 1881—ALI MUSJID, PEIWAR KOTAL, CHARASIA, AHMED KHEL, KABUL, and KANDAHAR. The ribbon is green with broad dark-red edges.

A decoration was also issued to the troops who took part in the famous march from Kabul to Kandahar. This is a five-pointed star, with small balls between the inner angles, and was made from the bronze guns taken at Kandahar. The ribbon is the single rainbow pattern, but unwatered.

Tibet 1903-4.

The medal issued for this expedition has a ribbon of dark red edged with green, with two white stripes: there is one clasp GYANTSE for those who took part in the operations about that place.

Indian troops have also earned medals in the following campaigns:—

Egypt 1882-85.

Medal ribbon—Three blue stripes and two white, with bars for TEL-EL-KEBIR, SUAKIN 1885, TOFREK. A bronze star was also presented by the Khedive of Egypt bearing the Sphinx and having a blue ribbon.

East and Central Africa 1897-98.

Medal ribbon—Half red half yellow. Bars issued to Indian troops were LUBWA'S, UGANDA 1897-98, "1898" for those who fought against the Ogaden Somalis, and UGANDA 1899.

China 1900.

The troops who took part in suppressing the Boxer rebellion received the same medal and ribbon as for the 1842 war but bearing a different head of Queen Victoria; one clasp was issued to Indian troops—RELIEF OF PEKIN.

South Africa 1899–1902.

This medal was given to men of the Indian cavalry who served with the remount department and to certain non-combatant followers. The ribbon has a broad centre stripe of orange, with dark blue and red stripes at the sides.

East Africa and Somaliland 1902–04.

Clasps of the Africa General Service Medal issued to Indian troops were—JUBALAND, SOMALILAND 1901, UGANDA 1900, B.C.A. 1899–1900 (the latter meaning British Central Africa), SOMALILAND 1902–04. The ribbon is yellow with two narrow green stripes between a broad black edging.

This medal was also issued for service in Somaliland, 1920, to men of the now 1st Battalion 4th Bombay Grenadiers.

The War of 1914–20.

1914 Star. A bright bronze decoration with a ribbon of red, white and blue, shaded and watered, was given to all who served in France and Flanders between 5th August, 1914, and midnight of 22nd-23rd November, 1914. To those who served within range of the enemy's mobile artillery between those dates a bronze clasp bearing the dates 5th August—22nd November, 1914, was also awarded.

1914–15 Star. The ribbon is similar to the 1914 star. This was awarded to all those who served overseas or on the North-west Frontier, India, between the 5th August, 1914, and 31st December, 1915, and who had not received the 1914 Star.

British War Medal. The ribbon is of deep yellow edged with light blue and white stripes with a thin black stripe between. This was given all who served in a theatre of war or on certain defence duties in India or overseas between 5th August, 1914, and 11th November, 1918.

Victory Medal. To avoid the interchange of allied war medals, each of the allied nations presented a medal struck in bronze, bearing the winged figure of victory, and having a double rainbow ribbon, to all ranks who served in a theatre of war only between 5th August, 1914, and 11th November, 1918. A bronze oak-leaf is worn on the ribbon by those who have been mentioned in despatches.

The General Service Medal 1923.

This was given to Indian troops for the military operations in Iraq, Persia and Kurdistan, 1924–29. Clasps issued were IRAQ, NORTH-WEST PERSIA, SOUTH PERSIA, KURDISTAN, in 1924; SOUTHERN DESERT IRAQ in 1929, and NORTHERN KURDISTAN in 1934. The ribbon is purple with an emerald green centre.

The following are the Meritorious Service Medals and Decorations awarded to the Indian Army for gallantry in the field:—

The Victoria Cross.

Instituted by Queen Victoria in 1856. The highest decoration awarded by the sovereign for a signal act of valour in the face of the enemy. Ribbon—Dark red, with a small replica of the cross thereon when worn alone. Indian soldiers were made eligible for this decoration by the King-Emperor's pronouncement at the Imperial Durbar at Delhi in December, 1912.

The Indian Order of Merit.

Instituted in 1837 for personal bravery and was originally the Indian soldier's Victoria Cross. Divided into three classes, a recipient can be advanced from one class to another. Ribbon—Dark blue, edged with red.

The Distinguished Service Order.

Instituted in 1866 and given to officers performing exceptional service in the field. Since 1917 has only been awarded

for acts of gallantry in the face of the enemy. The recipient must have first been mentioned in dispatches. Ribbon—Red, edged with blue.

East India Company's Meritorious Service Medal.

Instituted in 1848 and given for meritorious service, to the Company's troops. Ribbon—Red.

The Military Cross.

Instituted 31st December, 1914. Given for distinguished service in action to captains, subalterns and warrant officers. Ribbon—White, with a purple stripe.

The Indian Distinguished Service Medal.

Instituted 28th June, 1907, for distinguished service in the field of Indian officers and men. Ribbon—Red, with broad blue edges.

Order of British India.

Given for long and faithful service. Ribbon—Formerly blue, now red.

India Meritorious Service Medal.

Instituted 1888. Ribbon—Red, edged with white, with a white stripe down the centre.

"THE FIRST KHAKI UNIFORM"

A SOWAR, THE CORPS OF GUIDES, IN 1850

CHAPTER LXVII

"Khaki"

As the use of khaki for military dress developed in India a note on its history may properly find a place in this book. As will have been read, the inventor was Lieutenant (afterwards Lieut.-General Sir Harry) Lumsden, who raised "The Queen's Own Corps of Guides" in December, 1846.

Lumsden took the view that the border *banditti* could be best dealt with in a manner which they would well understand, if his own men wore their native style clothing, rather than adopt the semi-European dress then current in the Indian forces, and which the infantry battalions of the Punjab Frontier Force raised at the same time as the Guides were wearing.

No attempt at uniformity appears to have been made until the Corps had been in existence for more than a year, when were adopted a smock and wide pyjama trousers made of coarse home-spun cotton material, a cotton turban, and over all in the cold weather, jerkins of sheepskin and wadded cotton jackets. All were dyed locally with the product of a dwarf palm known as *mazari*, which possessed the property of colouring white cloth a drabbish grey; an indigenous method of producing camouflage clothing developed by the tribesmen as the result of constant vendettas and domestic wars. The leather jackets, however, dyed unsuccessfully with *mazari*, so mulberry juice was used instead, which produced a yellowy drab shade. This was the origin of drab or khaki uniform.

Camouflage clothing for those engaged in hostile operations where concealment afforded a prime factor to success was not a new military idea. Its first use on any scale was during the

colonizing wars in North America, two regiments then being dressed in short brown coats to make them less conspicuous in the woods. Then came the development of the rifleman dressed in dark green, operating as a scout and a sharpshooter to deal with the snipers who greatly harried the troops. Rifle regiments were subsequently formed, whose chief characteristic in tactical operations was concealment—hence no colours were carried to mark the rallying point, no polish or pipe-clay which would show them up. So thus for a time green became the principal colour for concealment in the armies of the world, including that of India.

Khaki, however, was best suited to the barren rocky mountains and plains of the Indian North-west. All the units of the Punjab Frontier Force save one—"Coke's Rifles" who stuck to indigo green—adopted khaki shortly after the Guides established it. Practically all the units hastily raised in the Punjab for service during the Great Mutiny, 1857–58, wore it at that time.

White uniform dyed dust colour was also then taken into use by British units, the first being the 52nd Foot, now the 2nd Oxfordshire and Buckinghamshire Light Infantry, at Sialkot on the 25th May, 1857. During the Siege of Lucknow, which began in July, 1857, archivists have recorded that the 32nd Foot, now the 1st Duke of Cornwall's Light Infantry, dyed their white uniforms an earth colour to make themselves less conspicuous. These two, however, despite what may be said to the contrary, were not the first British units to wear khaki, for the present 2nd Battalion Highland Light Infantry wore drill doublets so coloured, during the Kaffir War in South Africa in 1851–53. These, however, were isolated instances. And, what is more, khaki, except in a few instances, went out of general use in India as soon as the Great Mutiny was over.

It was re-introduced as a working dress in 1861 to the abhorence of the parade ground soldier, who saw it abolished in favour of white three years later.

White drill, dyed khaki on the spot, was used by all troops who took part in the Afghan War of 1878–80—the lessons taught by the red-coated battalions passing through the frontier defiles in the war of 1839 had been learnt.

After this war khaki became the official service dress in India. It was dyed regimentally, all kinds of pigments being used for the purpose—tea, coffee, mud, curry-powder. Each corps had its own recipe, some of which smelt high to heaven when dyeing operations were in progress.

Fast-dyed khakis invented by Spinners of Manchester were patented in 1884, but were not adopted in India for some years, the Queen's Own Corps of Guides continuing to dye their own until 1904.

Grey was the first British choice for active service, and soldiery so dressed were sent to Egypt to serve alongside khaki-clad troops from India during the campaigns of 1880–82.

For the re-conquest of the Sudan, 1897–98, all the troops were uniformed in khaki, and it was this campaign which really brought about its establishment as universal service dress for the British Army.

And so by the time of the second Boer War in 1899 the word khaki—the Persian for dust colour—had begun to find a place in English dictionaries.

The Americans started the adoption of khaki in the early nineteen-hundreds, Japan in 1905, the French following suit for their colonial forces about 1910, the Belgian and the Spanish armies in 1919. It is becoming more and more universal, and as a British War Minister a few short years ago said, when pressed to return the troops to "coloured" magnificence, "the uniform outlook gets drabber and drabber."

CHAPTER

Titles of Indian Regiments

Previous Titles.	Titles of 1903.
Cavalry	
1st (Duke of York's Own) Bengal Lancers (Skinner's Horse)	1st Duke of York's Own Lancers (Skinner's Horse).
3rd Bengal Cavalry (Skinner's Horse)	3rd Skinner's Horse
2nd Bengal Lancers	2nd Lancers (Gardner's Horse)
4th Bengal Lancers	4th Lancers
5th Bengal Cavalry	5th Cavalry
8th Bengal Lancers	8th Lancers
9th Bengal Lancers (Hodson's Horse)	9th Hodson's Horse
10th (Duke of Cambridge's Own) Bengal Lancers (Hodson's Horse)	10th Duke of Cambridge's Own Lancers (Hodson's Horse)
11th (Prince of Wales' Own) Bengal Lancers	11th Prince of Wales' Own Lancers
12th Bengal Cavalry	12th Cavalry
13th (Duke of Connaught's) Bengal Lancers	13th Duke of Connaught's Lancers
16th Bengal Lancers	16th Cavalry
3rd Madras Lancers	28th Light Cavalry
1st Madras Lancers	26th Light Cavalry
4th Lancers (Hyderabad Contingent)	30th Lancers (Gordon's Horse)

LXVIII

1903 and 1922

Titles during 1914–18.	Titles of 1922.
1st Duke of York's Own Lancers (Skinner's Horse)	1st Duke of York's Own Skinner's Horse
3rd Skinner's Horse . . .	
2nd Lancers (Gardner's Horse) .	2nd Lancers (Gardner's Horse)
4th Cavalry	
5th Cavalry	3rd Cavalry
8th Cavalry	
9th Hodson's Horse	4th Duke of Cambridge's Own Hodson's Horse.
10th Duke of Cambridge's Own Lancers (Hodson's Horse)	
11th (King Edward's Own Lancers) (Probyn's Horse)	5th King Edward's Own Probyn's Horse.
12th Cavalry	
13th Duke of Connaught's Lancers (Watson's Horse)	6th Duke of Connaught's Own Lancers.
16th Cavalry	
28th Light Cavalry	7th Light Cavalry.
26th King George's Own Light Cavalry	8th King George's Own Light Cavalry.
30th Lancers (Gordon's Horse) .	

Previous Titles.	Titles of 1903.
1st Lancers, Hyderabad Contingent	20th Deccan Horse . . .
2nd Lancers, Hyderabad Contingent	29th Lancers, Deccan Horse .
Queen's Own Corps of Guides	Queen's Own Corps of Guides .
1st Punjab Cavalry (Prince Albert Victor's Own)	21st Prince Albert Victor's Own Cavalry (Frontier Force)
3rd Punjab Cavalry	23rd Cavalry (Frontier Force) .
2nd Punjab Cavalry . . .	22nd Cavalry (Frontier Force) .
5th Punjab Cavalry	25th Cavalry (Frontier Force) .
1st (Duke of Connaught's Own) Bombay Lancers	31st Duke of Connaught's Own Lancers
2nd Bombay Lancers . . .	32nd Lancers . . . .
5th Bombay Cavalry (Scinde Horse)	35th Scinde Horse . . .
6th Bombay Cavalry (Jacob's Horse)	36th Jacob's Horse . . .
17th Bengal Lancers . . .	17th Cavalry . . . .
7th Bombay Lancers (Baluch Horse)	37th Lancers (Baluch Horse) .
2nd Madras Lancers	27th Light Cavalry . . .
3rd (Queen's Own) Bombay Light Cavalry	33rd Queen's Own Light Cavalry .
4th (Prince Albert Victor's Own) Bombay Cavalry (Poona Horse)	34th Prince Albert Victor's Own Poona Horse
6th (Prince of Wales') Bengal Cav.	6th Prince of Wales' Cavalry .
7th Bengal Lancers	7th Lancers . . . . .
18th Bengal Lancers . . .	18th Tiwana Lancers . . .
19th Bengal Lancers (Fane's Horse)	19th Lancers (Fane's Horse) .
14th Bengal Lancers (Murray's Jat Horse)	14th Murray's Jat Lancers . .
15th (Cureton's Multani) Bengal Lancers	15th Lancers (Cureton's Mooltanis)

Titles during 1914–18.	Titles of 1922.
20th Deccan Horse . . .	9th Royal Deccan Horse.
29th Lancers (Deccan Horse) .	
Queen Victoria's Own Corps of Guides (Frontier Force) (Lumsden's) Cavalry	10th Queen Victoria's Own Corps of Guides Cavalry (Frontier Force).
21st Prince Albert Victor's Own Cavalry (Frontier Force) (Daly's Horse)	11th Prince Albert Victor's Own Cavalry (Frontier Force).
23rd Cavalry (Frontier Force) .	
22nd Sam Browne's Cavalry (Frontier Force).	12th Cavalry (Frontier Force).
25th Cavalry (Frontier Force) .	
31st Duke of Connaught's Own Lancers	13th Duke of Connaught's Own Bombay Lancers.
32nd Lancers	
35th Scinde Horse	14th Prince of Wales' Own Scinde Horse.
36th Jacob's Horse	
17th Cavalry	15th Lancers
37th Lancers (Baluch Horse) .	
27th Light Cavalry . . .	16th Light Cavalry.
33rd Queen Victoria's Own Light Cavalry	17th Queen Victoria's Own Poona Horse.
34th Prince Albert Victor's Own Poona Horse	
6th King Edward's Own Cavalry	18th King Edward's Own Cavalry.
7th Hariana Lancers	
18th King George's Own Lancers	19th King George's Own Lancers.
19th Lancers (Fane's Horse) .	
14th Murray's Jat Lancers . .	20th Lancers.
15th Lancers (Cureton's Multanis)	

Previous Titles.	Titles of 1903.
1st Central India Horse . . .	38th Central India Horse . .
2nd Central India Horse . . .	39th Central India Horse . .
—	—
—	—
—	—
—	—
—	—
—	—
Sappers and Miners	
Bengal Sappers and Miners . .	1st Sappers and Miners . . .
Queen's Own Madras Sappers and Miners	2nd Queen's Own Sappers and Miners
Bombay Sappers and Miners . .	3rd Sappers and Miners . .
—	—
Artillery	
Kohat Mountain Battery. . .	21st Kohat Mountain Battery (Frontier Force)
Derajat Mountain Battery . .	22nd Derajat Mountain Battery (Frontier Force)
Peshawar Mountain Battery . .	23rd Peshawar Mountain Battery (Frontier Force)
Hazara Mountain Battery . .	24th Hazara Mountain Battery (Frontier Force)
Quetta Mountain Battery . .	25th Mountain Battery . .
Jullunder Mountain Battery . .	26th Jacob's Mountain Battery .
Gujrat Mountain Battery . .	27th Mountain Battery . .

Titles during 1914–18.	Titles of 1922.
38th King George's Own Central India Horse	21st King George's Own Central India Horse.
39th King George's Own Central India Horse	
40th Cavalry Regiment . . .	—
41st Cavalry Regiment . . .	—
42nd Cavalry Regiment . .	—
43rd Cavalry Regiment . .	—
44th Cavalry Regiment . . .	—
45th Cavalry Regiment . . .	—
1st King George's Own Sappers and Miners	King George's Own Bengal Sappers and Miners.
2nd Queen Victoria's Own Sappers and Miners	Queen Victoria's Own Madras Sappers and Miners.
3rd Sappers and Miners . .	Royal Bombay Sappers and Miners.
—	Burma Sappers and Miners.
21st Kohat Mountain Battery (Frontier Force)	101st Royal (Kohat) Pack Battery (Frontier Force).
22nd Derajat Mountain Battery (Frontier Force)	102nd (Derajat) Pack Battery (Frontier Force).
23rd Peshawar Mountain Battery (Frontier Force)	103rd (Peshawar) Pack Battery (Frontier Force).
24th Hazara Mountain Battery (Frontier Force)	104th (Hazara) Pack Battery (Frontier Force).
25th Mountain Battery . . .	105th (Bombay) Pack Battery.
26th Jacob's Mountain Battery	106th (Jacob's) Pack Battery.
27th Mountain Battery . . .	107th (Bengal) Pack Battery.

Previous Titles.	Titles of 1903.
Lahore Mountain Battery .	28th Mountain Battery . . .
Murree Mountain Battery .	29th Mountain Battery . . .
Abbottabad Mountain Battery .	30th Mountain Battery . . .
	31st Mountain Battery (1907) .
	32nd Mountain Battery (1907) .
—	—
—	—
—	—
—	—
—	—
—	—
—	—
—	—
—	—
—	—
—	—
—	—
—	—
—	—
Infantry	
1st Brahman Infantry .	1st Brahmans . . . . .
—	—
2nd (Queen's Own) Rajput Light Infantry	2nd Queen's Own Rajput Light Infantry
—	—
—	—

Titles during 1914–18.	Titles of 1922.
28th Mountain Battery . . .	108th (Lahore) Pack Battery.
29th Mountain Battery . . .	109th (Murree) Pack Battery.
30th Mountain Battery . . .	110th (Abbottabad) Pack Battery.
31st Mountain Battery . . .	111th (Dehra Dun) Pack Battery.
32nd Mountain Battery . .	112th (Poonch) Pack Battery.
33rd (Reserve) Mountain Battery	113th (Dardoni) Pack Battery.
34th (Reserve) Mountain Battery	114th (Rajputana) Pack Battery.
35th (Reserve) Mountain Battery	115th (Jhelum) Pack Battery.
36th (Reserve) Mountain Battery	116th (Zhob) Pack Battery.
37th (Reserve) Mountain Battery	117th (Rawalpindi) Pack Battery.
38th (Reserve) Mountain Battery	118th (Sohan) Pack Battery.
39th (Reserve) Mountain Battery	119th (Maymyo) Pack Battery.
40th (Reserve) Mountain Battery	—
41st (Reserve) Mountain Battery	—
42nd (Reserve) Mountain Battery	120th (Ambala) Pack Battery.
43rd (Reserve) Mountain Battery	121st (Nowshera) Pack Battery.
44th (Reserve) Mountain Battery	—
45th (Reserve) Mountain Battery	—
49th (Reserve) Mountain Battery	—
50th (Reserve) Mountain Battery	—
1st Battalion, 1st Brahmans	4th Battalion, 1st Punjab Regiment.
2nd Battalion, 1st Brahmans .	—
1st Battalion, 2nd Queen Victoria's Own Rajput Light Infantry	1st Battalion, 7th Rajput Regiment (Queen Victoria's Own Light Infantry).
2nd Battalion, 2nd Queen Victoria's Own Rajput Light Infantry	—
3rd Battalion, 2nd Queen Victoria's Own Rajput Light Infantry	—

Previous Titles.	Titles of 1903.
3rd Brahman Infantry . . .	3rd Brahmans . . . . .
	—
4th (Prince Albert Victor's) Rajput Infantry	4th Prince Albert Victor's Rajputs
	—
5th Bengal (Light) Infantry . .	5th Light Infantry . . . .
	—
6th Jat Light Infantry . .	6th Jat Light Infantry . . .
—	—
7th (Duke of Connaught's Own) Rajput Infantry	7th Duke of Connaught's Own Rajputs
—	—
8th Rajput Infantry . . .	8th Rajputs . . . . .
—	
Bhopal Battalion . . .	9th Bhopal Infantry . . .
—	—
—	—
—	—
10th Jat Infantry	10th Jats . . . . .
	—
11th Rajput Infantry . . .	11th Rajputs . . . .
	—
12th Bengal Pioneers (The Kelat-i-Ghilzie Regiment) . . .	12th Pioneers (The Kelat-i-Ghilzie Regiment)
	—
13th (Shekhawati) Rajput Infantry	13th Rajputs (The Shekhawati) Regiment

Titles during 1914–18.	Titles of 1922.
1st Battalion, 3rd Brahmans	—
2nd Battalion, 3rd Gaur Brahmans	—
1st Battalion, 4th Prince Albert Victor's Rajputs	2nd Battalion, 7th Rajput Regiment (Prince Albert Victor's).
2nd Battalion, 4th Prince Albert Victor's Rajputs	—
1st Battalion, 5th Light Infantry	—
2nd Battalion, 5th Light Infantry	—
1st Battalion, 6th Jat Light Infantry	1st Royal Battalion, 9th Jat Regiment (Light Infantry).
2nd Battalion, 6th Jat Light Infantry	10th Battalion, 9th Jat Regiment.
1st Battalion, 7th Duke of Connaught's Own Rajputs	3rd Battalion, 7th Rajput Regiment (Duke of Connaught's Own)
2nd Battalion, 7th Duke of Connaught's Own Rajputs	—
1st Battalion, 8th Rajputs . .	4th Battalion, 7th Rajput Regiment
2nd Battalion, 8th Rajputs . .	—
1st Battalion, 9th Bhopal Infantry	4th Battalion, 16th Punjab Regiment (Bhopal).
2nd Battalion, 9th (Delhi) Infantry	—
3rd Battalion, 9th Bhopal Infantry	—
4th Battalion, 9th Bhopal Infantry	—
1st Battalion, 10th Jats . .	3rd Battalion, 9th Jat Regiment.
2nd Battalion, 10th Jats . .	—
1st Battalion, 11th Rajputs . .	5th Battalion, 7th Rajput Regiment
2nd Battalion, 11th Rajputs .	—
1st Battalion, 12th Pioneers (The Kelat-i-Ghilzie Regiment)	2nd Battalion, 2nd Bombay Pioneers (Kelat-i-Ghilzie).
2nd Battalion, 12th Pioneers (The Kelat-i-Ghilzie Regiment)	—
13th Rajputs (The Shekhawati) Regiment	10th Battalion, 6th Rajputana Rifles (Shekhawati).

Previous Titles.	Titles of 1903.
14th (Ferozepore) Sikh Infantry .	14th Ferozepore Sikhs . . .
15th (Ludhiana) Sikh Infantry .	15th Ludhiana Sikhs . . .
—	—
16th (Lucknow) Rajput Infantry .	16th Rajputs (The Lucknow Regiment)
17th Musalman Rajput Infantry (The Loyal Regiment)	17th Infantry (The Loyal Regiment)
—	—
18th Musalman Rajput Infantry .	18th Infantry . . . . .
—	—
19th Punjab Infantry . . .	19th Punjabis . . . .
—	—
20th (Duke of Cambridge's Own) Punjab Infantry	20th Duke of Cambridge's Own Punjabis
21st Punjab Infantry . . .	21st Punjabis . . . . .
—	—
22nd Punjab Infantry . . .	22nd Punjabis . . . . .
—	—
23rd Punjab Pioneers . . .	23rd Sikh Pioneers . . . .
—	—
—	—
24th Punjab Infantry . . .	24th Punjabis . . .

Titles during 1914–18.	Titles of 1922.
14th King George's Own Ferozepore Sikhs	1st Battalion, 11th Sikh Regiment (King George's Own) (Ferozepore Sikhs).
1st Battalion, 15th Ludhiana Sikhs	2nd Battalion, 11th Sikh Regiment (Ludhiana Sikhs).
2nd Battalion, 15th Ludhiana Sikhs	—
16th Rajputs (The Lucknow Regiment)	10th Battalion, 7th Rajput Regiment (The Lucknow Regiment).
1st Battalion, 17th Infantry (The Loyal Regiment)	—
2nd Battalion, 17th Infantry (The Loyal Regiment)	—
1st Battalion, 18th Infantry .	4th Battalion, 9th Jat Regiment. [Since disbanded.
2nd Battalion, 18th Infantry .	—
1st Battalion, 19th Punjabis .	1st Battalion, 14th Punjab Regiment.
2nd Battalion, 19th Punjabis .	—
20th Duke of Cambridge's Own Infantry (Brownlow's Punjabis)	2nd Battalion, 14th Punjab Regiment (Duke of Cambridge's Own) (Brownlow's).
1st Battalion, 21st Punjabis . .	10th Battalion, 14th Punjab Regiment.
2nd Battalion, 21st Punjabis .	—
1st Battalion, 22nd Punjabis .	3rd Battalion, 14th Punjab Regiment.
2nd Battalion, 22nd Punjabis .	—
1st Battalion, 23rd Sikh Pioneers .	1st Battalion, 3rd Sikh Pioneers.
2nd Battalion, 23rd Sikh Pioneers	10th Battalion, 3rd Sikh Pioneers.
3rd Battalion, 23rd Sikh Pioneers	
24th Punjabis	4th Battalion, 14th Punjab Regiment.

Previous Titles.	Titles of 1903.
25th Punjab Infantry . . .	25th Punjabis . . .
—	—
26th Punjab Infantry . . .	26th Punjabis . . .
—	—
27th Punjab Infantry . . .	27th Punjabis . . .
—	—
28th Punjab Infantry . . .	28th Punjabis . . .
—	—
29th Punjab Infantry . . .	29th Punjabis . . .
30th Punjab Infantry . . .	30th Punjabis . . .
—	—
—	—
—	—
31st Punjab Infantry . . .	31st Punjabis . . .
32nd Punjab Pioneers . . .	32nd Sikh Pioneers . .
—	—
—	—
33rd Punjab Infantry . . .	33rd Punjabis . . .
—	—
34th Punjab Pioneers . . .	34th Sikh Pioneers . . .
—	—
—	—
35th Sikh Infantry	35th Sikhs . . . .
—	—

Titles during 1914–18.	Titles of 1922.
1st Battalion, 25th Punjabis	1st Battalion, 15th Punjab Regiment.
2nd Battalion, 25th Punjabis	—
1st Battalion, 26th Punjabis	2nd Battalion, 15th Punjab Regiment.
2nd Battalion, 26th Punjabis	—
1st Battalion, 27th Punjabis	3rd Battalion, 15th Punjab Regiment.
2nd Battalion, 27th Punjabis	—
1st Battalion, 28th Punjabis	4th Battalion, 15th Punjab Regiment.
2nd Battalion, 28th Punjabis	—
29th Punjabis	10th Battalion, 15th Punjab Regiment.
1st Battalion, 30th Punjabis	1st Battalion, 16th Punjab Regiment.
2nd Battalion, 30th Punjabis	—
3rd Battalion, 30th Punjabis	—
4th Battalion, 30th Punjabis	—
31st Punjabis	2nd Battalion, 16th Punjab Regiment.
1st Battalion, 32nd Sikh Pioneers	2nd Battalion, 3rd Sikh Pioneers.
2nd Battalion, 32nd Sikh Pioneers	—
3rd Battalion, 32nd Sikh Pioneers	—
1st Battalion, 33rd Punjabis	3rd Battalion, 16th Punjab Regiment.
2nd Battalion, 33rd Punjabis	—
1st Battalion, 34th Sikh Pioneers	3rd Royal Battalion, 3rd Sikh Pioneers.
2nd Battalion, 34th Sikh Pioneers	—
3rd Battalion, 34th Sikh Pioneers	—
1st Battalion, 35th Sikhs	10th Battalion, 11th Sikh Regiment.
2nd Battalion, 35th Sikhs	—

Previous Titles.	Titles of 1903.
36th Sikh Infantry	36th Sikhs
37th Dogra Infantry	37th Dogras
38th Dogra Infantry	38th Dogras
1st Battalion, 39th Garhwal Rifles	1st Battalion, 39th Garhwal Rifles
2nd Battalion, 39th Garhwal Rifles	2nd Battalion, 39th Garhwal Rifles
—	—
—	—
40th Punjab Infantry	40th Pathans
41st Dogra Infantry	41st Dogras
—	—
Deoli Irregular Force	42nd Deoli Regiment
—	—
Erinpura Irregular Force	43rd Erinpura Regiment
—	—
Merwara Battalion	44th Merwara Infantry
45th (Rattray's) Sikh Infantry	45th Rattray's Sikhs
46th Punjab Infantry	46th Punjabis
47th Sikh Infantry	47th Sikhs

Titles during 1914–18.	Titles of 1922.
36th Sikhs	4th Battalion, 11th Sikh Regiment.
37th Dogras	1st Battalion, 17th Dogra Regiment (Prince of Wales' Own).
38th Dogras	2nd Battalion, 17th Dogra Regiment
1st Battalion, 39th Garhwal Rifles	1st Battalion, 18th Royal Garhwal Rifles.
2nd Battalion, 39th Garhwal Rifles	2nd Battalion, 18th Royal Garhwal Rifles.
3rd Battalion, 39th Garhwal Rifles	3rd Battalion, 18th Royal Garhwal Rifles.
4th Battalion, 39th Garhwal Rifles	10th Battalion, 18th Royal Garhwal Rifles.
40th Pathans	5th Battalion, 14th Punjab Regiment.
1st Battalion, 41st Dogras	3rd Battalion, 17th Dogra Regiment.
2nd Battalion, 41st Dogras	10th Battalion, 17th Dogra Regiment.
1st Battalion, 42nd Deoli Regiment	—
2nd Battalion, 42nd Deoli Regt.	—
1st Battalion, 43rd Erinpura Regiment	—
2nd Battalion, 43rd Erinpura Regiment	—
44th Merwara Infantry	—
45th Rattray's Sikhs	3rd Battalion, 11th Sikh Regiment (Rattray's Sikhs).
46th Punjabis	10th Battalion, 16th Punjab Regiment.
47th Sikhs	5th Battalion, 11th Sikh Regiment (Duke of Connaught's Own).

Previous Titles.	Titles of 1903.
48th Bengal Pioneers	48th Pioneers
—	—
—	—
—	—
—	
1st Sikh Infantry	51st Sikhs (Frontier Force)
2nd (or Hill) Sikh Infantry	52nd Sikhs (Frontier Force)
3rd Sikh Infantry	53rd Sikhs (Frontier Force)
4th Sikh Infantry	54th Sikhs (Frontier Force)
—	—
Queen's Own Corps of Guides	Queen's Own Corps of Guides
—	—
—	—
1st Punjab Infantry	55th Coke's Rifles (Frontier Force)
—	—
2nd Punjab Infantry	56th Infantry (Frontier Force)

Titles during 1914–18.	Titles of 1922.
1st Battalion, 48th Pioneers	4th Battalion, 2nd Bombay Pioneers.
2nd Battalion, 48th Pioneers	—
49th Bengalis	—
1st Battalion, 50th Kumaon Rifles	1st Kumaon Rifles.
2nd Battalion, 50th Kumaon Rifles	2nd Kumaon Rifles. [Since disbanded.
51st Sikhs (Frontier Force)	1st Battalion, 12th Frontier Force Regiment (Prince of Wales' Own) (Sikhs).
52nd Sikhs (Frontier Force)	2nd Battalion, 12th Frontier Force Regiment (Sikhs).
53rd Sikhs (Frontier Force)	3rd Battalion, 12th Frontier Force Regiment (Sikhs)
1st Battalion, 54th Sikhs (Frontier Force)	4th Battalion, 12th Frontier Force Regiment (Sikhs).
2nd Battalion, 54th Sikhs (Frontier Force)	—
Queen Victoria's Own Corps of Guides (Frontier Force) (Lumsden's)	5th Battalion, 12th Frontier Force Regiment (Queen Victoria's Own Corps of Guides).
2nd Battalion, Queen Victoria's Own Corps of Guides (Frontier Force) (Lumsden's).	10th Battalion, 12th Frontier Force Regiment (Queen Victoria's Own Corps of Guides).
3rd Battalion, Queen Victoria's Own Corps of Guides (Frontier Force) (Lumsden's)	—
1st Battalion, 55th Coke's Rifles (Frontier Force)	1st Battalion, 13th Frontier Force Rifles (Coke's).
2nd Battalion, 55th Coke's Rifles (Frontier Force).	—
1st Battalion, 56th Punjabi Rifles (Frontier Force)	2nd Battalion, 13th Frontier Force Rifles.

Previous Titles.	Titles of 1903.
—	—
4th Punjab Infantry	57th Wilde's Rifles (Frontier Force)
5th Punjab Infantry	58th Vaughan's Rifles (Frontier Force)
6th Punjab Infantry	59th Scinde Rifles (Frontier Force)
1st Madras Pioneers	61st Pioneers
—	—
2nd Madras Infantry	62nd Punjabis
3rd (Palamcottah) Madras Light Infantry	63rd Palamcottah Light Infantry
—	—
4th Madras Pioneers	64th Pioneers
5th Madras Infantry	65th Carnatic Light Infantry
6th Madras Infantry	66th Punjabis
—	—
7th Madras Infantry	67th Punjabis
—	—
9th Madras Infantry	69th Punjabis
—	—
—	—
—	—
—	—

Titles during 1914–18.	Titles of 1922.
2nd Battalion, 56th Punjabi Rifles (Frontier Force)	10th Battalion, 13th Frontier Force Rifles.
57th Wilde's Rifles (Frontier Force)	4th Battalion, 13th Frontier Force Rifles (Wilde's).
58th Vaughan's Rifles (Frontier Force)	5th Battalion, 13th Frontier Force Rifles.
59th Scinde Rifles (Frontier Force)	6th Royal Battalion, 13th Frontier Force Rifles (Scinde).
1st Battalion, 61st King George's Own Pioneers	1st Battalion, 1st Madras Pioneers (King George's Own).
2nd Battalion, 61st King George's Own Pioneers	—
62nd Punjabis	1st Battalion, 1st Punjab Regiment.
1st Battalion, 63rd Palamcottah Light Infantry	—
2nd Battalion, 63rd Palamcottah Light Infantry	—
64th Pioneers	2nd Battalion, 1st Madras Pioneers.
1st Battalion, 66th Punjabis	2nd Battalion, 1st Punjab Regiment.
2nd Battalion, 66th Punjabis	—
1st Battalion, 67th Punjabis	1st Battalion, 2nd Punjab Regiment.
2nd Battalion, 67th Punjabis	10th Battalion, 2nd Punjab Regiment.
1st Battalion, 69th Punjabis	2nd Battalion, 2nd Punjab Regiment.
2nd Battalion, 69th Punjabis	
1st Battalion, 70th Burma Rifles	1st Battalion, 20th Burma Rifles.
2nd Battalion, 70th Burma Rifles	2nd Battalion, 20th Burma Rifles.
4th Battalion, 70th Burma Rifles	4th Battalion, 20th Burma Rifles (Chin). [Since disbanded.

Previous Titles.	Titles of 1903.
—	—
11th Coorg Infantry	71st Coorg Rifles
12th Burma Infantry	72nd Punjabis
—	—
13th Madras Infantry	73rd Carnatic Infantry
—	—
14th Madras Infantry	74th Punjabis
15th Madras Infantry	75th Carnatic Infantry
—	—
16th Madras Infantry	76th Punjabis
—	—
—	77th Moplah Rifles
—	78th Moplah Rifles
19th Madras Infantry	79th Carnatic Infantry
20th Madras Infantry	80th Carnatic Infantry
21st Madras Pioneers	81st Pioneers
—	—
22nd Madras Infantry	82nd Punjabis
23rd (Wallajahbad) Madras Light Infantry	83rd Wallajahbad Light Infantry
24th Madras Infantry	84th Punjabis
—	—

Titles during 1914–18.	Titles of 1922.
3rd (or 5th) Battalion, 70th Burma Rifles	10th Battalion, 20th Burma Rifles.
71st Punjabis	—
1st Battalion, 72nd Punjabis	3rd Battalion, 2nd Punjab Regiment.
2nd Battalion, 72nd Punjabis	—
1st Battalion, 73rd Carnatic Infantry	1st Battalion, 3rd Madras Regiment.
2nd Battalion, 73rd Malabar Infantry	—
74th Punjabis	4th Battalion, 2nd Punjab Regiment.
1st Battalion, 75th Carnatic Infantry	2nd Battalion, 3rd Madras Regiment.
2nd Battalion, 75th Carnatic Infantry	—
1st Battalion, 76th Punjabis	3rd Battalion, 1st Punjab Regiment
2nd Battalion, 76th Punjabis	—
—	—
79th Carnatic Infantry	3rd Battalion, 3rd Madras Regiment [Since disbanded.
80th Carnatic Infantry	—
1st Battalion, 81st Pioneers	10th Battalion, 1st Madras Pnrs.
2nd Battalion, 81st Pioneers	—
82nd Punjabis	5th Battalion, 1st Punjab Regiment
83rd Wallajahbad Light Infantry	4th Battalion, 3rd Madras Regiment (Wallajahbad Light Infantry). [Since disbanded.
84th Punjabis	10th Battalion, 1st Punjab Regiment.
85th Burman Rifles	3rd Battalion, 20th Burma Rifles (Kachin).

Previous Titles.	Titles of 1903.
26th Madras Infantry	86th Carnatic Infantry
27th Madras Infantry	87th Punjabis
28th Madras Infantry	88th Carnatic Infantry
29th Burma Infantry	89th Punjabis
	—
30th Burma Infantry	90th Punjabis
	—
31st Burma Light Infantry	91st Punjabis
	—
32nd Burma Infantry	92nd Punjabis
33rd Burma Infantry	93rd Burma Infantry
1st Infantry, Hyderabad Contingent	94th Russell's Infantry
	—
2nd Infantry, Hyderabad Contingent	95th Russell's Infantry
—	—
3rd Infantry, Hyderabad Contingent	96th Berar Infantry
	—

Titles during 1914–18.	Titles of 1922.
86th Carnatic Infantry	10th Battalion, 3rd Madras Regiment.
87th Punjabis	5th Battalion, 2nd Punjab Regiment
1st Battalion, 88th Carnatic Infantry	—
2nd Battalion, 88th Carnatic Infantry	—
1st Battalion, 89th Punjabis	1st Battalion, 8th Punjab Regiment.
2nd Battalion, 89th Punjabis	10th Battalion, 8th Punjab Regiment.
1st Battalion, 90th Punjabis	2nd Battalion, 8th Punjab Regiment.
2nd Battalion, 90th Punjabis	—
1st Battalion, 91st Punjabis (Light Infantry)	3rd Battalion, 8th Punjab Regiment.
2nd Battalion, 91st Punjabis (Light Infantry)	—
92nd Punjabis	4th Battalion, 8th Punjab Regiment (Prince of Wales' Own).
93rd Burma Infantry	5th Battalion, 8th Punjab Regiment (Burma).
1st Battalion, 94th Russell's Infantry	1st Battalion, 19th Hyderabad Regiment (Russell's).
2nd Battalion, 94th Russell's Infantry	—
1st Battalion, 95th Russell's Infantry	10th Battalion, 19th Hyderabad Regiment (Russell's).
2nd Battalion, 95th Russell's Infantry	—
1st Battalion, 96th Berar Infantry	2nd Battalion, 19th Hyderabad Regiment (Berar).
2nd Battalion, 96th Infantry	—

Previous Titles.	Titles of 1903.
4th Infantry, Hyderabad Contingent	97th Deccan Infantry
—	
5th Infantry, Hyderabad Contingent	98th Infantry
—	
6th Infantry, Hyderabad Contingent	99th Deccan Infantry
—	—
1st Bombay Grenadiers	101st Grenadiers
	—
2nd (Prince of Wales' Own) Bombay Grenadiers	102nd Prince of Wales' Own Grenadiers
	—
3rd Bombay Light Infantry	103rd Mahratta Light Infantry
—	—
—	—
4th Bombay Rifles	104th Wellesley's Rifles
5th Bombay Light Infantry	105th Mahratta Light Infantry
—	
	106th Hazara Pioneers (1904)
7th Bombay Pioneers	107th Pioneers
—	—
8th Bombay Infantry	108th Infantry

Titles during 1914–18	Titles of 1922.
1st Battalion, 97th Deccan Infantry	3rd Battalion, 19th Hyderabad Regiment
2nd Battalion, 97th Deccan Infantry	—
1st Battalion, 98th Infantry .	4th Battalion, 19th Hyderabad Regiment.
2nd Battalion, 98th Infantry .	—
1st Battalion, 99th Deccan Infantry	5th Battalion, 19th Hyderabad Regiment. [Since disbanded.
2nd Battalion, 99th Deccan Infantry	—
1st Battalion, The 101st Grenadiers	1st Battalion, 4th Bombay Grenadiers.
2nd Battalion, The 101st Grenadiers	—
1st Battalion, 102nd King Edward's Own Grenadiers	2nd Battalion, 4th Bombay Grenadiers (King Edward's Own).
2nd Battalion, 102nd King Edward's Own Grenadiers	—
1st Battalion, 103rd Mahratta Light Infantry	1st Battalion, 5th Mahratta Light Infantry.
2nd Battalion, 103rd Mahratta Light Infantry	—
3rd Battalion, 103rd (Kolhapur) Mahratta Light Infantry	—
104th Wellesley's Rifles .	1st Battalion, 6th Rajputana Rifles (Wellesley's).
105th Mahratta Light Infantry .	2nd Battalion, 5th Mahratta Light Infantry.
106th Hazara Pioneers	1st Battalion, 4th Hazara Pioneers.
1st Battalion, 107th Pioneers .	1st Battalion, 2nd Bombay Pioneers
2nd Battalion, 107th Pioneers .	—
108th Infantry.	3rd Battalion, 4th Bombay Grenadiers.

Previous Titles.	Titles of 1903.
9th Bombay Infantry . . .	109th Infantry . . . .
—	—
10th Bombay Light Infantry . .	110th Mahratta Light Infantry .
—	—
12th Bombay Infantry . . .	112th Infantry . . . .
—	—
13th Bombay Infantry . . .	113th Infantry . . . .
—	—
14th Bombay Infantry . . .	114th Mahrattas . . . .
16th Bombay Infantry . . .	116th Mahrattas . . . .
—	—
17th Bombay Infantry . . .	117th Mahrattas . . . .
—	—
19th Bombay Infantry . . .	119th Infantry (The Mooltan Regiment)
—	—
20th Bombay Infantry . . .	120th Rajputana Infantry . .
21st Bombay Infantry (Marine Battalion)	121st Pioneers . . . .
22nd Bombay Infantry . . .	122nd Rajputana Infantry .
23rd Bombay Rifles	123rd Outram's Rifles . . .
—	—

Titles during 1914–18.	Titles of 1922.
1st Battalion, 109th Infantry .	4th Battalion, 4th Bombay Grenadiers.
2nd Battalion, 109th Infantry	—
110th Mahratta Light Infantry .	3rd Battalion, 5th Mahratta Light Infantry
111th Mahars	—
1st Battalion, 112th Infantry .	5th Battalion, 4th Bombay Grenadiers. [Since disbanded.
2nd Battalion, 112th Infantry .	—
1st Battalion, 113th Infantry .	10th Battalion, 4th Bombay Grenadiers.
2nd Battalion, 113th Infantry .	—
114th Mahrattas	10th Battalion, 5th Mahratta Light Infantry
1st Battalion, 116th Mahrattas .	4th Battalion, 5th Mahratta Light Infantry
2nd Battalion, 116th Mahrattas .	—
1st Battalion, 117th Mahrattas .	5th Royal Battalion, 5th Mahratta Light Infantry.
2nd Battalion, 117th Mahrattas .	
1st Battalion, 119th Infantry (The Mooltan Regiment)	2nd Battalion, 9th Jat Regiment (Mooltan Battalion).
2nd Battalion, 119th Infantry (The Mooltan Regiment)	—
120th Rajputana Infantry . .	2nd Battalion, 6th Rajputana Rifles (Prince of Wales' Own).
121st Pioneers	10th Battalion, 2nd Bombay Pioneers (Marine Battalion).
122nd Rajputana Infantry . .	3rd Battalion, 6th Rajputana Rifles.
1st Battalion, 123rd Outram's Rifles	4th Battalion, 6th Rajputana Rifles (Outram's).
2nd Battalion, 123rd Outram's Rifles	—

Previous Titles.	Titles of 1903.
24th (Duchess of Connaught's Own) Baluchistan Infantry	124th Duchess of Connaught's Own Baluchistan Infantry
—	—
—	—
25th Bombay Rifles	125th Napier's Rifles . . .
—	—
26th Baluchistan Infantry . .	126th Baluchistan Infantry . .
27th Baluch Light Infantry . .	127th Baluch Light Infantry .
—	—
28th Bombay Pioneers . . .	128th Pioneers . . . .
—	—
29th (Duke of Connaught's Own) Baluch Infantry	129th Duke of Connaught's Own Baluchis
—	—
30th Baluch Infantry . . .	130th Baluchis . . . .
—	—
—	—

Titles during 1914–18.	Titles of 1922.
1st Battalion, 124th Duchess of Connaught's Baluchistan Infantry	1st Battalion, 10th Baluch Regiment (Duchess of Cannaught's Own).
2nd Battalion, 124th Duchess of Connaught's Baluchistan Infantry	10th Battalion, 10th Baluch Regiment.
3rd Battalion, 124th Duchess of Connaught's Baluchistan Infantry	—
1st Battalion, 125th Napier's Rifles	5th Battalion, 6th Rajputana Rifles (Napier's).
2nd Battalion, 125th Napier's Rifles	—
126th Baluchistan Infantry .	2nd Battalion, 10th Baluch Regiment.
1st Battalion, 127th Queen Mary's Own Baluch Light Infantry	3rd Battalion, 10th Baluch Regiment (Queen Mary's Own).
2nd Battalion, 127th Queen Mary's Own Baluch Light Infantry	—
1st Battalion, 128th Pioneers .	3rd Battalion, 2nd Bombay Pioneers
2nd Battalion, 128th Pioneers .	—
1st Battalion, 129th Duke of Connaught's Own Baluchis	4th Battalion, 10th Baluch Regiment (Duke of Connaught's Own)
2nd Battalion, 129th Duke of Connaught's Own Baluchis	—
130th (King George's Own) Baluchis (Jacob's Rifles)	5th Battalion, 10th Baluch Regiment (King George's Own) (Jacob's Rifles).
1st Battalion, 131st United Provinces Regiment	—
2nd Battalion, 131st United Provinces Regiment	—

Previous Titles.	Titles of 1903.
—	—
—	—
—	—
—	—
—	—
—	—
—	—
—	—
—	—
—	—
—	—
—	—
—	—
—	—
—	—

Titles during 1914-18.	Titles of 1922.
1st Battalion, 133rd Regiment .	—
1st Battalion, 140th Patiala Infantry	—
1st Battalion, 141st Bikanir Infantry	—
1st Battalion, 142nd Jodhpur Infantry	—
1st Battalion, 143rd Narsingh (Dholpur) Infantry	—
1st Battalion, 144th Bharatpur Infantry	—
1st Battalion, 145th Alwar (Jai Paltan) Infantry	—
1st Battalion, 150th Indian Infantry	—
2nd Battalion, 150th Indian Infantry	—
3rd Battalion, 150th Indian Infantry	—
1st Battalion, 151st Sikh Infantry	—
2nd Battalion, 151st Indian Infantry	—
3rd Battalion, 151st Punjabi Rifles	—
1st Battalion, 152nd Punjabis	—
2nd Battalion, 152nd Punjabis .	—
3rd Battalion, 152nd Punjabis .	—
1st Battalion, 153rd Punjabis .	—
2nd Battalion, 153rd Punjabis .	—
3rd Battalion, 153rd Rifles .	—
1st Battalion, 154th Indian Infantry	—
2nd Battalion, 154th Indian Infantry	—

Previous Titles.	Titles of 1903.
Gurkhas.	
1st Battalion, 1st Gurkha Rifles .	1st Battalion, 1st Gurkha Rifles (The Malaun Regiment).
2nd Battalion, 1st Gurkha Rifles .	2nd Battalion, 1st Gurkha Rifles (The Malaun Regiment)
—	—
1st Battalion, 2nd Gurkha Rifles (Prince of Wales' Own) (The Sirmoor Rifles)	1st Battalion, 2nd Prince of Wales' Own Gurkha Rifles (The Sirmoor Rifles)
2nd Battalion, 2nd Gurkha Rifles (Prince of Wales' Own) (The Sirmoor Rifles)	2nd Battalion, 2nd Prince of Wales' Own Gurkha Rifles (The Sirmoor Rifles)
1st Battalion, 3rd Gurkha Rifles .	1st Battalion, 3rd Gurkha Rifles .
2nd Battalion, 3rd Gurkha Rifles .	2nd Battalion, 3rd Gurkha Rifles.
—	
—	

Titles during 1914–18.	Titles of 1922.
3rd Battalion, 154th Indian Inf.	—
1st Battalion, 155th Indian Pioneers	—
2nd Battalion, 155th Indian Inf.	—
1st Battalion, 156th Indian Inf.	—
1st Battalion, 1st King George's Own Gurkha Rifles (The Malaun Regiment)	1st Battalion, 1st King George's Own Gurkha Rifles (The Malaun Regiment).
2nd Battalion, 1st King George's Own Gurkha Rifles (The Malaun Regiment)	2nd Battalion, 1st King George's Own Gurkha Rifles (The Malaun Regiment).
3rd Battalion, 1st King George's Own Gurkha Rifles (The Malaun Regiment)	—
1st Battalion, 2nd King Edward's Own Gurkha Rifles (The Sirmoor Rifles)	1st Battalion, 2nd King Edward's Own Gurkha Rifles (The Sirmoor Rifles).
2nd Battalion, 2nd King Edward's Own Gurkha Rifles (The Sirmoor Rifles)	2nd Battalion, 2nd King Edward's Own Gurkha Rifles (The Sirmoor Rifles).
3rd Battalion, 2nd King Edward's Own Gurkha Rifles (The Sirmoor Rifles)	—
1st Battalion, 3rd Queen Alexandra's Own Gurkha Rifles	1st Battalion, 3rd Queen Alexandra's Own Gurkha Rifles.
2nd Battalion, 3rd Queen Alexandra's Own Gurkha Rifles	2nd Battalion, 3rd Queen Alexandra's Own Gurkha Rifles.
3rd Battalion, 3rd Queen Alexandra's Own Gurkha Rifles	—
4th Battalion, 3rd Queen Alexandra's Own Gurkha Rifles	—

Previous Titles.	Titles of 1903.
1st Battalion, 4th Gurkha Rifles	**1st Battalion, 4th Gurkha Rifles**
2nd Battalion, 4th Gurkha Rifles	**2nd Battalion, 4th Gurkha Rifles.**
1st Battalion, 5th Gurkha Rifles	**1st Battalion, 5th Gurkha Rifles**
2nd Battalion, 5th Gurkha Rifles	**2nd Battalion, 5th Gurkha Rifles.**
—	
42nd Gurkha Rifles	**6th Gurkha Rifles**
—	**2nd Battalion, 6th Gurkha Rifles (1904)**
—	—
8th Gurkha Rifles	**2nd Battalion, 10th Gurkha Rifles (till 1907)**
—	**2nd Battalion, 7th Gurkha Rifles (1907)**
—	—
44th Gurkha Rifles	**8th Gurkha Rifles**
43rd Gurkha Rifles	**7th Gurkha Rifles**
9th Gurkha Rifles	**1st Battalion, 9th Gurkha Rifles (1904)**
—	**2nd Battalion, 9th Gurkha Rifles (1904)**
—	—
10th Gurkha Rifles	**1st Battalion, 10th Gurkha Rifles.**
—	**2nd Battalion, 10th Gurkha Rifles (1908)**
—	—
—	—
—	—
—	—

Titles during 1914–18.	Titles of 1922.
1st Battalion, 4th Gurkha Rifles	1st Battalion, 4th Gurkha Rifles.
2nd Battalion, 4th Gurkha Rifles.	2nd Battalion, 4th Gurkha Rifles.
1st Battalion, 5th Gurkha Rifles (Frontier Force)	1st Battalion, 5th Royal Gurkha Rifles (Frontier Force)
2nd Battalion, 5th Gurkha Rifles (Frontier Force)	2nd Battalion, 5th Royal Gurkha Rifles (Frontier Force).
3rd Battalion, 5th Gurkha Rifles (Frontier Force)	—
1st Battalion, 6th Gurkha Rifles	1st Battalion, 6th Gurkha Rifles.
2nd Battalion, 6th Gurkha Rifles	2nd Battalion, 6th Gurkha Rifles
3rd Battalion, 6th Gurkha Rifles	
1st Battalion, 7th Gurkha Rifles	1st Battalion, 7th Gurkha Rifles.
2nd Battalion, 7th Gurkha Rifles.	2nd Battalion, 7th Gurkha Rifles.
3rd Battalion, 7th Gurkha Rifles	—
1st Battalion, 8th Gurkha Rifles	1st Battalion, 8th Gurkha Rifles.
2nd Battalion, 8th Gurkha Rifles.	2nd Battalion, 8th Gurkha Rifles.
3rd Battalion, 8th Gurkha Rifles	—
1st Battalion, 9th Gurkha Rifles	1st Battalion, 9th Gurkha Rifles.
2nd Battalion, 9th Gurkha Rifles.	2nd Battalion, 9th Gurkha Rifles.
3rd Battalion, 9th Gurkha Rifles	—
1st Battalion, 10th Gurkha Rifles	1st Battalion, 10th Gurkha Rifles.
2nd Battalion, 10th Gurkha Rifles	2nd Battalion, 10th Gurkha Rifles.
1st Battalion, 11th Gurkha Rifles	—
2nd Battalion, 11th Gurkha Rifles	—
3rd Battalion, 11th Gurkha Rifles	—
4th Battalion, 11th Gurkha Rifles	—

CHAPTER LXIX

Commanders-in-Chief of the Army in India 1748-1939

Name.	Assumed Command.
1. Major Stringer Lawrence	January, 1748
2. Colonel John Adlercron	1754.
3. Colonel Robert Clive	December, 1756.
4. Major John Caillaud	25th February, 1760.
5. Major John Carnac	31st December, 1760.
6. Lieut.-Colonel Eyre Coote	April, 1761.
7. Major Thomas Adams	1763.
8. Major John Carnac	January, 1764.
9. Major Hector Munro	July, 1764.
10. Brigadier-General John Carnac	January, 1765.
11. Major-General Robert Lord Clive	April, 1765.
12. Colonel Richard Smith	29th January, 1767.
13. Brigadier-General Sir Robert Barker	24th March, 1770.
14. Colonel Alexander Chapman	18th January, 1774.
15. Lieut.-General Sir John Clavering, K.B.	2nd November, 1774.
16. Lieut.-General Sir Eyre Coote, K.B.	25th March, 1779.
17. Lieut.-General Robert Sloper	21st July, 1785.
18. Lieut.-General Charles Earl Cornwallis, K.G.	12th September, 1786.
19. Major-General Sir Robert Abercromby, K.B.	28th October, 1793.

Name.	Assumed Command.
20. Lieut.-General Sir Alured Clarke, K.B.	16th March, 1797.
21. Lieut.-General Gerard Lake (afterwards Lord Lake)	13th March, 1801.
22. General Charles Marquis Cornwallis, K.G.	30th July, 1805.
23. General Gerard Lord Lake . .	5th October, 1805.
24. Lieut.-General George Hewitt .	17th October, 1807.
25. Lieut.-General Sir George Nugent .	14th January, 1812.
26. General Francis, Earl of Moira (afterwards Marquis of Hastings)	4th October, 1813.
27. Lieut.-General the Hon'ble Sir Edward Paget, G.C.B.	13th January, 1823.
28. General Shapleton, Lord Combermere, G.C.B.	7th October, 1825.
29. General George, Earl of Dalhousie, G.C.B.	1st January, 1830.
30. General Sir Edward Barnes, G.C.B.	10th January, 1832.
31. General Lord William H. C. Bentinck, G.C.B.	15th October, 1833.
32. Lieut.-General the Hon'ble Sir Henry Fane, G.C.B.	5th September, 1835.
33. Major-General Sir Jasper Nicolls, K.C.B.	7th December, 1839.
34. General Sir Hugh Gough, Bart, G.C.B. (afterwards Lord Gough)	8th August, 1843.
35. General Sir Charles James Napier, G.C.B.	7th May, 1849.
36. General Sir William Maynard Gomm, K.C.B.	6th December, 1850.
37. General the Hon'ble George Anson	23rd January, 1856.

Name.	Assumed Command.
38. Lieut.-General Sir Patrick Grant, K.C.B.	17th June, 1857 (Officiating).
39. General Sir Colin Campbell, G.C.B. (afterwards Lord Clyde)	13th August, 1857.
40. General Sir Hugh H. Rose, G.C.B.	4th June, 1860.
41. General Sir William Rose Mansfield, K.C.B.	23rd March, 1865.
42. General Lord Napier of Magdala, G.C.B., G.C.S.I.	9th April, 1870.
43- General Sir Fred. P. Haines, K.C.B. .	10th April, 1876
44. General Sir Donald M. Stewart, G.C.B., C.I.E.	8th April, 1881.
45. General Sir Fred. S. Roberts, V.C., G.C.B., C.I.E.	28th November, 1885.
46. General Sir Geo. S. White, V.C., G.C.I.E., K.C.B.	8th April, 1893.
47. Lieut.-General Sir Charles Edward Nairne, K.C.B.	20th March, 1898 (Provisional).
48. General Sir W. S. A. Lockhart, G.C.B., K.C.S.I.	4th November, 1898.
49. General Sir A. P. Palmer, K.C.B. .	19th March, 1900.
50. General Viscount Kitchener of Khartoum, G.C.B., O.M., G.C.M.G.	28th November, 1902.
51. General Sir O'Moore Creagh, V.C., G.C.B.	10th September, 1909.
52. General Sir B. Duff, G.C.B., G.C.S.I. K.C.V.O., C.I.E.	8th March, 1914.
53. General Sir C. C. Monro G.C.S.I., G.C.M.G., K.C.B.	1st October, 1916.
54. General Lord Rawlinson of Trent, G.C.B., G.C.V.O., K.C.M.G., A.-D.-C.	21st November, 1920.

Name.	Assumed Command.
55. Field Marshal Lord Birdwood of Anzac and of Totnes, G.C.B., G.C.S.I., G.C.M.G., G.C.V.O., C.I.E., D.S.O., D.C.L., LL.D.	6th August, 1925.
56. Field Marshal Sir Philip W. Chetwode, G.C.B., O.M., G.C.S.I., K.C.M.G., D.S.O., D.C.L.	30th November, 1930.
57. General Sir Robert A. Cassells, G.C.B., G.C.S.I., D.S.O.	30th November, 1935.

(*India Office photo*)

"INDIA ARMS"

STUDENTS OF A LAHORE COLLEGE LEARNING THE RUDIMENTS OF MILITARY TRAINING BEFORE JOINING THE NEW TERRITORIAL FORCE

CHAPTER LXX

Order of Precedence of Units of the Indian Army

Cavalry.

Governor-General's Body-Guard.
Governor's Body-Guard, Madras.
Governor's Body-Guard, Bombay.
Governor's Body-Guard, Bengal.
16th Light Cavalry.
7th Light Cavalry.
8th K. G. V's O. Light Cavalry.
Skinner's Horse (1st D. Y. O. Cavalry).
2nd Royal Lancers (Gardner's Horse).
13th D. C. O. Lancers.
The Poona Horse (17th Q. V. O. Cavalry).
The Scinde Horse (14th P. W. O. Cavalry).
3rd Cavalry.
18th K. E. VII's O. Cavalry.
The Guides Cavalry (10th Q. V. Own F. F.).
P. A. V. O. Cavalry (11th F. F.).
Sam Browne's Cavalry (12th F. F.). (Training Regiment).
Hodson's Horse (4th D. C. O. Lancers).
Probyn's Horse (5th K. E. O. Lancers).
6th D. C. O. Lancers (Watson's Horse).
20th Lancers (Training Regiment) 15th Lancers (Training Regiment).

19th K. G. V's O. Lancers.
The Royal Deccan Horse (9th Horse).
The Central India Horse (21st K.G.V's O. Horse).

Sappers and Miners.

Madras, Bengal, R. Bombay.

Indian Signal Corps.

Infantry.

1st Bn., 1st Punjab Regt.
2nd Bn., 1st Punjab Regt.
1st, 10th, 2nd, 3rd Bns., 2nd Punjab Regt.
4th Bn., 2nd Punjab Regt.
3rd Bn., 1st Punjab Regt.
1st Bn., 4th Bombay Grenadiers.
5th Bn., 1st Punjab Regt.
10th Bn., 1st Punjab Regt.
2nd Bn., 4th Bombay Grenadiers.
1st Bn., 5th Mahratta Light Infantry.
1st Bn., 6th Rajputana Rifles.
2nd Bn., 5th Mahratta Light Infantry.
3rd Bn., 5th Mahratta Light Infantry.
1st Bn., 7th Rajput Regt.
5th Bn., 2nd Punjab Regt.
1st, 10th Bns., 8th Punjab Regt.
2nd Bn., 7th Rajput Regt.
2nd, 3rd, 4th, 5th Bns., 8th Punjab Regt.
10th Bn., 4th/9th Regiments.
10th, 4th Bns., 5th Mahratta Light Infantry.
5th R. Bn., 5th Mahratta Light Infantry.
1st R. Bn., 9th Jat Regt.

3rd, 4th Bns., 7th Rajput Regt.
2nd Bn., 9th Jat Regt.
2nd Bn., 6th Rajputana Rifles.
3rd, 4th Bns., 6th Rajputana Rifles.
1st, 10th Bns., 10th Baluch Regt.
5th Bn., 6th Rajputana Rifles.
9th Gurkha Rifles.
3rd Bn., 9th Jat Regt.
1st Gurkha Rifles.
5th Bn., 7th Rajput Regt.
2nd Bn., 10th Baluch Regt.
10th Bn., 6th Rajputana Rifles.
3rd Bn., 10th Baluch Regt.
1st Bn., 2nd R. Bn., 11th Sikh Regt.
2nd Gurkha Rifles.
3rd Gurkha Rifles.
5th, 10th, 1st, 2nd Bns., 3rd R. Bn., 4th Bn., 12th Frontier Force Regt.
4th Bn., 10th Baluch Regt.
1st, 2nd, 10th, 4th, 5th Bns., 13th Frontier Force Rifles.
6th R. Bn., 13th Frontier Force Rifles.
4th P. W. O. Gurkha Rifles.
10th Bn., 7th Rajput Regt.
1st, 2nd, 10th, 3rd Bns., 14th Punjab Regt.
4th Bn., 14th Punjab Regt.
1st, 2nd, 3rd, 4th, 10th Bns., 15th Punjab Regt.
1st, 2nd Bns., 16th Punjab Regt.
5th R. Gurkha Rifles.
5th Bn., 10th Baluch Regt.
3rd Bn., 16th Punjab Regt.
10th, 4th Bns., 11th Sikh Regt.
1st, 2nd Bns., 17th Dogra Regt.
18th R. Garhwal Rifles.
5th Bn., 14th Punjab Regt.
6th Gurkha Rifles.

3rd, 10th Bns., 17th Dogra Regt.
7th Gurkha Rifles.
8th Gurkha Rifles.
3rd Bn., 11th Sikh Regt.
10th Bn., 16th Punjab Regt.
5th Bn., 11th Sikh Regt.
1st, 10th, 2nd and 4th Bns., 19th Hyderabad Regt.
Escort to the British Envoy (Nepal).
4th Bn., 16th Punjab Regt.
10th Gurkha Rifles.
Kumaon Rifles.

All infantry regiments of the late Punjab Frontier Force are considered light infantry.